PROBLEMS & SOLUTIONS IN THEORETICAL & MATHEMATICAL PHYSICS

PROBLEMS & SOLUTIONS IN THEORETICAL & MATHEMATICAL PHYSICS

Second Edition

Volume II: Advanced Level

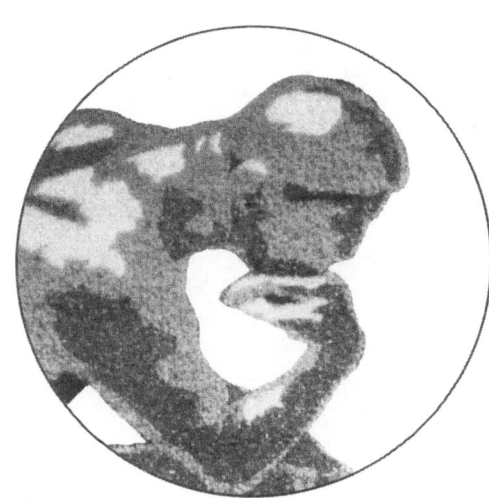

Willi-Hans Steeb

Rand Afrikaans University, South Africa

World Scientific
New Jersey • London • Singapore • Hong Kong

Published by

World Scientific Publishing Co. Pte. Ltd.

5 Toh Tuck Link, Singapore 596224

USA office: Suite 202, 1060 Main Street, River Edge, NJ 07661

UK office: 57 Shelton Street, Covent Garden, London WC2H 9HE

British Library Cataloguing-in-Publication Data
A catalogue record for this book is available from the British Library.

PROBLEMS AND SOLUTIONS IN THEORETICAL AND MATHEMATICAL PHYSICS
Vol. 2: Advanced Level (Second Edition)

ISBN 981-238-988-1
ISBN 981-238-987-3 (pbk)

This book is printed on acid-free paper.

Printed in Singapore by Uto-Print

Preface

The purpose of this book is to supply a collection of problems together with their detailed solution which will prove to be valuable to students as well as to research workers in the fields of mathematics, physics, engineering and other sciences. The topics range in difficulty from elementary to advanced. Almost all problems are solved in detail and most of the problems are self-contained. All relevant definitions are given. Students can learn important principles and strategies required for problem solving. Teachers will also find this text useful as a supplement, since important concepts and techniques are developed in the problems. The material was tested in my lectures given around the world.

The book is divided into two volumes. Volume I presents the introductory problems for undergraduate and advanced undergraduate students. In Volume II the more advanced problems together with their detailed solutions are collected, to meet the needs of graduate students and researchers. Problems included cover most of the new fields in theoretical and mathematical physics such as Lax representation, Bäcklund transformation, soliton equations, Lie algebra valued differential forms, Hirota technique, Painlevé test, the Bethe ansatz, the Yang–Baxter relation, wavelets, chaos, fractals, complexity, etc.

In the reference section some other books are listed having useful problems for students in theoretical physics and mathematical physics. Some or related problems to those given in Volume I can be found in these books. In Volume II references are given to books and original articles where some of the advanced problems can be found.

I wish to express my gratitude to Catharine Thompson and Lance Hoffman for a critical reading of the manuscripts. Finally, I appreciate the help of the Lady from Madeira.

Any useful suggestions and comments are welcome.

Email addresses of the author:

`steeb_wh@yahoo.com`

`whs@na.rau.ac.za`

`Willi-Hans.Steeb@fhso.ch`

Home pages of the author:

`http://issc.rau.ac.za`

`http://zeus.rau.ac.za/steeb/steeb.html`

`http://www.fhso.ch/~steeb`

Contents

Notation

$\emptyset$	empty set
$\mathbf{N}$	natural numbers
$\mathbf{Z}$	integers
$\mathbf{Q}$	rational numbers
$\mathbf{R}$	real numbers
$\mathbf{R}^+$	nonnegative real numbers
$\mathbf{C}$	complex numbers
$\mathbf{R}^n$	n-dimensional Euclidean space
$\mathbf{C}^n$	n-dimensional complex linear space
i	$:= \sqrt{-1}$
$\Re z$	real part of the complex number z
$\Im z$	imaginary part of the complex number z
$\mathbf{x} \in \mathbf{R}^n$	element $\mathbf{x}$ of $\mathbf{R}^n$
$A \subset B$	subset A of set B
$A \cap B$	the intersection of the sets A and B
$A \cup B$	the union of the sets A and B
$f \circ g$	composition of two mappings $(f \circ g)(x) = f(g(x))$
u	dependent variable
t	independent variable (time variable)
x	independent variable (space variable)
$\mathbf{x}^T = (x_1, x_2, \ldots, x_n)$	vector of independent variables, ^{T} means transpose
$\mathbf{u}^T = (u_1, u_2, \ldots, u_n)$	vector of dependent variables, ^{T} means transpose
$\|\cdot\|$	norm
$\mathbf{x} \cdot \mathbf{y}$	scalar product (inner product)

$\mathbf{x} \times \mathbf{y}$	vector product
$\otimes$	Kronecker product, tensor product
det	determinant of a square matrix
tr	trace of a square matrix
I	unit matrix
$[\,,\,]$	commutator
δ_{jk}	Kronecker delta with $\delta_{jk} = 1$ for $j = k$ and $\delta_{jk} = 0$ for $j \neq k$
sgn(x)	the sign of x, 1 if $x > 0$, -1 if $x < 0$, 0 if $x = 0$
λ	eigenvalue
ϵ	real parameter
$\wedge$	Grassmann product (exterior product, wedge product)
H	Hamilton function
$\hat{H}$	Hamilton operator

Chapter 1

Lax Representations in Classical Mechanics

Problem 1. Let

$$\frac{d\mathbf{u}}{dt} = \mathbf{V}(\mathbf{u}), \quad \mathbf{u} \equiv (u_1, u_2, \ldots, u_m)^T \tag{1}$$

be an autonomous system of first-order ordinary differential equations. Assume that the functions $V_k : \mathbf{R}^m \to \mathbf{R}$ are smooth. Assume that this system can be written in the form (the so-called *Lax representation*)

$$\frac{dL}{dt} = [A, L](t) \equiv [A(t), L(t)] \tag{2}$$

where A and L are $n \times n$ matrices and $[A, L] \equiv AL - LA$. The $n \times n$ matrices L and A are called a *Lax pair*.

(i) Show that

$$\frac{dL^k}{dt} = [A, L^k](t). \tag{3}$$

(ii) Show that $\mathrm{tr}(L^k)$ $(k = 1, 2, \ldots)$ are first integrals, where $\mathrm{tr}(\cdot)$ denotes the trace.

(iii) Assume that L^{-1} exists. Show that $\mathrm{tr}(L^{-1})$ is a first integral.

(iv) Show that the solution of the matrix differential Eq. (2) is given by

$$L(t) = e^{tA} L e^{-tA} \tag{4}$$

where $L = L(t = 0)$.

Solution.   (i) The formula (3) is true by assumption for $k = 1$. If it is true for k, then

$$\frac{dL^{k+1}}{dt} = \frac{d}{dt}(L^k L) = \frac{dL^k}{dt}L + L^k\frac{dL}{dt}$$

$$= ([A, L^k]L + L^k[A, L])(t) \quad \text{by the induction hypothesis}$$

$$= (AL^k L - L^k AL + L^k AL - L^k LA)(t) = (AL^{k+1} - L^{k+1}A)(t)$$

$$= [A, L^{k+1}](t).$$

(ii) We have

$$\frac{d}{dt}\text{tr}\, L^k = \text{tr}\left(\frac{dL^k}{ft}\right) = \text{tr}([A, L^k]) = \text{tr}(AL^k - L^k A)$$

$$= \text{tr}(AL^k) - \text{tr}(L^k A) = 0$$

since $\text{tr}(XY) = \text{tr}(YX)$ for arbitrary $n \times n$ matrices X and Y. Consequently, $\text{tr}\, L^k$ $(k = 1, 2, \ldots)$ are first integrals of system (1).

(iii) From $L^{-1}L = I$ where I is the unit matrix it follows that

$$\frac{dL^{-1}}{dt} = -L^{-1}\frac{dL}{dt}L^{-1}.$$

Then

$$\frac{d}{dt}\text{tr}\, L^{-1} = \text{tr}\left(\frac{d}{dt}L^{-1}\right) = -\text{tr}\left(L^{-1}\frac{dL}{dt}L^{-1}\right)$$

$$= -\text{tr}(L^{-1}[A, L]L^{-1}) = -\text{tr}(L^{-1}ALL^{-1}) + \text{tr}(L^{-1}LAL^{-1})$$

$$= -\text{tr}(L^{-1}A) + \text{tr}(AL^{-1}) = 0.$$

(iv) From (4) it follows that

$$\frac{dL}{dt} = e^{tA}ALe^{-tA} - e^{tA}LAe^{-tA} = e^{tA}(AL - LA)e^{-tA}$$

$$= e^{tA}[A, L]e^{-tA} = [A, L](t).$$

Problem 2.   Let $H : \mathbf{R}^4 \to \mathbf{R}$ be the Hamilton function

$$H(\mathbf{p}, \mathbf{q}) = \frac{1}{2}(p_1^2 + p_2^2) + e^{q_2 - q_1}. \tag{1}$$

(i) Find the equations of motion. The Hamilton function (1) is a first integral, i.e. $dH/dt = 0$. Find the second first integral on inspection of the equations of motion.

(ii) Define

$$a := \frac{1}{2}e^{(q_2 - q_1)/2},$$

$$b_1 := \frac{1}{2}p_1,$$

$$b_2 := \frac{1}{2}p_2.$$

Give the equations of motion for a, b_1 and b_2.

(iii) Show that the equations of motion for a, b_1 and b_2 can be written in Lax form, i.e.,

$$\frac{dL}{dt} = [A, L](t)$$

where

$$L := \begin{pmatrix} b_1 & a \\ a & b_2 \end{pmatrix}, \quad A := \begin{pmatrix} 0 & a \\ -a & 0 \end{pmatrix}.$$

Solution.   (i) The *Hamilton equations of motion* are given by

$$\frac{dq_1}{dt} = \frac{\partial H}{\partial p_1} = p_1, \tag{2a}$$

$$\frac{dq_2}{dt} = \frac{\partial H}{\partial p_2} = p_2, \tag{2b}$$

$$\frac{dp_1}{dt} = -\frac{\partial H}{\partial q_1} = e^{q_2 - q_1}, \tag{2c}$$

$$\frac{dp_2}{dt} = -\frac{\partial H}{\partial q_2} = -e^{q_2 - q_1}. \tag{2d}$$

Adding (2c) and (2d) we find the first integral

$$I(\mathbf{p},\mathbf{q}) = p_1 + p_2$$

since

$$\frac{dI}{dt} = \frac{dp_1}{dt} + \frac{dp_2}{dt} = e^{q_2 - q_1} - e^{q_2 - q_1} = 0.$$

Obviously, H is the other first integral.

(ii) From

$$a := \frac{1}{2}e^{(q_2-q_1)/2},$$

we obtain

$$\frac{da}{dt} = \frac{1}{4}e^{(q_2-q_1)/2}\left(\frac{dq_2}{dt} - \frac{dq_1}{dt}\right) = \frac{1}{4}e^{(q_2-q_1)/2}(p_2 - p_1)$$

$$= \frac{1}{2}e^{(q_2-q_1)/2}(b_2 - b_1).$$

Thus

$$\frac{da}{dt} = a(b_2 - b_1).$$

From

$$b_1 := \frac{1}{2}p_1,$$

it follows that

$$\frac{db_1}{dt} = \frac{1}{2}\frac{dp_1}{dt} = \frac{1}{2}e^{q_2-q_1} = 2a^2.$$

Analogously,

$$\frac{db_2}{dt} = \frac{1}{2}\frac{dp_2}{dt} = -\frac{1}{2}e^{q_2-q_1} = -2a^2.$$

To summarize, the equations of motion for a, b_1 and b_2 are given by

$$\frac{da}{dt} = a(b_2 - b_1),$$

$$\frac{db_1}{dt} = 2a^2,$$

$$\frac{db_2}{dt} = -2a^2.$$

(iii) The Lax representation is given by L and A since

$$[A, L] \equiv AL - LA = \begin{pmatrix} 0 & a \\ -a & 0 \end{pmatrix}\begin{pmatrix} b_1 & a \\ a & b_2 \end{pmatrix} - \begin{pmatrix} b_1 & a \\ a & b_2 \end{pmatrix}\begin{pmatrix} 0 & a \\ -a & 0 \end{pmatrix}$$

$$= \begin{pmatrix} 2a^2 & a(b_2 - b_1) \\ a(b_2 - b_1) & -2a^2 \end{pmatrix}$$

and

$$\frac{dL}{dt} = \begin{pmatrix} \dfrac{db_1}{dt} & \dfrac{da}{dt} \\ \dfrac{da}{dt} & \dfrac{db_2}{dt} \end{pmatrix}.$$

The first integrals are given by

$$I_1(a, b_1, b_2) = \text{tr}\, L = b_1 + b_2,$$

$$I_2(a, b_1, b_2) = \text{tr}\, L^2 = 2a^2 + b_1^2 + b_2^2.$$

Problem 3.   Consider the nonlinear system of ordinary differential equations

$$\frac{du_1}{dt} = (\lambda_3 - \lambda_2)u_2 u_3, \tag{1a}$$

$$\frac{du_2}{dt} = (\lambda_1 - \lambda_3)u_3 u_1, \tag{1b}$$

$$\frac{du_3}{dt} = (\lambda_2 - \lambda_1)u_1 u_2 \tag{1c}$$

where $\lambda_j \in \mathbf{R}$. The dynamical system (1) describes *Euler's rigid body motion*.

(i) Show that the first integrals are given by

$$I_1(\mathbf{u}) = u_1^2 + u_2^2 + u_3^2, \quad I_2(\mathbf{u}) = \lambda_1 u_1^2 + \lambda_2 u_2^2 + \lambda_3 u_3^2. \tag{2}$$

(ii) A Lax representation is given by

$$\frac{dL}{dt} = [L, \lambda L](t) \tag{3}$$

where

$$L := \begin{pmatrix} 0 & -u_3 & u_2 \\ u_3 & 0 & -u_1 \\ -u_2 & u_1 & 0 \end{pmatrix}, \quad \lambda L := \begin{pmatrix} 0 & -\lambda_3 u_3 & \lambda_2 u_2 \\ \lambda_3 u_3 & 0 & -\lambda_1 u_1 \\ -\lambda_2 u_2 & \lambda_1 u_1 & 0 \end{pmatrix}.$$

Show that $\text{tr}(L^k)$ ($k = 1, 2, \ldots$) provides only one first integral of system (1).

(iii) Instead of (3) we consider now

$$\frac{d(L + Ay)}{dt} = [L + Ay, \lambda L + By](t) \tag{4}$$

where y is a dummy variable and A and B are time-independent diagonal matrices, i.e, $A = \text{diag}(A_1, A_2, A_3)$ and $B = \text{diag}(B_1, B_2, B_3)$ with $A_j, B_j \in \mathbf{R}$. Equation (4) decomposes into various powers of y, namely

$$y^0 : \frac{dL}{dt} = [L, \lambda L](t), \tag{5a}$$

$$y^1 : 0 = [L, B] + [A, \lambda L], \tag{5b}$$

$$y^2 : [A, B] = 0. \tag{5c}$$

Equation (5c) is satisfied identically since A and B are diagonal matrices. Equation (5b) leads to

$$\lambda_i = \frac{B_j - B_k}{A_j - A_k} \tag{6}$$

where (i, j, k) are permutations of $(1, 2, 3)$. Equation (6) can be satisfied by setting

$$B_j = A_j^2, \quad \lambda_i = A_j + A_k.$$

Consequently the original Lax pair L, λL satisfies the *extended Lax pair*

$$L + Ay, \quad \lambda L + By.$$

Show that $\text{tr}((L + Ay)^2)$ and $\text{tr}((L + Ay)^3)$ provide the first integrals (2).

Solution.   (i) Straightforward calculation yields

$$\frac{d}{dt} I_1(\mathbf{u}) = 2u_1 \frac{du_1}{dt} + 2u_2 \frac{du_2}{dt} + 2u_3 \frac{du_3}{dt}.$$

Thus

$$\frac{d}{dt} I_1(\mathbf{u}) = 2u_1(\lambda_3 - \lambda_2)u_2 u_3 + 2u_2(\lambda_1 - \lambda_3)u_3 u_1 + 2u_3(\lambda_2 - \lambda_1)u_1 u_2 = 0.$$

Analogously, $dI_2(\mathbf{u})/dt = 0$.

(ii) We obtain $\text{tr } L = 0$ and

$$\text{tr } L^2 = -2(u_1^2 + u_2^2 + u_3^2) = -2I_1(\mathbf{u}).$$

Since L does not depend on λ we cannot find I_2.

(iii) Straightforward calculation yields

$$\text{tr}((L + Ay)^2) = -2I_1(\mathbf{u}) + C_1,$$

$$\text{tr}((L + Ay)^3) = -3yI_2(\mathbf{u}) + C_2$$

where C_1 and C_2 are constants. Thus the extended Lax pair provides both first integrals.

Problem 4.   The Lie algebra $sl(2, \mathbf{R})$ is defined by the commutator rules

$$[X_0, X_+] = 2X_+, \quad [X_0, X_-] = -2X_-, \quad [X_+, X_-] = X_0$$

where X_0, X_+, X_- denote the generators. Let

$$\{A(\mathbf{p}, \mathbf{q}), B(\mathbf{p}, \mathbf{q})\} := \sum_{j=1}^{N} \left(\frac{\partial A}{\partial q_j} \frac{\partial B}{\partial p_j} - \frac{\partial B}{\partial q_j} \frac{\partial A}{\partial p_j} \right)$$

be the *Poisson bracket*. Let

$$t_+ := \frac{1}{2} \sum_{k=1}^{N} p_k^2,$$

$$t_- := -\frac{1}{2} \sum_{k=1}^{N} q_k^2,$$

$$t_0 := -\sum_{k=1}^{N} p_k q_k.$$

(i) Show that the functions $-t_+, -t_-, -t_0$ form a basis of a Lie algebra under the Poisson bracket and that the Lie algebra is isomorphic to $sl(2, \mathbf{R})$.

(ii) Let

$$H(\mathbf{p}, \mathbf{q}) = \frac{1}{2} \sum_{k=1}^{N} p_k^2 + U(\mathbf{q}) \equiv t_+ + U(\mathbf{q}),$$

$$t_- = -\frac{1}{2} \sum_{k=1}^{N} q_k^2, \quad t_0 = -\sum_{k=1}^{N} p_k q_k.$$

Find the condition on U such that $\{-H, -t_-, -t_0\}$ forms a basis of a Lie algebra which is isomorphic to $sl(2, \mathbf{R})$.

(iii) The Hamilton system

$$H(\mathbf{p}, \mathbf{q}) = \frac{1}{2} \sum_{k=1}^{N} p_k^2 + \sum_{j=2}^{N} \sum_{k=1}^{j-1} \frac{a^2}{(q_j - q_k)^2}$$

admits a Lax representation with

$$L := \begin{pmatrix} p_1 & \dfrac{ia}{(q_1 - q_2)} & \cdots & \dfrac{ia}{(q_1 - q_N)} \\ \dfrac{ia}{(q_2 - q_1)} & p_2 & \cdots & \dfrac{ia}{(q_2 - q_N)} \\ \vdots & & & \vdots \\ \dfrac{ia}{(q_N - q_1)} & & \cdots & p_N \end{pmatrix}$$

where a is a nonzero real constant. Let $N = 3$. Find the first integrals.

Solution.   (i) Straightforward calculation yields

$$\{-t_0, -t_+\} = \{t_0, t_+\} = -2t_+ ,$$

$$\{-t_0, -t_-\} = \{t_0, t_-\} = 2t_- ,$$

$$\{-t_+, -t_-\} = \{t_+, t_-\} = -t_0 .$$

Consider the map

$$X_0 \to -t_0 , \quad X_- \to -t_- , \quad X_+ \to -t_+ .$$

Then the two Lie algebras are isomorphic.

(ii) Since

$$\{-t_0, -H\} = -\sum_{j=1}^{N} \left(p_j^2 - \frac{\partial U}{\partial q_j} q_j \right) ,$$

$$\{-H, -t_-\} = \sum_{j=1}^{N} p_j q_j = -t_0 ,$$

we find from the condition

$$\{-t_0, -H\} = -2H$$

that the potential U satisfies the linear partial differential equation

$$\sum_{j=1}^{N} q_j \frac{\partial U}{\partial q_j} = -2U .$$

This is a homogeneous equation (of rank -2) for the potential U. This equation admits the two solutions

$$U(\mathbf{q}) = \sum_{j=2}^{N} \sum_{k=1}^{j-1} \frac{a^2}{(q_j - q_k)^2}$$

and

$$U(\mathbf{q}) = \frac{a^2}{q_1^2 + q_2^2 + \cdots + q_N^2} .$$

(iii) The first integrals are the invariants of L. From L we find (with $N = 3$) that

$$\mathrm{tr}(L) = p_1 + p_2 + p_3$$

and

$$\mathrm{tr}(L^2) = p_1^2 + p_2^2 + p_3^2 + 2a^2 \left(\frac{1}{(q_1 - q_2)^2} + \frac{1}{(q_3 - q_1)^2} + \frac{1}{(q_2 - q_3)^2} \right) .$$

Obviously,

$$2H = \mathrm{tr}\, L^2 .$$

The calculation of $\mathrm{tr}(L^3)$ is very lengthy. To find the third first integral we use

$$\det(L) = \lambda_1 \lambda_2 \lambda_3 = p_1 p_2 p_3 - a^2 \left(\frac{p_1}{(q_2 - q_3)^2} + \frac{p_2}{(q_3 - q_1)^2} + \frac{p_3}{(q_1 - q_2)^2} \right)$$

where λ_1, λ_2 and λ_3 denote the eigenvalues of L.

Problem 5.   Assume that

$$L := \begin{pmatrix} p_1 & if_{12} & 0 & \cdots & 0 & 0 & -if_{N1} \\ -if_{12} & p_2 & if_{23} & \cdots & 0 & 0 & 0 \\ 0 & -if_{23} & p_3 & \cdots & & & \\ \vdots & & & & & & if_{N-1,N} \\ if_{N1} & 0 & 0 & \cdots & 0 & -if_{N-1,N} & p_N \end{pmatrix}$$

where $f_{n,n+1} := f(q_n - q_{n+1})$ with $f(\mathbf{q})$ a certain given smooth real function. Assume that the Hamilton function H is given by

$$H(\mathbf{p}, \mathbf{q}) = \frac{1}{2} \operatorname{tr} L^2 .$$

Let

$$A := i \begin{pmatrix} h_1 & g_{12} & 0 & \cdots & 0 & g_{N1} \\ g_{12} & h_2 & g_{23} & \cdots & 0 & 0 \\ 0 & g_{23} & h_3 & & & \\ & & & & & \\ 0 & \cdots & & & & g_{N-1,N} \\ g_{N1} & 0 & & \cdots & g_{N-1,N} & h_N \end{pmatrix}$$

where $g(\mathbf{q})$ and $h_k(\mathbf{p}, \mathbf{q})$ are real-valued smooth functions. Find the condition on f, g and h such that L, A are a Lax pair of the Hamilton function H.

Solution.   From the Hamilton function we obtain

$$H(\mathbf{p}, \mathbf{q}) = \frac{1}{2} \operatorname{tr} L^2 = \frac{1}{2} \sum_{k=1}^{N} p_k^2 + \sum_{k=1}^{N} f_{k,k+1}^2$$

with $N + 1 \equiv 1$, i.e. modulo N. Then the Hamilton equations of motion are given by

$$\frac{dq_k}{dt} = \frac{\partial H}{\partial p_k} = p_k , \tag{1a}$$

$$\frac{dp_k}{dt} = -\frac{\partial H}{\partial q_k} = 2(-f_{k,k+1}' f_{k,k+1} + f_{k-1,k}' f_{k-1,k}) \tag{1b}$$

where $k = 1, \ldots, N$ and f' means differentiation with respect to the argument. For f we obtain

$$\frac{df_{k,k+1}}{dt} = (p_k - p_{k+1}) f_{k,k+1}'$$

where we used $f_{k,k+1} = f(q_k - q_{k+1})$. From the requirement that L, A are a Lax pair, i.e. L, A satisfies $dL/dt = [L, A](t)$ we obtain

$$\frac{dp_k}{dt} = 2(f_{k-1,k} g_{k-1,k} - f_{k,k+1} g_{k,k+1}) , \tag{2a}$$

$$\frac{df_{k,k+1}}{dt} = g_{k,k+1}(p_k - p_{k+1}) + i f_{k,k+1}(h_{k+1} - h_k) \quad \text{modulo } N , \tag{2b}$$

$$0 = f_{k,k+1} g_{k,k+2} - g_{k,k+1} f_{k,k+2} . \tag{2c}$$

The choice $h_k = 0$ and

$$g_{k,k+1} = f'_{k,k+1},$$ (3a)

$$0 = f_{k,k+1}g_{k,k+2} - g_{k,k+1}f_{k,k+2}$$ (3b)

provides the consistency of system (2) and system (1). Inserting (3a) into (3b) yields

$$\frac{f'_{k,k+1}}{f_{k,k+1}} = \frac{f'_{k,k+2}}{f_{k,k+2}}.$$

Consequently,

$$\frac{f'_{k,k+1}}{f_{k,k+1}} = c \equiv \text{const}.$$ (4)

Therefore, the solution of (4) is given by

$$f_{k,k+1} \equiv f(q_k - q_{k+1}) = A\exp(c(q_k - q_{k+1}))$$

where A is a constant.

Chapter 2

Kronecker and Tensor Products

Problem 1. Let A be an $m \times n$ matrix and let B be a $p \times q$ matrix. The *Kronecker product* of A and B is defined as

$$A \otimes B := \begin{pmatrix} a_{11}B & a_{12}B & \cdots & a_{1n}B \\ a_{21}B & a_{22}B & \cdots & a_{2n}B \\ \vdots & \vdots & \ddots & \vdots \\ a_{m1}B & a_{m2}B & \cdots & a_{mn}B \end{pmatrix}. \tag{1}$$

Thus $A \otimes B$ is an $mp \times nq$ matrix. Now let A be an $m \times m$ matrix and let B be an $n \times n$ matrix. Show that

$$\operatorname{tr}(A \otimes B) \equiv (\operatorname{tr} A)(\operatorname{tr} B). \tag{2}$$

Solution. From definition (1) we find

$$\operatorname{diag}(a_{jj}B) = (a_{jj}b_{11}, a_{jj}b_{22}, \ldots, a_{jj}b_{nn}).$$

Therefore

$$\operatorname{tr}(A \otimes B) = \sum_{j=1}^{m} \sum_{k=1}^{n} a_{jj} b_{kk}.$$

Since

$$\operatorname{tr} A = \sum_{j=1}^{m} a_{jj}, \quad \operatorname{tr} B = \sum_{k=1}^{n} b_{kk},$$

we find (2).

Problem 2. Let P and Q be two $n \times n$ matrices and I the $n \times n$ unit matrix. Assume that

$$P^2 = P$$

and

$$Q^2 = Q.$$

(i) Show that

$$(P \otimes Q)^2 = P \otimes Q. \tag{1}$$

(ii) Show that

$$(P \otimes (I - P))^2 = P \otimes (I - P).$$

(iii) Assume that $QP = 0$. Calculate $(Q \otimes Q)(P \otimes P)$.

Solution. (i)

$$(P \otimes Q)(P \otimes Q) = (P^2 \otimes Q^2) = P \otimes Q.$$

(ii)

$$(P \otimes (I - P))(P \otimes (I - P)) = (P^2 \otimes (I - P)^2) = P \otimes (I - P - P + P^2).$$

Thus

$$(P \otimes (I - P))(P \otimes (I - P)) = P \otimes (I - P - P + P) = P \otimes (I - P).$$

(iii) We find

$$(Q \otimes Q)(P \otimes P) = (QP) \otimes (QP) = 0.$$

Remark. *Equation* (1) *can be extended to N-factors. Let* $P_1, \ldots, P_N$ *be matrices with* $P_j^2 = P_j$, *where* $j = 1, 2, \ldots, N$. *Then*

$$(P_1 \otimes P_2 \otimes \cdots \otimes P_N)^2 = P_1 \otimes P_2 \otimes \cdots \otimes P_N.$$

Problem 3. (i) Let A be an $n \times n$ matrix and B be an $m \times m$ matrix. I_n is the $n \times n$ unit matrix and I_m the $m \times m$ unit matrix. Calculate

$$[A \otimes I_m, I_n \otimes B]$$

where $[\, , \,]$ denotes the commutator.

(ii) Let A, B, C, D be $n \times n$ matrices. Assume that

$$[A, B] = 0, \quad [C, D] = 0. \tag{1}$$

Calculate

$$[A \otimes C, B \otimes D].$$

Solution.   (i) Straightforward calculation yields

$$[A \otimes I_m, I_n \otimes B] = (A \otimes I_m)(I_n \otimes B) - (I_n \otimes B)(A \otimes I_m)$$

$$= A \otimes B - A \otimes B$$

$$= 0.$$

Consequently, the matrices $A \otimes I_m$ and $I_n \otimes B$ commute.

(ii) We find

$$[A \otimes C, B \otimes D] = (A \otimes C)(B \otimes D) - (B \otimes D)(A \otimes C)$$

$$= (AB) \otimes (CD) - (BA) \otimes (DC)$$

$$= (AB) \otimes (CD) - (AB) \otimes (CD)$$

$$= 0$$

where we have used (1).

Problem 4.   (i) Let A be an $m \times m$ matrix with eigenvalues

$$\lambda_1, \lambda_2, \ldots, \lambda_m$$

and the corresponding eigenvectors

$$\mathbf{a}_1, \mathbf{a}_2, \ldots, \mathbf{a}_m.$$

Let B be an $n \times n$ matrix with eigenvalues

$$\mu_1, \mu_2, \ldots, \mu_n$$

and the corresponding eigenvectors

$$\mathbf{b}_1, \mathbf{b}_2, \ldots, \mathbf{b}_n.$$

Show that the matrix $A \otimes B$ has the eigenvalues $\lambda_j \mu_k$ with the corresponding eigenvectors $\mathbf{a}_j \otimes \mathbf{b}_k$, where $1 \leq j \leq m$ and $1 \leq k \leq n$.

(ii) Let

$$X := \begin{pmatrix} 0 & -i \\ i & 0 \end{pmatrix}, \quad Y := \begin{pmatrix} 1 & 0 \\ 0 & -1 \end{pmatrix}.$$

Find the eigenvalues of $X \otimes Y$.

Solution. (i) From the eigenvalue equations

$$A\mathbf{a}_j = \lambda_j \mathbf{a}_j \quad 1 \le j \le m$$

and

$$B\mathbf{b}_k = \mu_k \mathbf{b}_k \quad 1 \le k \le n,$$

we find

$$(A\mathbf{a}_j) \otimes (B\mathbf{b}_k) = \lambda_j \mu_k (\mathbf{a}_j \otimes \mathbf{b}_k).$$

Since

$$(A\mathbf{a}_j) \otimes (B\mathbf{b}_k) \equiv (A \otimes B)(\mathbf{a}_j \otimes \mathbf{b}_k),$$

we obtain

$$(A \otimes B)(\mathbf{a}_j \otimes \mathbf{b}_k) = \lambda_j \mu_k (\mathbf{a}_j \otimes \mathbf{b}_k).$$

Equation (1) is an eigenvalue equation. Consequently,

$$\mathbf{a}_j \otimes \mathbf{b}_k$$

is an eigenvector of $A \otimes B$ with eigenvalue $\lambda_j \mu_k$.

(ii) Both matrices are hermitian and unitary. Thus it follows that the eigenvalues are given by $\{1, -1\}$. Since the eigenvalues of X and Y are given by $\{1, -1\}$, we find that the eigenvalues of $X \otimes Y$ are given by $\{1, 1, -1, -1\}$.

Problem 5. (i) Let A be an $m \times m$ matrix and B be an $n \times n$ matrix. Let

$$A\mathbf{a}_j = \lambda_j \mathbf{a}_j, \quad j = 1, 2, \dots, m,$$

$$B\mathbf{b}_k = \mu_k \mathbf{b}_k, \quad k = 1, 2, \dots, n$$

be their respective eigenvalue equations. Find the eigenvalues and eigenvectors of

$$A \otimes I_n + I_m \otimes B$$

where I_n is the $n \times n$ unit matrix and I_m is the $m \times m$ unit matrix.

(ii) Let

$$X := \begin{pmatrix} 0 & -i \\ i & 0 \end{pmatrix}, \quad Y := \begin{pmatrix} 1 & 0 \\ 0 & -1 \end{pmatrix}.$$

Find the eigenvalues of

$$X \otimes I_2 + I_2 \otimes Y.$$

Solution.   (i) We have

$$(A \otimes I_n + I_m \otimes B)(\mathbf{a} \otimes \mathbf{b}) = (A \otimes I_n)(\mathbf{a} \otimes \mathbf{b}) + (I_m \otimes B)(\mathbf{a} \otimes \mathbf{b})$$
$$= (A\mathbf{a}) \otimes (I_n\mathbf{b}) + (I_m\mathbf{a}) \otimes (B\mathbf{b})$$
$$= (\lambda\mathbf{a}) \otimes \mathbf{b} + \mathbf{a} \otimes (\mu\mathbf{b})$$
$$= (\lambda + \mu)(\mathbf{a} \otimes \mathbf{b}).$$

Consequently, the eigenvectors and eigenvalues of $A \otimes I_n + I_m \otimes B$ are given by

$$\mathbf{a}_j \otimes \mathbf{b}_k$$

and

$$\lambda_j + \mu_k$$

respectively, where $j = 1, 2, \ldots, m$ and $k = 1, 2, \ldots, n$.

(ii) Since the eigenvalues of X and Y are given by $\{1, -1\}$, we find that the eigenvalues of $X \otimes I_2 + I_2 \otimes Y$ take the form $\{2, 0, 0, -2\}$.

Problem 6.   Let A, B be two arbitrary $n \times n$ matrices. Let I be the $n \times n$ unit matrix. Prove that

$$\operatorname{tr} e^{A \otimes I + I \otimes B} \equiv (\operatorname{tr} e^A)(\operatorname{tr} e^B). \tag{1}$$

Solution.   Since

$$[A \otimes I, I \otimes B] = 0,$$

we have

$$\operatorname{tr} e^{(A \otimes I + I \otimes B)} = \operatorname{tr}(e^{A \otimes I} e^{I \otimes B}).$$

Now

$$e^{A \otimes I} = \sum_{k=0}^{\infty} \frac{(A \otimes I)^k}{k!} , \quad e^{I \otimes B} = \sum_{k=0}^{\infty} \frac{(I \otimes B)^k}{k!} .$$

An arbitrary term in the expansion of $e^{A \otimes I} e^{I \otimes B}$ is given by

$$\frac{1}{n!} \frac{1}{m!} (A \otimes I)^n (I \otimes B)^m .$$

Now we have

$$\frac{1}{n!} \frac{1}{m!} (A \otimes I)^n (I \otimes B)^m \equiv \frac{1}{n!} \frac{1}{m!} (A^n \otimes I^n)(I^m \otimes B^m)$$

$$\equiv \frac{1}{n!} \frac{1}{m!} (A^n \otimes I)(I \otimes B^m) \equiv \frac{1}{n!} \frac{1}{m!} (A^n \otimes B^m) .$$

Therefore

$$\operatorname{tr} \left(\frac{1}{n!} \cdot \frac{1}{m!} (A \otimes I)^n (I \otimes B)^m \right) = \frac{1}{n!} \frac{1}{m!} (\operatorname{tr} A^n)(\operatorname{tr} B^m) .$$

Since

$$(\operatorname{tr} e^A)(\operatorname{tr} e^B) \equiv \operatorname{tr} \left(\sum_{k=0}^{\infty} \frac{A^k}{k!} \right) \operatorname{tr} \left(\sum_{k=0}^{\infty} \frac{B^k}{k!} \right) = \left(\sum_{k=0}^{\infty} \frac{1}{k!} \operatorname{tr} A^k \right) \left(\sum_{k=0}^{\infty} \frac{1}{k!} \operatorname{tr} B^k \right),$$

we have proved the identity (1).

Problem 7.   Let

$$A_j := \overbrace{I \otimes \cdots \otimes I \otimes A \otimes I \otimes \cdots \otimes I}^{N\text{-factors}}$$

where A is at the jth place. Let

$$X := A_1 + A_2 + \cdots + A_N$$

where A is an $n \times n$ matrix and I is the $n \times n$ unit matrix. Calculate

$$\frac{\operatorname{tr}(A_j e^X)}{\operatorname{tr} e^X} .$$

Solution.   Since $[A_j, A_k] = 0$ for $j = 1, 2, \ldots, N$ and $k = 1, 2, \ldots, N$ we have

$$e^X \equiv e^{A_1 + A_2 + \cdots + A_N} \equiv e^{A_1} e^{A_2} \cdots e^{A_N} .$$

Now

$$\overbrace{e^{A_j} = I \otimes \cdots \otimes I \otimes e^A \otimes I \otimes \cdots \otimes I}^{N\text{-factors}}$$

where e^A is at the jth place. It follows that

$$e^{A_1} e^{A_2} \cdots e^{A_N} = e^A \otimes e^A \otimes \cdots \otimes e^A .$$

Therefore

$$\operatorname{tr} e^X \equiv (\operatorname{tr} e^A)^N .$$

Now

$$A_j e^X = (I \otimes \cdots \otimes I \otimes A \otimes I \otimes \cdots \otimes I)(e^A \otimes e^A \otimes \cdots \otimes e^A)$$

or

$$A_j e^X = (e^A \otimes \cdots \otimes e^A \otimes A e^A \otimes e^A \otimes \cdots \otimes e^A) .$$

Thus

$$\operatorname{tr}(A_j e^X) = (\operatorname{tr} e^A)^{N-1}(\operatorname{tr}(A e^A)) .$$

Consequently

$$\frac{\operatorname{tr}(A_j e^X)}{\operatorname{tr} e^X} = \frac{\operatorname{tr}(A e^A)}{\operatorname{tr} e^A} .$$

Problem 8.   Let

$$\mathbf{u}_1 = \begin{pmatrix} 1 \\ 0 \end{pmatrix} , \quad \mathbf{u}_2 = \begin{pmatrix} 0 \\ 1 \end{pmatrix} , \quad \mathbf{v}_1 = \frac{1}{\sqrt{2}} \begin{pmatrix} 1 \\ 1 \end{pmatrix} , \quad \mathbf{v}_2 = \frac{1}{\sqrt{2}} \begin{pmatrix} 1 \\ -1 \end{pmatrix} .$$

Calculate

$$\mathbf{u}_1 \otimes \mathbf{u}_1 , \quad \mathbf{u}_1 \otimes \mathbf{u}_2 , \quad \mathbf{u}_2 \otimes \mathbf{u}_1 , \quad \mathbf{u}_2 \otimes \mathbf{u}_2 ,$$

$$\mathbf{v}_1 \otimes \mathbf{v}_1 , \quad \mathbf{v}_1 \otimes \mathbf{v}_2 , \quad \mathbf{v}_2 \otimes \mathbf{v}_1 , \quad \mathbf{v}_2 \otimes \mathbf{v}_2 ,$$

$$\mathbf{u}_1^T \otimes \mathbf{u}_1 , \quad \mathbf{u}_1^T \otimes \mathbf{u}_2 , \quad \mathbf{u}_2^T \otimes \mathbf{u}_1 , \quad \mathbf{u}_2^T \otimes \mathbf{u}_2 ,$$

$$\mathbf{v}_1^T \otimes \mathbf{v}_1 , \quad \mathbf{v}_1^T \otimes \mathbf{v}_2 , \quad \mathbf{v}_2^T \otimes \mathbf{v}_1 , \quad \mathbf{v}_2^T \otimes \mathbf{v}_2 .$$

Discuss!

Solution. (a) Using the definition of the Kronecker product we find

$$\mathbf{u}_1 \otimes \mathbf{u}_1 = \begin{pmatrix} 1 \\ 0 \\ 0 \\ 0 \end{pmatrix}, \quad \mathbf{u}_1 \otimes \mathbf{u}_2 = \begin{pmatrix} 0 \\ 1 \\ 0 \\ 0 \end{pmatrix},$$

$$\mathbf{u}_2 \otimes \mathbf{u}_1 = \begin{pmatrix} 0 \\ 0 \\ 1 \\ 0 \end{pmatrix}, \quad \mathbf{u}_2 \otimes \mathbf{u}_2 = \begin{pmatrix} 0 \\ 0 \\ 0 \\ 1 \end{pmatrix}. \tag{1}$$

The set $\{\mathbf{u}_1, \mathbf{u}_2\}$ is the standard basis of $\mathbf{R}^2$. From (1) we find that the set

$$\{\mathbf{u}_1 \otimes \mathbf{u}_1, \ \mathbf{u}_1 \otimes \mathbf{u}_2, \ \mathbf{u}_2 \otimes \mathbf{u}_1, \ \mathbf{u}_2 \otimes \mathbf{u}_2\} \tag{2}$$

is the standard basis in $\mathbf{R}^4$.

(b) Using the definition of the Kronecker product we obtain

$$\mathbf{v}_1 \otimes \mathbf{v}_1 = \frac{1}{2} \begin{pmatrix} 1 \\ 1 \\ 1 \\ 1 \end{pmatrix}, \quad \mathbf{v}_1 \otimes \mathbf{v}_2 = \frac{1}{2} \begin{pmatrix} 1 \\ -1 \\ 1 \\ -1 \end{pmatrix},$$

$$\mathbf{v}_2 \otimes \mathbf{v}_1 = \frac{1}{2} \begin{pmatrix} 1 \\ 1 \\ -1 \\ -1 \end{pmatrix}, \quad \mathbf{v}_2 \otimes \mathbf{v}_2 = \frac{1}{2} \begin{pmatrix} 1 \\ -1 \\ -1 \\ 1 \end{pmatrix}.$$

The set $\{\mathbf{v}_1, \mathbf{v}_2\}$ is a basis of $\mathbf{R}^2$. From (2) we find that the set

$$\{\mathbf{v}_1 \otimes \mathbf{v}_1, \ \mathbf{v}_1 \otimes \mathbf{v}_2, \ \mathbf{v}_2 \otimes \mathbf{v}_1, \ \mathbf{v}_2 \otimes \mathbf{v}_2\}$$

is a basis in $\mathbf{R}^4$.

(c) We find

$$\mathbf{u}_1^T \otimes \mathbf{u}_1 = (1, 0) \otimes \begin{pmatrix} 1 \\ 0 \end{pmatrix} = \begin{pmatrix} 1 & 0 \\ 0 & 0 \end{pmatrix},$$

$$\mathbf{u}_1^T \otimes \mathbf{u}_2 = (1, 0) \otimes \begin{pmatrix} 0 \\ 1 \end{pmatrix} = \begin{pmatrix} 0 & 0 \\ 1 & 0 \end{pmatrix},$$

$$\mathbf{u}_2^T \otimes \mathbf{u}_1 = (0, 1) \otimes \begin{pmatrix} 1 \\ 0 \end{pmatrix} = \begin{pmatrix} 0 & 1 \\ 0 & 0 \end{pmatrix},$$

$$\mathbf{u}_2^T \otimes \mathbf{u}_2 = (0,1) \otimes \begin{pmatrix} 0 \\ 1 \end{pmatrix} = \begin{pmatrix} 0 & 0 \\ 0 & 1 \end{pmatrix}.$$

Consequently,

$$\mathbf{u}_1^T \otimes \mathbf{u}_1 + \mathbf{u}_2^T \otimes \mathbf{u}_2$$

is the 2×2 unit matrix.

(d) We obtain

$$\mathbf{v}_1^T \otimes \mathbf{v}_1 = \frac{1}{2}(1,1) \otimes \begin{pmatrix} 1 \\ 1 \end{pmatrix} = \frac{1}{2}\begin{pmatrix} 1 & 1 \\ 1 & 1 \end{pmatrix},$$

$$\mathbf{v}_1^T \otimes \mathbf{v}_2 = \frac{1}{2}(1,1) \otimes \begin{pmatrix} 1 \\ -1 \end{pmatrix} = \frac{1}{2}\begin{pmatrix} 1 & 1 \\ -1 & -1 \end{pmatrix},$$

$$\mathbf{v}_2^T \otimes \mathbf{v}_1 = \frac{1}{2}(1,-1) \otimes \begin{pmatrix} 1 \\ 1 \end{pmatrix} = \frac{1}{2}\begin{pmatrix} 1 & -1 \\ 1 & -1 \end{pmatrix},$$

$$\mathbf{v}_2^T \otimes \mathbf{v}_2 = \frac{1}{2}(1,-1) \otimes \begin{pmatrix} 1 \\ -1 \end{pmatrix} = \frac{1}{2}\begin{pmatrix} 1 & -1 \\ -1 & 1 \end{pmatrix}.$$

Consequently,

$$\mathbf{v}_1^T \otimes \mathbf{v}_1 + \mathbf{v}_2^T \otimes \mathbf{v}_2$$

is the 2×2 unit matrix.

Problem 9.   Let A be a symmetric 4×4 matrix over $\mathbf{R}$. Assume that the eigenvalues are given by $\lambda_1 = 0$, $\lambda_2 = 1$, $\lambda_3 = 2$ and $\lambda_4 = 3$ with the corresponding normalized eigenvectors

$$\mathbf{u}_1 = \frac{1}{\sqrt{2}} \begin{pmatrix} 1 \\ 0 \\ 0 \\ 1 \end{pmatrix}, \quad \mathbf{u}_2 = \frac{1}{\sqrt{2}} \begin{pmatrix} 1 \\ 0 \\ 0 \\ -1 \end{pmatrix},$$

$$\mathbf{u}_3 = \frac{1}{\sqrt{2}} \begin{pmatrix} 0 \\ 1 \\ 1 \\ 0 \end{pmatrix}, \quad \mathbf{u}_4 = \frac{1}{\sqrt{2}} \begin{pmatrix} 0 \\ 1 \\ -1 \\ 0 \end{pmatrix}.$$

Find the matrix A with the help of the *spectral theorem*.

Solution. The eigenvectors to different eigenvalues are orthogonal. Since the normalized eigenvectors are pairwise orthogonal we find from the spectral theorem

$$A = \sum_{j=1}^{4} \lambda_j \mathbf{u}_j^T \otimes \mathbf{u}_j.$$

Thus

$$A = \lambda_1 \mathbf{u}_1^T \otimes \mathbf{u}_1 + \lambda_2 \mathbf{u}_2^T \otimes \mathbf{u}_2 + \lambda_3 \mathbf{u}_3^T \otimes \mathbf{u}_3 + \lambda_4 \mathbf{u}_4^T \otimes \mathbf{u}_4.$$

Inserting the eigenvalues yields

$$A = \mathbf{u}_2^T \otimes \mathbf{u}_2 + 2\mathbf{u}_3^T \otimes \mathbf{u}_3 + 3\mathbf{u}_4^T \otimes \mathbf{u}_4.$$

Using the definition of the Kronecker product we find that

$$A = \frac{1}{2} \begin{pmatrix} 1 & 0 & 0 & -1 \\ 0 & 5 & -1 & 0 \\ 0 & -1 & 5 & 0 \\ -1 & 0 & 0 & 1 \end{pmatrix}.$$

Problem 10. Let U be a 4×4 unitary matrix. Assume that U can be written as

$$U = V \otimes W$$

where V and W are 2×2 unitary matrices. Show that U can be written as

$$U = \exp(iA \otimes I_2 + iI_2 \otimes B) \tag{1}$$

where A and B are 2×2 hermitian matrices and I_2 is the 2×2 unit matrix.

Solution. Any unitary matrix Z can be written as

$$Z = \exp(iX)$$

where X is a hermitian matrix. Thus

$$V = \exp(iA),$$
$$W = \exp(iB)$$

where A and B are 2×2 hermitian matrices. Consequently

$$U = V \otimes W = \exp(iA) \otimes \exp(iB).$$

Using the identity

$$\exp(iA) \otimes \exp(iB) \equiv \exp(iA \otimes I_2) \exp(iI_2 \otimes B)$$

and

$$[A \otimes I_2, I_2 \otimes B] = 0$$

we find Eq. (1).

Problem 11.   Let A_i and B_j be $n \times n$ matrices over **C**, where $i, j = 1, 2, \ldots, p$. We define

$$r_{12} := \sum_{i=1}^{p} A_i \otimes B_i \otimes I, \quad r_{13} := \sum_{j=1}^{p} A_j \otimes I \otimes B_j, \quad r_{23} := \sum_{k=1}^{p} I \otimes A_k \otimes B_k$$

where I is the $n \times n$ unit matrix. Find

$$[r_{12}, r_{13}], \quad [r_{12}, r_{23}], \quad [r_{13}, r_{23}]$$

where $[,]$ denotes the commutator.

Solution.   Since

$$(A_i \otimes B_i \otimes I)(A_j \otimes I \otimes B_j) - (A_j \otimes I \otimes B_j)(A_i \otimes B_i \otimes I)$$

$$= (A_i A_j) \otimes B_i \otimes B_j - (A_j A_i) \otimes B_i \otimes B_j$$

$$= (A_i A_j - A_j A_i) \otimes B_i \otimes B_j = [A_i, A_j] \otimes B_i \otimes B_j,$$

thus

$$[r_{12}, r_{13}] = \sum_{i=1}^{p} \sum_{j=1}^{p} [A_i, A_j] \otimes B_i \otimes B_j.$$

Analogously

$$[r_{12}, r_{23}] = \sum_{i=1}^{p} \sum_{k=1}^{p} A_i \otimes [B_i, A_k] \otimes B_k,$$

$$[r_{13}, r_{23}] = \sum_{j=1}^{p} \sum_{k=1}^{p} A_j \otimes A_k \otimes [B_j, B_k].$$

Remark. *Assume that A_i and B_j are elements of a Lie algebra. The* Classical Yang–Baxter Equation *(CYBE)* *is given by*

$$[r_{12}, r_{13}] + [r_{12}, r_{23}] + [r_{13}, r_{23}] = 0.$$

Problem 12. Let $X = (x_{ij})$ be an $m \times n$ matrix. We define the vector

$$v(X) := (x_{11}, x_{12}, \ldots, x_{1n}, x_{21}, \ldots, x_{2n}, \ldots, x_{mn})^T \tag{1}$$

where ^{T} denotes transpose. Thus $v(X)$ is a column vector. Let C be a $p \times m$ matrix and let D be a $q \times n$ matrix. Then we have

$$v(CXD^T) = (C \otimes D)v(X). \tag{2}$$

Show that

$$D \otimes C = P(C \otimes D)Q \tag{3}$$

where P and Q are permutation matrices depending only on p, q and m, n respectively.

Solution. Consider the equation

$$CXD^T = E. \tag{4}$$

Using (2) we find

$$(C \otimes D)v(X) = v(E). \tag{5}$$

Similarly, application of (2) to the transpose of (4) yields

$$(D \otimes C)v(X^T) = v(E^T). \tag{6}$$

In view of (1) it is clear that

$$v(E^T) = Pv(E), \quad v(X) = Qv(X^T) \tag{7}$$

for suitable permutation matrices P and Q which depend only on the dimensions of E and X, respectively. Equation (6), together with that resulting from substitution of (7) into (5), verifies (3).

Problem 13. The starting point in the construction of the *Yang–Baxter equation* is the 2×2 matrix

$$T := \begin{pmatrix} a & b \\ c & d \end{pmatrix}$$

where a, b, c, and d are $n \times n$ linear operators of an algebra over the complex numbers. In other words, T is an operator-valued matrix. Let I be the 2×2 identity matrix. We define the 4×4 matrices

$$T_1 = T \otimes I, \quad T_2 = I \otimes T$$

where $\otimes$ denotes the Kronecker product. Thus T_1 and T_2 are operator-valued 4×4 matrices. Applying the rules for the Kronecker product we find

$$T_1 = T \otimes I = \begin{pmatrix} a & 0 & b & 0 \\ 0 & a & 0 & b \\ c & 0 & d & 0 \\ 0 & c & 0 & d \end{pmatrix}, \quad T_2 = I \otimes T = \begin{pmatrix} a & b & 0 & 0 \\ c & d & 0 & 0 \\ 0 & 0 & a & b \\ 0 & 0 & c & d \end{pmatrix}.$$

The Yang–Baxter equation is given by

$$R_q T_1 T_2 = T_2 T_1 R_q \tag{1}$$

where R_q is a 4×4 matrix and q a nonzero complex number. Let

$$R_q := \begin{pmatrix} 1 & 0 & 0 & 0 \\ 0 & -1 & 0 & 0 \\ 0 & 1+q & q & 0 \\ 0 & 0 & 0 & 1 \end{pmatrix}.$$

Find the condition on a, b, c and d such that condition (1) is satisfied.

Solution.   Altogether we find 16 relations which have to be satisfied. Equation (1) gives rise to the relations of the algebra elements a, b, c and d

$$ab = q^{-1}ba, \quad dc = qcd, \quad bc = -qcb$$

$$bd = -db, \quad ac = -ca, \quad [a,d] = (1 + q^{-1})bc$$

where all the commutative relations have been omitted and $[\,,\,]$ denotes the commutator.

Problem 14.   Calculate the eigenvalues of the Hamilton operator

$$\hat{H} := \lambda(S_x \otimes \hat{L}_x + S_y \otimes \hat{L}_y + S_z \otimes \hat{L}_z) \tag{1}$$

where we .consider a subspace G_1 of the Hilbert space $L_2(S^2)$ with

$$S^2 := \{(x, y, z) : x^2 + y^2 + z^2 = 1\}.$$

Here $\otimes$ denotes the *tensor product*. The linear operators $\hat{L}_x$, $\hat{L}_y$, $\hat{L}_z$ act in the subspace G_1. A basis of subspace G_1 is

$$Y_{1,0} = \sqrt{\frac{3}{4\pi}} \cos\theta \, ,$$

$$Y_{1,1} = \sqrt{\frac{3}{8\pi}} \sin\theta e^{i\phi} \, ,$$

$$Y_{1,-1} = \sqrt{\frac{3}{8\pi}} \sin\theta e^{-i\phi} \, .$$

The operators (matrices) S_x, S_y, S_z act in the Hilbert space $\mathbf{C}^2$ with the basis

$$\begin{pmatrix} 1 \\ 0 \end{pmatrix}, \quad \begin{pmatrix} 0 \\ 1 \end{pmatrix} \, .$$

The matrices S_z, S_+ and S_- are given by

$$S_z := \frac{1}{2}\hbar \begin{pmatrix} 1 & 0 \\ 0 & -1 \end{pmatrix}, \quad S_+ := \hbar \begin{pmatrix} 0 & 1 \\ 0 & 0 \end{pmatrix}, \quad S_- := \hbar \begin{pmatrix} 0 & 0 \\ 1 & 0 \end{pmatrix}$$

where $S_\pm := S_x \pm iS_y$. The operators $\hat{L}_x$, $\hat{L}_y$ and $\hat{L}_z$ take the form

$$\hat{L}_z := -i\hbar \frac{\partial}{\partial\phi} \, ,$$

$$\hat{L}_+ := \hbar e^{i\phi} \left(\frac{\partial}{\partial\theta} + i \cot\theta \frac{\partial}{\partial\phi} \right) \, ,$$

$$\hat{L}_- := \hbar e^{-i\phi} \left(-\frac{\partial}{\partial\theta} + i \cot\theta \frac{\partial}{\partial\phi} \right)$$

where

$$\hat{L}_\pm := \hat{L}_x \pm i\hat{L}_y \, . \tag{2}$$

Give an interpretation of the Hamilton operator $\hat{H}$ and of the subspace G_1.

Remark. *In some textbooks we find the notation* $\hat{H} = \lambda \mathbf{S} \cdot \mathbf{L}$.

Solution. Using (2) the Hamilton operator (1) can be written as

$$\hat{H} = \lambda(S_z \otimes \hat{L}_z) + \frac{\lambda}{2}(S_+ \otimes \hat{L}_- + S_- \otimes \hat{L}_+) \, .$$

In the tensor product space $\mathbf{C}^2 \otimes G_1$, a basis is given by

$$|1\rangle = \begin{pmatrix} 1 \\ 0 \end{pmatrix} \otimes Y_{1,0}, \quad |2\rangle = \begin{pmatrix} 1 \\ 0 \end{pmatrix} \otimes Y_{1,-1}, \quad |3\rangle = \begin{pmatrix} 1 \\ 0 \end{pmatrix} \otimes Y_{1,1},$$

$$|4\rangle = \begin{pmatrix} 0 \\ 1 \end{pmatrix} \otimes Y_{1,0}, \quad |5\rangle = \begin{pmatrix} 0 \\ 1 \end{pmatrix} \otimes Y_{1,-1}, \quad |6\rangle = \begin{pmatrix} 0 \\ 1 \end{pmatrix} \otimes Y_{1,1}.$$

In the following we use

$$\hat{L}_+ Y_{1,1} = 0, \quad \hat{L}_+ Y_{1,0} = \hbar\sqrt{2} Y_{1,1}, \quad \hat{L}_+ Y_{1,-1} = \hbar\sqrt{2} Y_{1,0},$$

$$\hat{L}_- Y_{1,1} = \hbar\sqrt{2} Y_{1,0}, \quad \hat{L}_- Y_{1,0} = \hbar\sqrt{2} Y_{1,-1}, \quad \hat{L}_- Y_{1,-1} = 0,$$

$$\hat{L}_z Y_{1,1} = \hbar Y_{1,1}, \quad \hat{L}_z Y_{1,0} = 0, \quad \hat{L}_z Y_{1,-1} = -\hbar Y_{1,-1}.$$

For the state $|1\rangle$ we find

$$\hat{H}|1\rangle = \left[\lambda(S_z \otimes \hat{L}_z) + \frac{\lambda}{2}(S_+ \otimes \hat{L}_- + S_- \otimes \hat{L}_+) \right] \begin{pmatrix} 1 \\ 0 \end{pmatrix} \otimes Y_{1,0}.$$

Thus

$$\hat{H}|1\rangle$$
$$= \lambda \left[S_z \begin{pmatrix} 1 \\ 0 \end{pmatrix} \otimes \hat{L}_z Y_{1,0} \right] + \frac{\lambda}{2} \left[S_+ \begin{pmatrix} 1 \\ 0 \end{pmatrix} \otimes \hat{L}_- Y_{1,0} + S_- \begin{pmatrix} 1 \\ 0 \end{pmatrix} \otimes \hat{L}_+ Y_{1,0} \right].$$

Finally

$$\hat{H}|1\rangle = \frac{\lambda}{2} S_- \begin{pmatrix} 1 \\ 0 \end{pmatrix} \otimes \hat{L}_+ Y_{1,0} = \frac{\lambda}{\sqrt{2}} \hbar^2 |6\rangle.$$

Analogously, we find

$$\hat{H}|2\rangle = -\frac{\lambda\hbar^2}{2}|2\rangle + \frac{\lambda\hbar^2}{\sqrt{2}}|4\rangle,$$

$$\hat{H}|3\rangle = \frac{\lambda\hbar^2}{2}|3\rangle,$$

$$\hat{H}|4\rangle = \frac{\lambda\hbar^2}{2}|2\rangle,$$

$$\hat{H}|5\rangle = \frac{\lambda\hbar^2}{2}|5\rangle,$$

$$\hat{H}|6\rangle = -\frac{\lambda\hbar^2}{2}|6\rangle + \frac{\lambda\hbar^2}{\sqrt{2}}|1\rangle.$$

Hence the states $|3\rangle$ and $|5\rangle$ are eigenstates with the eigenvalues $E_{1,2} = \lambda\hbar^2/2$. The states $|1\rangle$ and $|6\rangle$ form a two-dimensional subspace. The matrix representation is given by

$$\begin{pmatrix} 0 & \dfrac{\lambda\hbar^2}{\sqrt{2}} \\ \dfrac{\lambda\hbar^2}{\sqrt{2}} & -\dfrac{\lambda\hbar^2}{2} \end{pmatrix}.$$

The eigenvalues are

$$E_{3,4} = -\frac{\lambda\hbar^2}{2} \pm \frac{3\lambda\hbar^2}{4}.$$

Analogously, the states $|2\rangle$ and $|4\rangle$ form a two-dimensional subspace. The matrix representation is given by

$$\begin{pmatrix} -\dfrac{\lambda\hbar^2}{2} & \dfrac{\lambda\hbar^2}{\sqrt{2}} \\ \dfrac{\lambda\hbar^2}{\sqrt{2}} & 0 \end{pmatrix}.$$

The eigenvalues are

$$E_{5,6} = -\frac{\lambda\hbar^2}{2} \pm \frac{3\lambda\hbar^2}{4}.$$

Remark. *The Hamilton operator* (1) *describes the* spin-orbit coupling.

Chapter 3

Nambu Mechanics

Problem 1. In *Nambu mechanics* the equations of motion are given by

$$\frac{du_j}{dt} = \frac{\partial(u_j, I_1, \ldots, I_{n-1})}{\partial(u_1, u_2, \ldots, u_n)}, \quad j = 1, 2, \ldots, n \tag{1}$$

where $\partial(u_j, I_1, \ldots, I_{n-1})/\partial(u_1, u_2, \ldots, u_n)$ denotes the Jacobian and $I_k : \mathbf{R}^n \to \mathbf{R}$ $(k = 1, \ldots, n-1)$ are $n-1$ smooth functions. Show that I_k $(k = 1, \ldots, n-1)$ are first integrals.

Remark. *Equation (1) can also be written as using summation convention*

$$\frac{du_j}{dt} = \varepsilon_{jl_1 \cdots l_{n-1}}(\partial_{l_1} I_1) \cdots (\partial_{l_{n-1}} I_{n-1}), \quad j = 1, 2, \ldots, n \tag{2}$$

where $\varepsilon_{jl_1 \cdots l_{n-1}}$ is the generalized Levi-Civita symbol and $\partial_j \equiv \partial/\partial u_j$.

Solution. The proof that $I_1, \ldots, I_{n-1}$ are first integrals of system (1) (or system (2)) is as follows. We have (summation convention is used)

$$\frac{dI_i}{dt} = \frac{\partial I_i}{\partial u_j}\frac{du_j}{dt} = \varepsilon_{jl_1 \cdots l_{n-1}}(\partial_j l_i)(\partial_{l_1} I_1) \cdots (\partial_{l_{n-1}} I_{n-1})$$

$$= \frac{\partial(I_i, I_1, \ldots, I_{n-1})}{\partial(u_1, \ldots, u_n)}.$$

Thus $dI_i/dt = 0$ as the Jacobian matrix has two equal rows and is therefore singular.

Remark. *If the first integrals are polynomials, then the dynamical system (1) is algebraically completely integrable.*

Problem 2.   Let

$$\frac{du_j}{dt} = \frac{\partial(u_j, I_1, \ldots, I_{n-1})}{\partial(u_1, u_2, \ldots, u_n)}, \quad j = 1, 2, \ldots, n \tag{1}$$

where $\partial(u_j, I_1, \ldots, I_{n-1})/\partial(u_1, u_2, \ldots, u_n)$ denotes the Jacobian and $I_k :$ $\mathbf{R}^n \to \mathbf{R}$ $(k = 1, \ldots, n-1)$ are $n-1$ smooth functions. From problem (1) we know that I_k $(k = 1, \ldots, n-1)$ are first integrals. Assume that the dynamical system admits $n-1$ polynomial first integrals. Do the $n-1$ polynomial first integrals uniquely determine the dynamical system?

Remark.   *If $I_1, \ldots, I_{n-1}$ are first integrals, then $f(I_1, \ldots, I_{n-1})$ is also a first integral, where f is a smooth function.*

Solution.   We restrict ourselves to the case $n = 3$. Then the equations of motion are given by

$$\frac{du_1}{dt} = \frac{\partial I_1}{\partial u_2}\frac{\partial I_2}{\partial u_3} - \frac{\partial I_1}{\partial u_3}\frac{\partial I_2}{\partial u_2}, \tag{2a}$$

$$\frac{du_2}{dt} = \frac{\partial I_1}{\partial u_3}\frac{\partial I_2}{\partial u_1} - \frac{\partial I_1}{\partial u_1}\frac{\partial I_2}{\partial u_3}, \tag{2b}$$

$$\frac{du_3}{dt} = \frac{\partial I_1}{\partial u_1}\frac{\partial I_2}{\partial u_2} - \frac{\partial I_1}{\partial u_2}\frac{\partial I_2}{\partial u_1}. \tag{2c}$$

If I_1 and I_2 are quadratic polynomials, i.e.

$$I_1(\mathbf{u}) = \frac{a_{11}u_1^2}{2} + a_{12}u_1u_2 + \cdots + \frac{a_{33}u_3^2}{2} + a_1u_1 + a_2u_2 + a_3u_3,$$

$$I_2(\mathbf{u}) = \frac{b_{11}u_1^2}{2} + b_{12}u_1u_2 + \cdots + \frac{b_{33}u_3^2}{2} + b_1u_1 + b_2u_2 + b_3u_3,$$

then the equations of motion are given by

$$\frac{du_1}{dt} = (a_{12}b_{13} - a_{13}b_{12})u_1^2 + (a_{12}b_{23} + a_{22}b_{13} - a_{13}b_{22} - a_{23}b_{12})u_1u_2$$

$$+ (a_{12}b_{33} + a_{23}b_{13} - a_{13}b_{23} - a_{33}b_{12})u_1u_3 + (a_{22}b_{23} - a_{23}b_{22})u_2^2$$

$$+ (a_{22}b_{33} + a_{33}b_{22})u_2u_3 + (a_{23}b_{33} - a_{33}b_{23})u_3^2$$

$$+ (a_{12}b_3 + a_2b_{13} - a_{13}b_2 - a_3b_{12})u_1$$

$$+ (a_{22}b_3 + a_2b_{23} - a_{23}b_2 - a_3b_{22})u_2$$

$$+ (a_{23}b_3 + a_2b_{33} - a_{33}b_2 - a_3b_{23})u_3 + a_2b_3 - a_3b_2, \tag{3a}$$

$$\frac{du_2}{dt} = (a_{13}b_{11} - a_{11}b_{13})u_1^2 + (a_{13}b_{12} + a_{23}b_{11} + a_{11}b_{23} - a_{12}b_{13})u_1 u_2$$

$$+ (a_{33}b_{11} - a_{11}b_{33})u_1 u_3 + (a_{23}b_{12} - a_{12}b_{23})u_2^2$$

$$+ (a_{23}b_{13} + a_{33}b_{12} - a_{12}b_{33} - a_{13}b_{23})u_2 u_3$$

$$+ (a_{33}b_{13} - a_{13}b_{33})u_3^2 + (a_{13}b_1 + a_3 b_{11} - a_{11}b_3 - a_1 b_{13})u_1$$

$$+ (a_{23}b_1 + a_3 b_{12} - a_{12}b_3 - a_1 b_{23})u_2$$

$$+ (a_{33}b_1 + a_3 b_{13} - a_{13}b_3 - a_1 b_{33})u_3 + a_3 b_1 - a_1 b_3 , \qquad (3b)$$

$$\frac{du_3}{dt} = (a_{11}b_{12} - a_{12}b_{11})u_1^2 + (a_{11}b_{22} - a_{22}b_{11})u_1 u_2$$

$$+ (a_{11}b_{23} + a_{13}b_{12} - a_{12}b_{13} - a_{23}b_{11})u_1 u_3$$

$$+ (a_{12}b_{22} - a_{22}b_{12})u_2^2 + (a_{12}b_{23} + a_{13}b_{22} - a_{22}b_{13} - a_{23}b_{12})u_2 u_3$$

$$+ (a_{13}b_{23} - a_{23}b_{13})u_3^2 + (a_{11}b_2 + a_1 b_{12} - a_{12}b_1 - a_2 b_{11})u_1$$

$$+ (a_{12}b_2 + a_1 b_{22} - a_{22}b_1 - a_2 b_{12})u_2$$

$$+ (a_{13}b_2 + a_1 b_{23} - a_{23}b_1 - a_2 b_{13})u_3 + a_1 b_2 - a_2 b_1 . \qquad (3c)$$

We consider

$$\frac{du_1}{dt} = u_2 , \qquad (4a)$$

$$\frac{du_2}{dt} = 4\frac{u_3^2}{u_1} , \qquad (4b)$$

$$\frac{du_3}{dt} = -\frac{u_2 u_3}{u_1} . \qquad (4c)$$

On inspection we find that

$$I_1(\mathbf{u}) = \frac{u_2^2}{2} + 2u_3^2 ,$$

$$I_2(\mathbf{u}) = u_1 u_3$$

are first integrals. If I_1 and I_2 are first integrals, then $f(I_1, I_2)$ and $g(I_1, I_2)$ (f and g are smooth functions) are also first integrals, provided that $\partial(f,g)/\partial(I_1, I_2) \neq 0$. Inserting f and g into system (2) yields

$$\frac{du_1}{dt} = u_1 u_2 \left(\frac{\partial f}{\partial I_1} \frac{\partial g}{\partial I_2} - \frac{\partial f}{\partial I_2} \frac{\partial g}{\partial I_1} \right),$$

$$\frac{du_2}{dt} = 4u_3^2 \left(\frac{\partial f}{\partial I_1} \frac{\partial g}{\partial I_2} - \frac{\partial f}{\partial I_2} \frac{\partial g}{\partial I_1} \right),$$

$$\frac{du_3}{dt} = -u_2 u_3 \left(\frac{\partial f}{\partial I_1} \frac{\partial g}{\partial I_2} - \frac{\partial f}{\partial I_2} \frac{\partial g}{\partial I_1} \right).$$

In order to reconstruct system (4) from the first integrals I_1 and I_2 we have to impose the following condition

$$\frac{\partial f}{\partial I_1} \frac{\partial g}{\partial I_2} - \frac{\partial f}{\partial I_2} \frac{\partial g}{\partial I_1} = \frac{1}{u_1}. \tag{5}$$

We see that condition (5) cannot be satisfied. Thus the answer to our question is not affirmative.

Remark.   *If $f(I_1, I_2) = I_1$ and $g(I_1, I_2) = I_2$ then we obtain the system*

$$\frac{du_1}{dt} = u_1 u_2, \tag{6a}$$

$$\frac{du_2}{dt} = 4u_3^2, \tag{6b}$$

$$\frac{du_3}{dt} = -u_2 u_3. \tag{6c}$$

System (6) is a special case of system (3).

Problem 3.   Let

$$\Omega_1 \frac{du_1}{dt} = (\Omega_3 - \Omega_2)u_2 u_3, \tag{1a}$$

$$\Omega_2 \frac{du_2}{dt} = (\Omega_1 - \Omega_3)u_3 u_1, \tag{1b}$$

$$\Omega_3 \frac{du_3}{dt} = (\Omega_2 - \Omega_1)u_1 u_2 \tag{1c}$$

where Ω_1, Ω_2 and Ω_3 are positive constants. This dynamical system describes *Euler's rigid body motion*. Show that this system can be derived within Nambu mechanics.

Solution.   Using system (3) of Problem 2 the equations of motion can be derived within Nambu mechanics from the first integrals

$$I_1(\mathbf{u}) = \frac{1}{2}(\Omega_1 u_1^2 + \Omega_2 u_2^2 + \Omega_3 u_3^2),$$

$$I_2(\mathbf{u}) = \frac{1}{2}\left(\frac{\Omega_1}{\Omega_2\Omega_3}u_1^2 + \frac{\Omega_2}{\Omega_3\Omega_1}u_2^2 + \frac{\Omega_3}{\Omega_1\Omega_2}u_3^2\right).$$

The first integral I_1 represents the total kinetic energy.

Remark.   *The general solution of system (1) can be expressed in terms of Jacobi elliptic functions.*

Problem 4.   The nonrelativistic motion of a charged particle with mass m and charge q in a constant electric field $\mathbf{E} = (E_1, E_2, E_3)$ and constant magnetic field $\mathbf{B} = (B_1, B_2, B_3)$ is given by

$$\frac{dv_1}{dt} = \frac{q}{m}E_1 + \frac{q}{m}(B_3v_2 - B_2v_3),$$

$$\frac{dv_2}{dt} = \frac{q}{m}E_2 + \frac{q}{m}(B_1v_3 - B_3v_1),$$

$$\frac{dv_3}{dt} = \frac{q}{m}E_3 + \frac{q}{m}(B_2v_1 - B_1v_2)$$

where $\mathbf{v} = (v_1, v_2, v_3)$ denotes the velocity of the particle. If $B_1 = B_2 = 0$, $B_3 \neq 0$ and $E_2 = E_3 = 0$, $E_1 \neq 0$, then we obtain

$$\frac{dv_1}{dt} = \frac{q}{m}E_1 + \frac{q}{m}B_3v_2, \tag{1a}$$

$$\frac{dv_2}{dt} = -\frac{q}{m}B_3v_1, \tag{1b}$$

$$\frac{dv_3}{dt} = 0. \tag{1c}$$

Show that system (1) can be derived within Nambu mechanics.

Solution.   Using (3) of Problem 2 we find that the dynamical system (1) can be derived from the first integrals

$$I_1(\mathbf{v}) = -v_3,$$

$$I_2(\mathbf{v}) = \frac{1}{2}B_3^2v_1^2 + \frac{1}{2}B_3^2v_2^2 + E_1B_3v_2.$$

Chapter 4

Gateaux and Fréchet Derivatives

Problem 1.   Let W_1 be some topological vector space and f is a (non-linear) mapping $f : W \to W_1$. We call f *Gateaux differentiable* in $u \in W$ if there exists a mapping $\theta \in L(W, W_1)$ such that for all $v \in W$

$$\lim_{\epsilon \to 0} \frac{1}{\epsilon}(f(u + \epsilon v) - f(u) - \epsilon \theta v) = 0$$

in the topology of W_1. The linear mapping $\theta \in L(W, W_1)$ is the called *Gateaux derivative* of f in u and is written as $\theta = f'(u)$.

(i) Let

$$\frac{du_1}{dt} - \sigma(u_2 - u_1) = 0\,,$$

$$\frac{du_2}{dt} + u_1 u_3 - r u_1 + u_2 = 0\,, \tag{1}$$

$$\frac{du_3}{dt} - u_1 u_2 + b u_3 = 0$$

where σ, b and r are positive constants. The left-hand side defines a map f. Calculate the Gateaux derivative.

Remark.   *Equation* (1) *is called the* Lorenz model. *The equation* $\theta \mathbf{v} = \mathbf{0}$ *is called the corresponding* variational equation (*or* linearized equation).

Calculate the variational equation of the Lorenz model.

(ii) Consider the map

$$f(u) := \frac{\partial u}{\partial t} - 6u\frac{\partial u}{\partial x} - \frac{\partial^3 u}{\partial x^3}.$$
(2)

Calculate the Gateaux derivative. Calculate the variational equation.

Remark.   *The nonlinear partial differential equation $f(u) = 0$, i.e.*

$$\frac{\partial u}{\partial t} = 6u\frac{\partial u}{\partial x} + \frac{\partial^3 u}{\partial x^3}$$
(3)

is called the Korteweg–de Vries equation.

Solution.   (i) Since

$$f(\mathbf{u} + \epsilon\mathbf{v}) - f(\mathbf{u}) = \begin{pmatrix} \frac{\epsilon dv_1}{dt} - \epsilon\sigma(v_2 - v_1) \\ \frac{\epsilon dv_2}{dt} + \epsilon u_1 v_3 + \epsilon u_3 v_1 + \epsilon^2 v_1 v_3 - \epsilon r v_1 + \epsilon v_2 \\ \frac{\epsilon dv_3}{dt} - \epsilon u_1 v_2 - \epsilon u_2 v_1 - \epsilon^2 v_1 v_2 + \epsilon b v_3 \end{pmatrix},$$

we obtain

$$\theta\mathbf{v} = \begin{pmatrix} \frac{dv_1}{dt} - \sigma(v_2 - v_1) \\ \frac{dv_2}{dt} + u_1 v_3 + u_3 v_1 - r v_1 + v_2 \\ \frac{dv_3}{dt} - u_1 v_2 - u_2 v_1 + b v_3 \end{pmatrix}.$$

Then the variational equation $\theta\mathbf{v} = \mathbf{0}$ takes the form

$$\frac{dv_1}{dt} = \sigma(v_2 - v_1),$$

$$\frac{dv_2}{dt} = -u_1 v_3 - u_3 v_1 + r v_1 - v_2,$$
(4)

$$\frac{dv_3}{dt} = u_1 v_2 + u_2 v_1 - b v_3.$$

Remark.   *To solve system (4) we have first to solve system (1).*

(ii) From (2) we obtain

$$f(u + \epsilon v) = \frac{\partial u}{\partial t} + \epsilon\frac{\partial v}{\partial t} - 6u\frac{\partial u}{\partial x} - 6\epsilon v\frac{\partial u}{\partial x} - 6\epsilon u\frac{\partial v}{\partial x} - 6\epsilon^2 v\frac{\partial v}{\partial x} - \frac{\partial^3 u}{\partial x^3} - \epsilon\frac{\partial^3 v}{\partial x^3}.$$

Therefore

$$f(u + \epsilon v) - f(u) = \epsilon \frac{\partial v}{\partial t} - 6\epsilon v \frac{\partial u}{\partial x} - 6\epsilon u \frac{\partial v}{\partial x} - 6\epsilon^2 v \frac{\partial v}{\partial x} - \epsilon \frac{\partial^3 v}{\partial x^3}.$$

It follows that

$$\lim_{\epsilon \to 0} \frac{1}{\epsilon} (f(u + \epsilon v) - f(u)) = \frac{\partial v}{\partial t} - 6 \frac{\partial u}{\partial x} v - 6u \frac{\partial v}{\partial x} - \frac{\partial^3 v}{\partial x^3}.$$

Therefore

$$\theta v \equiv f'(u)v = \frac{\partial v}{\partial t} - 6 \frac{\partial u}{\partial x} v - 6u \frac{\partial v}{\partial x} - \frac{\partial^3 v}{\partial x^3}. \tag{5}$$

From (5) we find the variational equation $\theta v = 0$ as

$$\frac{\partial v}{\partial t} = 6 \frac{\partial u}{\partial x} v + 6u \frac{\partial v}{\partial x} + \frac{\partial^3 v}{\partial x^3}.$$

This is a linear partial differential equation in v. A solution u from (3) has to be inserted. An example is the trivial solution $u(x, t) = 0$. Then we have

$$\frac{\partial v}{\partial t} = \frac{\partial^3 v}{\partial x^3}.$$

Using this equation we can study the stability of the solution $u(x, t) = 0$.

Problem 2. The difference equation

$$u_{t+1} = 4u_t(1 - u_t) \tag{1}$$

is called the *logistic equation*, where $t = 0, 1, 2, \ldots$ and $u_0 \in [0, 1]$. It can be shown that $u_t \in [0, 1]$ for all $t \in \mathbf{N}$. Equation (1) can also be written as the map $f : [0, 1] \to [0, 1]$

$$f(u) = 4u(1 - u). \tag{2}$$

Then u_t is the tth iterate of f. Calculate the Gateaux derivative of the map f. Find the variational equation of (1).

Solution. Since

$$f(u + \epsilon v) - f(u) = 4(u + \epsilon v)(1 - u - \epsilon v) - 4u(1 - u),$$

we obtain

$$f(u + \epsilon v) - f(u) = 4\epsilon v - 8\epsilon uv - 4\epsilon^2 v^2.$$

Therefore

$$\frac{d}{d\epsilon}(f(u + \epsilon v) - f(u))\bigg|_{\epsilon=0} = 4v - 8uv = 4(1 - 2u)v.$$

Thus we find that the Gateaux derivative of the map (2) is given by

$$\theta v = 4v - 8uv = 4(1 - 2u)v.$$

The variational equation of the nonlinear difference equation (1) is then

$$v_{t+1} = 4v_t - 8u_t v_t = 4(1 - 2u_t)v_t.$$

Remark.   *The stability of fixed points and periodic solutions is studied with the help of the variational equation (see Chapter 5).*

Problem 3.   The Gateaux derivative of an operator-valued function $R(u)$ is defined as

$$R'(u)[v]w := \frac{\partial[R(u + \epsilon v)w]}{\partial\epsilon}\bigg|_{\epsilon=0} \tag{1}$$

where ϵ is a real parameter. We say that $R'(u)[v]w$ is the derivative of $R(u)$ evaluated at v and then applied to w, where w, v and u are smooth functions of $x_1, \ldots, x_n$ and t. Let $n = 1$ and

$$R(u) := \frac{\partial}{\partial x} + u + \frac{\partial u}{\partial x}D_x^{-1}$$

where u is a smooth function of x and t and

$$D_x^{-1}f(x) := \int^x f(s)ds.$$

Calculate the Gateaux derivative of $R(u)$.

Solution.   From (1) we find

$$R(u + \epsilon v)w = \frac{\partial w}{\partial x} + (u + \epsilon v)w + \frac{\partial}{\partial x}(u + \epsilon v)D_x^{-1}w$$

or

$$R(u + \epsilon v)w = \frac{\partial w}{\partial x} + uw + \frac{\partial u}{\partial x}D_x^{-1}w + \epsilon vw + \epsilon\frac{\partial v}{\partial x}D_x^{-1}w.$$

Therefore

$$R'(u)[v]w = vw + \frac{\partial v}{\partial x}D_x^{-1}w \equiv \left(v + \frac{\partial v}{\partial x}D_x^{-1}\right)w.$$

Problem 4. Let $f, g : W \to W$ be two maps, where W is a topological vector space ($u \in W$). Assume that the Gateaux derivative of f and g exists, i.e.

$$f'(u)[v] := \frac{\partial f(u + \epsilon v)}{\partial \epsilon}\bigg|_{\epsilon=0}, \quad g'(u)[v] := \frac{\partial g(u + \epsilon v)}{\partial \epsilon}\bigg|_{\epsilon=0}. \tag{1}$$

The *Lie product* (or *commutator*) of f and g is defined by

$$[f, g] := f'(u)[g] - g'(u)[f]. \tag{2}$$

Let

$$f(u) := \frac{\partial u}{\partial t} - u\frac{\partial u}{\partial x} - \frac{\partial^3 u}{\partial x^3}, \quad g(u) := \frac{\partial u}{\partial t} - \frac{\partial^2 u}{\partial x^2}. \tag{3}$$

Calculate $[f, g]$.

Solution. Applying the definition (1) we find

$$f'(u)[g] \equiv \frac{\partial}{\partial \epsilon}\left[\frac{\partial}{\partial t}(u + \epsilon g(u))\right]_{\epsilon=0}$$

$$- \frac{\partial}{\partial \epsilon}\left[(u + \epsilon g(u))\frac{\partial}{\partial x}(u + \epsilon g(u))\right]_{\epsilon=0}$$

$$- \frac{\partial}{\partial \epsilon}\left[\frac{\partial^3}{\partial x^3}(u + \epsilon g(u))\right]_{\epsilon=0}.$$

Consequently,

$$f'(u)[g] = \frac{\partial g(u)}{\partial t} - g(u)\frac{\partial u}{\partial x} - u\frac{\partial g(u)}{\partial x} - \frac{\partial^3 g(u)}{\partial x^3}.$$

For the second term on the right-hand side of (2), we find

$$g'(u)[f] = \frac{\partial}{\partial \epsilon}\left[\frac{\partial}{\partial t}(u + \epsilon f(u))\right]_{\epsilon=0} - \frac{\partial}{\partial \epsilon}\left[\frac{\partial^2}{\partial x^2}(u + \epsilon f(u))\right]_{\epsilon=0}.$$

Consequently

$$g'(u)[f] = \frac{\partial f(u)}{\partial t} - \frac{\partial^2 f(u)}{\partial x^2}.$$

Hence

$$[f, g] = \frac{\partial}{\partial t}(g(u) - f(u)) - g(u)\frac{\partial u}{\partial x} - u\frac{\partial g(u)}{\partial x} - \frac{\partial^3 g(u)}{\partial x^3} + \frac{\partial^2 f(u)}{\partial x^2}.$$

Inserting f and g given by (3) we find

$$[f, g] = \frac{\partial}{\partial t}\left(-\frac{\partial^2 u}{\partial x^2} + u\frac{\partial u}{\partial x} + \frac{\partial^3 u}{\partial x^3}\right) - \left(\frac{\partial u}{\partial t} - \frac{\partial^2 u}{\partial x^2}\right)\frac{\partial u}{\partial x}$$

$$- u\left(\frac{\partial^2 u}{\partial t \partial x} - \frac{\partial^3 u}{\partial x^3}\right) - \frac{\partial^4 u}{\partial t \partial x^3} + \frac{\partial^5 u}{\partial x^5} + \frac{\partial^3 u}{\partial x^2 \partial t}$$

$$- 3\frac{\partial u}{\partial x}\frac{\partial^2 u}{\partial x^2} - u\frac{\partial^3 u}{\partial x^3} - \frac{\partial^5 u}{\partial x^5}.$$

It follows that

$$[f, g] = \frac{\partial u}{\partial t}\frac{\partial u}{\partial x} + u\frac{\partial^2 u}{\partial x \partial t} - \frac{\partial u}{\partial t}\frac{\partial u}{\partial x} + \frac{\partial^2 u}{\partial x^2}\frac{\partial u}{\partial x} - u\frac{\partial^2 u}{\partial t \partial x}$$

$$+ u\frac{\partial^3 u}{\partial x^3} - 3\frac{\partial u}{\partial x}\frac{\partial^2 u}{\partial x^2} - u\frac{\partial^3 u}{\partial x^3}.$$

Finally

$$[f, g] = -2\frac{\partial u}{\partial x}\frac{\partial^2 u}{\partial x^2}.$$

Problem 5. Let

$$\frac{\partial u}{\partial t} = A(u) \tag{1}$$

be an evolution equation, where A is a differential expression, i.e. A depends on u and its partial derivatives with respect to $x_1, \ldots, x_n$ and on the variables $x_1, \ldots, x_n$ and t. A function

$$\sigma\left(x_1, \ldots, x_n, t, u(\mathbf{x}, t), \frac{\partial u}{\partial x_1}, \ldots\right)$$

is called a *symmetry generator* of (1) if

$$\frac{\partial \sigma}{\partial t} = A'(u)[\sigma] \tag{2}$$

where $A'(u)[\sigma]$ is the Gateaux derivative of A, i.e.

$$A'(u)[\sigma] = \frac{\partial}{\partial \epsilon}A(u + \epsilon\sigma)|_{\epsilon=0}.$$

Let $n = 1$ and

$$\frac{\partial u}{\partial t} = 6u\frac{\partial u}{\partial x} + \frac{\partial^3 u}{\partial x^3}. \tag{3}$$

Show that

$$\sigma = 3t\frac{\partial u}{\partial x} + \frac{1}{2} \tag{4}$$

is a symmetry generator.

Solution.   First we calculate the Gateaux derivative of A, i.e.

$$A'(u)[\sigma] = \frac{\partial}{\partial \epsilon}A(u + \epsilon\sigma)|_{\epsilon=0}$$

$$= \frac{\partial}{\partial \epsilon}\left(6(u + \epsilon\sigma)\left(\frac{\partial(u + \epsilon\sigma)}{\partial x}\right) + \frac{\partial^3}{\partial x^3}(u + \epsilon\sigma)\right)\Bigg|_{\epsilon=0}.$$

Therefore

$$A'(u)[\sigma] = 6\frac{\partial u}{\partial x}\sigma + 6u\frac{\partial \sigma}{\partial x} + \frac{\partial^3 \sigma}{\partial x^3}. \tag{5}$$

From (4) we find that the left-hand side of (2) is given by

$$\frac{\partial \sigma}{\partial t} = 3\frac{\partial u}{\partial x} + 3t\frac{\partial^2 u}{\partial x\partial t}. \tag{6}$$

Inserting σ from (4) into (5) gives

$$A'(u)[\sigma] = 6\frac{\partial u}{\partial x}\left(3t\frac{\partial u}{\partial x} + \frac{1}{2}\right) + 18ut\frac{\partial^2 u}{\partial x^2} + 3t\frac{\partial^4 u}{\partial x^4}$$

or

$$A'(u)[\sigma] = 18tu\frac{\partial^2 u}{\partial x^2} + 18t\left(\frac{\partial u}{\partial x}\right)^2 + 3\frac{\partial u}{\partial x} + 3t\frac{\partial^4 u}{\partial x^4}.$$

From (3) we obtain

$$\frac{\partial^2 u}{\partial x\partial t} = 6\left(\frac{\partial u}{\partial x}\right)^2 + 6u\frac{\partial^2 u}{\partial x^2} + \frac{\partial^4 u}{\partial x^4}. \tag{7}$$

Inserting the right-hand side of (7) into (6) leads to

$$\frac{\partial \sigma}{\partial t} = 3\frac{\partial u}{\partial x} + 18t\left(\frac{\partial u}{\partial x}\right)^2 + 18tu\frac{\partial^2 u}{\partial x^2} + 3t\frac{\partial^4 u}{\partial x^4}.$$

This proves (2).

Problem 6.   Let

$$\frac{\partial u}{\partial t} = A(u)$$

be an evolution equation, where A is a nonlinear differential expression, i.e. A depends on u and its partial derivatives with respect to $x_1, \ldots, x_n$ and on the variables $x_1, \ldots, x_n$ and t. An operator $R(u)$ is called a *recursion operator* if

$$R'(u)[A(u)]v = [A'(u)v, R(u)v] \tag{1}$$

where

$$R'(u)[A(u)]v := \frac{\partial}{\partial \epsilon}(R(u + \epsilon A(u))v)|_{\epsilon=0} \tag{2}$$

and the commutator is defined as

$$[A'(u)v, R(u)v] := \frac{\partial}{\partial \epsilon}A'(u)(v + \epsilon R(u)v)|_{\epsilon=0} - \frac{\partial}{\partial \epsilon}R(u)(v + \epsilon A'(u)v)|_{\epsilon=0}. \tag{3}$$

Let $n = 1$ and

$$\frac{\partial u}{\partial t} = \frac{\partial^2 u}{\partial x^2} + 2u\frac{\partial u}{\partial x}, \tag{4}$$

i.e.

$$A(u) \equiv \frac{\partial^2 u}{\partial x^2} + 2u\frac{\partial u}{\partial x}.$$

Show that

$$R(u) = \frac{\partial}{\partial x} + u + \frac{\partial u}{\partial x}D_x^{-1}$$

is a recursion operator of (4), where

$$D_x^{-1}f(x) = \int^x f(s)ds. \tag{5}$$

Solution.   First we calculate the left-hand side of (1). When we apply the operator $R(u)$ to v, we obtain

$$R(u)v = \frac{\partial v}{\partial x} + uv + \frac{\partial u}{\partial x}D_x^{-1}v.$$

From definition (2) we find that the left-hand side is given by

$$R'(u)[A(u)]v = \frac{\partial}{\partial \epsilon} \left(\frac{\partial}{\partial x} + u + \epsilon A(u) + \frac{\partial(u + \epsilon A(u))}{\partial x} D_x^{-1} \right) v$$

$$= A(u)v + \frac{\partial A(u)}{\partial x} D_x^{-1} v$$

$$= \left(\frac{\partial^2 u}{\partial x^2} + 2u \frac{\partial u}{\partial x} \right) v + \left(\frac{\partial^3 u}{\partial x^3} + 2 \left(\frac{\partial u}{\partial x} \right)^2 + 2u \frac{\partial^2 u}{\partial x^2} \right) D_x^{-1} v.$$

$$(6)$$

To calculate the commutator we first have to find the Gateaux derivative of A. Since

$$A'(u)[v] := \left. \frac{\partial A(u + \epsilon v)}{\partial \epsilon} \right|_{\epsilon=0},$$

we find

$$A'(u)[v] = \left. \frac{\partial}{\partial \epsilon} \left(\frac{\partial^2}{\partial x^2}(u + \epsilon v) + 2(u + \epsilon v) \frac{\partial}{\partial x}(u + \epsilon v) \right) \right|_{\epsilon=0}$$

$$= \frac{\partial^2 v}{\partial x^2} + 2u \frac{\partial v}{\partial x} + 2 \frac{\partial u}{\partial x} v.$$

Therefore

$$A'(u)v = \frac{\partial^2 v}{\partial x^2} + 2u \frac{\partial v}{\partial x} + 2 \frac{\partial u}{\partial x} v.$$

For the first term of the commutator (3), we find

$$\frac{\partial}{\partial \epsilon} A'(u)(v + \epsilon R(u)v)|_{\epsilon=0}$$

$$= \frac{\partial}{\partial \epsilon} \left[\frac{\partial^2}{\partial x^2}(v + \epsilon R(u)v) + 2u \frac{\partial}{\partial x}(v + \epsilon R(u)v) \right.$$

$$\left. + 2 \frac{\partial u}{\partial x}(v + \epsilon R(u)v) \right]_{\epsilon=0}$$

$$= \frac{\partial^3 v}{\partial x^3} + 3 \frac{\partial^2 u}{\partial x^2} v + 5 \frac{\partial u}{\partial x} \frac{\partial v}{\partial x} + 3u \frac{\partial^2 v}{\partial x^2} + 6u \frac{\partial u}{\partial x} v + 2u^2 \frac{\partial v}{\partial x}$$

$$+ \left(\frac{\partial^3 u}{\partial x^3} + 2 \left(\frac{\partial u}{\partial x} \right)^2 + 2u \frac{\partial^2 u}{\partial x^2} \right) D_x^{-1} v.$$

$$(7)$$

For the second term of the commutator we obtain

$$\frac{\partial}{\partial\epsilon}R(u)(v+\epsilon A'(u)v)|_{\epsilon=0} = \frac{\partial}{\partial\epsilon}\left[\frac{\partial}{\partial x}(v+\epsilon A'(u)v)\right.$$

$$\left. + u(v+\epsilon A'(u)v) + \frac{\partial u}{\partial x}D_x^{-1}(v+\epsilon A'(u)v)\right]_{\epsilon=0}$$

$$= \frac{\partial A'(u)v}{\partial x} + uA'(u)v + \frac{\partial u}{\partial x}D_x^{-1}A'(u)v .$$

Using the identity $D_x^{-1}(v\partial u/\partial x + u\partial v/\partial x) = uv$, where D_x^{-1} is defined by (5) we find

$$\frac{\partial}{\partial\epsilon}R(u)(v+\epsilon A'(u)v)|_{\epsilon=0}$$

$$= \frac{\partial^3 v}{\partial x^3} + 2\frac{\partial^2 u}{\partial x^2}v + 5\frac{\partial u}{\partial x}\frac{\partial v}{\partial x} + 3u\frac{\partial^2 v}{\partial x^2} + 4u\frac{\partial u}{\partial x}v + 2u^2\frac{\partial v}{\partial x} . \qquad (8)$$

Combining (7) and (8) we find that the commutator $[A'(u)v, R(u)v]$ is given by the right-hand side of (6) which proves (1).

Problem 7.   Let f be a continuously differentiable real function on the interval $I = [a, b]$, i.e.

$$f : C^1[a, b] \to \mathbf{R} .$$

The norm is defined as

$$\|f\| := \max_{a\le t\le b}(|f(t)|, |f'(t)|) \qquad (1)$$

where $f'(t) \equiv df/dt$. Let $\Phi(t, y, z)$ be a given real function for $t \in I$ and y, z arbitrary. Φ is assumed to possess partial derivatives up to second order with respect to all arguments. Find the *Fréchet derivative* of the functional

$$Tf(t) := \int_a^b \Phi(t, f, f')dt .$$

Solution.   Taylor's theorem yields

$$T(f + g) - T(f) = \int_a^b [\Phi(t, f+g, f'+g') - \Phi(t, f, f')]dt$$

$$= \int_a^b [g\Phi_f + g'\Phi_{f'} + g^2\tilde{\Phi}_{ff} + gg'\tilde{\Phi}_{ff'} + g'^2\tilde{\Phi}_{f'f'}]dt$$

where g is continuously differentiable in I and

$$\frac{\partial \Phi}{\partial f} \equiv \Phi_f := \left. \frac{\partial \Phi(t, y, z)}{\partial y} \right|_{y \to f(t), z \to f'(t)} ,$$

$$\frac{\partial \Phi}{\partial f'} \equiv \Phi_{f'} := \left. \frac{\partial \Phi(t, y, z)}{\partial z} \right|_{y \to f(t), z \to f'(t)} .$$

The tilde indicates that the function has to be taken at some intermediate argument. With the norm (1) we have

$$|g|, |g'| \le \|g\| .$$

Therefore the contributions of the second derivatives under the integral sign may be estimated by $\|g\| \epsilon(\|g\|)$ where ϵ is a null function. Thus the Fréchet derivative of $T(f)$ becomes

$$T'_{(f)} g = \int_a^b \left[\frac{\partial \Phi}{\partial f} g(t) + \frac{\partial \Phi}{\partial f'} g'(t) \right] dt .$$

Remark. *If $f(a) = f_a$ and $f(b) = f_b$, then $g(a) = g(b) = 0$. Partial integration in this case yields*

$$\int_a^b \frac{\partial \Phi}{\partial f'} g'(t) dt = - \int_a^b \left(\frac{d}{dt} \frac{\partial \Phi}{\partial f'} \right) g(t) dt .$$

Consequently, the Fréchet derivative of T is given by

$$T'_{(f)} g = \int_a^b \left[\frac{\partial \Phi}{\partial f} - \frac{d}{dt} \frac{\partial \Phi}{\partial f'} \right] g(t) dt .$$

Remark. *In variational calculus we find that an extremum of Tf with the function $f \in C^2[a, b]$ and $f(a) = f_a$, $f(b) = f_b$ is given by*

$$\frac{\partial \Phi}{\partial f} - \frac{d}{dt} \frac{\partial \Phi}{\partial f'} = 0 . \tag{2}$$

Equation (2) is called the Euler–Lagrange equation.

Remark. *If the Fréchet derivative of a mapping exists, then the Gateaux derivative of this mapping also exists.*

Using the Gateaux derivative we start from

$$I(t, f(t), f'(t)) = \int_a^b F(t, f(t), f'(t)) dt .$$

Then

$$I(t, f(t) + \epsilon g(t), f'(t) + \epsilon g'(t)) = \int_a^b F(t, f(t) + \epsilon g(t), f'(t) + \epsilon g'(t))dt \,.$$

We assume that that F admits a Taylor expansion around f and f'. Then

$$\frac{d}{d\epsilon} I(t, f(t) + \epsilon g(t), f'(t) + \epsilon g'(t))\Big|_{\epsilon = 0} = \int_a^b \left(\frac{\partial F}{\partial f} g + \frac{\partial F}{\partial f'} g' \right) dt \,.$$

From the boundary conditions $g(a) = g(b) = 0$ and integration by parts, we obtain

$$\int_a^b \left(\frac{\partial F}{\partial f} - \frac{d}{dt} \frac{\partial F}{\partial f'} \right) g(t)dt = 0 \,.$$

Problem 8.   The Euler–Lagrange equation can be written as

$$\frac{d}{dt} \frac{\partial F(t, f(t), f'(t))}{\partial f'} - \frac{\partial F}{\partial f} = 0 \,. \tag{1}$$

Show that if F does not depend explicitly on t the Euler–Lagrange equation can be written as

$$\frac{d}{dt} \left(f' \frac{\partial F}{\partial f'} - F \right) = 0 \,. \tag{2}$$

Solution.   From (1) we find

$$f'' \frac{\partial^2 F}{\partial f'^2} + f' \frac{\partial^2 F}{\partial f \partial f'} + \frac{\partial^2 F}{\partial f' \partial t} - \frac{\partial F}{\partial f} = 0 \,. \tag{3}$$

Since F does not depend explicitly on t, Eq. (3) simplifies to

$$f'' \frac{\partial^2 F}{\partial f'^2} + f' \frac{\partial^2 F}{\partial f \partial f'} - \frac{\partial F}{\partial f} = 0 \,.$$

On the other hand, from (2) we obtain

$$f' \left(f'' \frac{\partial^2 F}{\partial f'^2} + f' \frac{\partial^2 F}{\partial f \partial f'} - \frac{\partial F}{\partial f} \right) = 0$$

where we assume that $f'(t) \neq 0$. Thus this proves that the Euler–Lagrange equation can be written in the form (2). From (2) it follows that

$$f' \frac{\partial F}{\partial f'} - F = \text{const} \,.$$

Problem 9.   Show that if $M \in C[a, b]$ and

$$\int_a^b M(x)h(x)dx = 0$$

for all $h \in \mathcal{H}$, then it is necessary that

$$M(x) = 0 \quad \text{for all } x \in [a, b]$$

where

$$\mathcal{H} := \{h : h \in C^1[a, b], \ h(a) = h(b) = 0\}.$$

Solution.   We prove it by contradiction. Suppose that there exists an $x_0 \in (a, b)$ such that $M(x_0) \neq 0$. Then we may assume without loss of generality that $M(x_0) > 0$. Since $M \in C[a, b]$ there exists a neighbourhood $N^\delta(x_0) \subset (a, b)$ such that

$$M(x) > 0$$

for all $x \in N^\delta(x_0)$. Let

$$h(x) = \begin{cases} (x - x_0 - \delta)^2(x - x_0 + \delta)^2 & \text{for } x \in N^\delta(x_0), \\ 0 & \text{for } x \notin N^\delta(x_0). \end{cases}$$

Clearly, $h \in \mathcal{H}$. With this choice of h we find

$$\int_a^b M(x)h(x)dx = \int_{x_0-\delta}^{x_0+\delta} M(x)(x - x_0 - \delta)^2(x - x_0 + \delta)^2 dx > 0$$

with contradicts our hypothesis. Thus $M(x) = 0$ for all $x \in (a, b)$. That $M(x)$ also has to vanish at the endpoints of the interval follows from the continuity of M.

Problem 10.   Let $\rho \in C[a, b]$. Consider the function

$$J(u(x)) = \int_a^b \rho(x)\sqrt{1 + u'(x)^2}dx$$

which is defined for all $u \in C^1[a, b]$. Find the Gateaux derivative.

Solution.   From

$$J(u + \epsilon v) = \int_a^b \rho(x)\sqrt{1 + (u + \epsilon v)'(x)^2}dx,$$

it follows that

$$\frac{\partial}{\partial \epsilon} J(u + \epsilon v) = \int_a^b \frac{\rho(x)(u + \epsilon v)'(x)v'(x)}{\sqrt{1 + (u + \epsilon v)'(x)^2}} dx \ .$$

This is justified by the continuity of this last integrand on $[a, b] \times \mathbf{R}$. Setting $\epsilon = 0$ we obtain

$$\left. \frac{\partial}{\partial \epsilon} J(u + \epsilon v) \right|_{\epsilon=0} = \int_a^b \frac{\rho(x)u'(x)v'(x)}{\sqrt{1 + u'(x)^2}} dx \ .$$

Chapter 5

Stability and Bifurcations

Problem 1. The nonlinear difference equation

$$u_{t+1} = 4u_t(1 - u_t)$$

is called the *logistic equation*, where $t = 0, 1, 2, \ldots$ and $u_0 \in [0, 1]$.

(i) Find the fixed points.

(ii) Study the stability of the fixed points.

Solution. (i) Since $u_0 \in [0, 1]$ we find $u_t \in [0, 1]$ for $t = 1, 2, \ldots$. The *fixed points* are given by the solutions of the equations

$$u^* = 4u^*(1 - u^*).$$

Solving this algebraic equation gives the two fixed points

$$u_1^* = 0, \quad u_2^* = \frac{3}{4}.$$

Remark. *The fixed points are time-independent solutions.*

(ii) The variational equation is given by

$$v_{t+1} = (4 - 8u_t)v_t. \tag{1}$$

Inserting the fixed point $u_1^* = 0$ into (1) yields

$$v_{t+1} = 4v_t.$$

For $v_0 \neq 0$ the quantity $|v_t|$ is growing and therefore the fixed point $u_1^* = 0$ is unstable. Inserting the fixed point $u_2^* = 3/4$ into (1) yields

$$v_{t+1} = -2v_t .$$

Consequently, if $v_0 \neq 0$, then $|v_t|$ is growing and therefore the fixed point $u_2^* = 3/4$ is also unstable.

Problem 2.   Consider the logistic equation $(1 \leq r \leq 4)$

$$u_{t+1} = ru_t(1 - u_t), \quad t = 0, 1, 2, \ldots, \quad u_0 \in [0, 1]. \tag{1}$$

Show that $r = 3$ is a bifurcation point.

Solution.   The fixed points (time-independent solutions) are given by the solution of the equation

$$u^* = ru^*(1 - u^*) .$$

We find

$$u_1^* = 0, \quad u_2^* = \frac{r-1}{r} .$$

From

$$f_r(u) = ru(1 - u) ,$$

we find

$$\frac{df_r}{du} = r(1 - 2u) .$$

Thus the variational equation is given by

$$v_{t+1} = r(1 - 2u_t)v_t \tag{2}$$

where u_t is a solution of (1). Inserting the fixed point u_2^* into (2) gives

$$v_{t+1} = (2 - r)v_t .$$

Thus, if $r > 3$, then $|2 - r| > 1$ and the fixed point u_2^* is unstable. If $r < 3$, then $|2 - r| < 1$ and the fixed point u_2^* is stable, i.e. v_t tends to zero as $t \to \infty$. Thus at $r = 3$ we have a qualitative change in the behaviour of the solution of (1).

Problem 3. Let

$$\frac{du_j}{dt} = F_j(\mathbf{u}), \quad j = 1, 2, \ldots, n \tag{1}$$

be an autonomous system of ordinary differential equations of first order, where F_j are smooth functions with $F_j : \mathbf{R}^n \to \mathbf{R}$. Then the variational equation is given by

$$\frac{dv_j}{dt} = \sum_{k=1}^{n} \frac{\partial F_j}{\partial u_k}(\mathbf{u}(t))v_k, \quad j = 1, 2, \ldots, n.$$

A vector $\mathbf{u}^*$ is called a *fixed point* or (*stationary point*) or *steady-state solution* of (1) if

$$F_j(\mathbf{u}^*) = 0, \quad j = 1, 2, \ldots, n.$$

Find the fixed points of the equation

$$\frac{du_1}{dt} = \sigma(u_2 - u_1),$$

$$\frac{du_2}{dt} = -u_1 u_3 + r u_1 - u_2, \tag{2}$$

$$\frac{du_3}{dt} = u_1 u_2 - b u_3$$

where b, σ and r are positive constants. Study the stability of the fixed points.

Remark. *System (2) is the so-called Lorenz model.*

Solution. The variational equation of system (2) can be written in matrix form as

$$\begin{pmatrix} \dfrac{dv_1}{dt} \\ \dfrac{dv_2}{dt} \\ \dfrac{dv_3}{dt} \end{pmatrix} = \begin{pmatrix} -\sigma & \sigma & 0 \\ r - u_3(t) & -1 & -u_1(t) \\ u_2(t) & u_1(t) & -b \end{pmatrix} \begin{pmatrix} v_1 \\ v_2 \\ v_3 \end{pmatrix}. \tag{3}$$

The system of equations

$$\sigma(u_2^* - u_1^*) = 0, \quad -u_1^* u_3^* + r u_1^* - u_2^* = 0, \quad u_1^* u_2^* - b u_3^* = 0$$

determine the fixed points. For all $r > 0$ the system (2) possesses the fixed point

$$u_1^* = u_2^* = u_3^* = 0 \,.$$

Inserting this solution into the variational Eq. (3) leads to a linear system of differential equations with constant coefficients, i.e.

$$\begin{pmatrix} \dfrac{dv_1}{dt} \\ \dfrac{dv_2}{dt} \\ \dfrac{dv_3}{dt} \end{pmatrix} = \begin{pmatrix} -\sigma & \sigma & 0 \\ r & -1 & 0 \\ 0 & 0 & -b \end{pmatrix} \begin{pmatrix} v_1 \\ v_2 \\ v_3 \end{pmatrix} \,. \tag{4}$$

From the matrix of the right-hand side of (4) we obtain the characteristic equation

$$(\lambda + b)(\lambda^2 + (\sigma + 1)\lambda + \sigma(1 - r)) = 0 \,.$$

This equation has three real roots when $r > 0$. All are negative when $r < 1$. One becomes positive when $r > 1$. Consequently, when $r > 1$ the system becomes unstable. For $r > 1$ the system (2) admits two additional fixed points, namely

$$u_1^* = u_2^* = \sqrt{b(r - 1)} \,, \quad u_3^* = r - 1$$

and

$$u_1^* = u_2^* = -\sqrt{b(r - 1)} \,, \quad u_3^* = r - 1 \,.$$

For either of these solutions, the characteristic equation of the matrix in (3) is given by

$$\lambda^3 + (\sigma + b + 1)\lambda^2 + (r + \sigma)b\lambda + 2\sigma b(r - 1) = 0 \,.$$

This equation possesses one real negative root and two complex-conjugate roots when $r > 1$. The complex conjugate roots are purely imaginary if the product of the coefficients of λ^2 and λ equals the constant term, i.e.

$$r = \frac{\sigma(\sigma + b + 3)}{\sigma - b - 1} \,. \tag{5}$$

Thus for r larger then the right-hand side of (5) the two additional fixed points are unstable.

Remark. *The Lorenz model* (2) *shows so-called* chaotic *behaviour for certain parameter values, for example* $b = 8/3$, $\sigma = 10$ *and* $r = 28$.

Problem 4. Consider the autonomous system of ordinary differential equations

$$\frac{du_1}{dt} = \mu u_1 + u_2 - u_1(u_1^2 + u_2^2)\,, \tag{1a}$$

$$\frac{du_2}{dt} = -u_1 + \mu u_2 - u_2(u_1^2 + u_2^2)\,, \tag{1b}$$

$$\frac{du_3}{dt} = u_3(1 - u_1^2 + u_2^2 - u_3^2) \tag{1c}$$

where $\mu \in \mathbf{R}$ is a bifurcation parameter.

(i) Find the fixed points.

(ii) Study the stability of the fixed points.

(iii) Does *Hopf bifurcation* occur for this system?

Solution. (i) The fixed points are determined by the equations

$$0 = \mu u_1^* + u_2^* - u_1^*(u_1^{*2} + u_2^{*2})\,,$$

$$0 = -u_1^* + \mu u_2^* - u_2^*(u_1^{*2} + u_2^{*2})\,,$$

$$0 = u_3^*(1 - u_1^{*2} - u_2^{*2} - u_3^{*2})\,.$$

We find the fixed points

$$\mathbf{e}_0 = (0,0,0)\,, \quad \mathbf{e}_1 = (0,0,1)\,, \quad \mathbf{e}_2 = (0,0,-1)\,.$$

(ii) Let us now study the stability of the fixed points. The linearized equation (variational equation) of Eq. (1) is given by

$$\frac{dv_1}{dt} = (\mu - 3u_1^2 - u_2^2)v_1 + (1 - 2u_1 u_2)v_2\,, \tag{2a}$$

$$\frac{dv_2}{dt} = (-1 - 2u_1 u_2)v_1 + (\mu - u_1^2 - 3u_2^2)v_2\,, \tag{2b}$$

$$\frac{dv_3}{dt} = -2u_1 u_3 v_1 - 2u_2 u_3 v_2 + (1 - u_1^2 - u_2^2 - 3u_3^2)v_3\,. \tag{2c}$$

From the linearized equation we obtain for the fixed point e_0 the matrix

$$A_0 \begin{pmatrix} \mu & 1 & 0 \\ -1 & \mu & 0 \\ 0 & 0 & 1 \end{pmatrix}.$$

The eigenvalues of A_0 are

$$\lambda_1^0 = 1, \quad \lambda_{2,3}^0 = \mu \pm i.$$

From the linearized equation we obtain for the fixed point e_1 the matrix

$$A_1 = \begin{pmatrix} \mu & 1 & 0 \\ -1 & \mu & 0 \\ 0 & 0 & -2 \end{pmatrix}.$$

The eigenvalues of A_1 are

$$\lambda_1^1 = -2, \quad \lambda_{2,3}^1 = \mu \pm i.$$

From the linearized equation we obtain for the fixed point e_2 the matrix

$$A_2 = \begin{pmatrix} \mu & 1 & 0 \\ -1 & \mu & 0 \\ 0 & 0 & -2 \end{pmatrix}.$$

The eigenvalues of A_2 are

$$\lambda_1^2 = -2, \quad \lambda_{2,3}^2 = \mu \pm i.$$

Thus the fixed point e_0 is unstable for all $\mu \in \mathbf{R}$. The fixed points e_1 and e_2 are asymptotically stable for $\mu < 0$ and asymptotically unstable for $\mu > 0$.

(iii) The condition of the *Hopf bifurcation* is satisfied. This means

$$\Re(\lambda_{2,3}(\mu = 0)) = 0,$$

$$\Re\left(\frac{d\lambda_{2,3}}{d\mu}\bigg|_{\mu=0}\right) = 1 \neq 0$$

for all three fixed points, where $\Re$ denotes the real part. Thus in the neighbourhood of the fixed points we find periodic solutions. For e_0 and $\mu > 0$ we obtain

$$v_1^0(\mu, t) = \sqrt{\mu} \sin t, \quad v_2^0(\mu, t) = \sqrt{\mu} \cos t, \quad v_3^0(\mu, t) = 0.$$

For e_1 and $1 \geq \mu > 0$ we find

$$v_1^1(\mu, t) = \sqrt{\mu} \sin t, \quad v_2^1(\mu, t) = \sqrt{\mu} \cos t, \quad v_3^1(\mu, t) = \sqrt{1 - \mu}.$$

For e_2 and $1 \geq \mu > 0$ we obtain

$$v_1^2(\mu, t) = \sqrt{\mu} \sin t, \quad v_2^2(\mu, t) = \sqrt{\mu} \cos t, \quad v_3^2(\mu, t) = -\sqrt{1 - \mu}.$$

The matrix of the variational Eq. (2) takes the form for e_0

$$B_0 = \begin{pmatrix} -2v_1^2 & 1 - 2v_1v_2 & 0 \\ -1 - 2v_1v_2 & -2v_2^2 & 0 \\ 0 & 0 & 1 - \mu \end{pmatrix}$$

where $v_1(\mu, t) = \sqrt{\mu} \sin t$ and $v_2(\mu, t) = \sqrt{\mu} \cos t$. The eigenvalues are

$$1 - \mu, \quad -\mu \pm \sqrt{\mu^2 - 1}.$$

For $e_{1,2}$ and $0 < \mu \leq 1$ we find

$$B_{1,2} = \begin{pmatrix} -2v_1^2 & 1 - 2v_1v_2 & 0 \\ -1 - 2v_1v_2 & -2v_2^2 & 0 \\ \mp 2v_1\sqrt{1 - \mu} & \mp 2v_2\sqrt{1 - \mu} & 2(\mu - 1) \end{pmatrix}.$$

The eigenvalues are

$$2(\mu - 1), \quad -\mu \pm \sqrt{\mu^2 - 1}.$$

The bifurcation phenomena can be understood as follows. Let

$$r^2 := u_1^2 + u_2^2, \quad R^2 := u_1^2 + u_2^2 + u_3^2.$$

Then from (1) it follows that

$$\frac{d}{dt}(r^2) = 2r^2(\mu - r^2), \tag{3a}$$

$$\frac{d}{dt}(R^2) = 2r^2(\mu - r^2) + 2(R^2 - r^2)(1 - R^2). \tag{3b}$$

This system admits for $\mu > 0$ the time-independent solution

$$R^2(\mu, t) = r^2(\mu, t) = \mu. \tag{4}$$

Linearizing system (3) around this solution we obtain the matrix

$$A(\mu) = \begin{pmatrix} -2\mu & 0 \\ -2\mu & 2(1-\mu) \end{pmatrix}.$$

The eigenvalues of $A(\mu)$ are given by

$$\lambda = -2\mu, \quad \lambda = 2(1-\mu).$$

The matrix $A(\mu)$ is singular for $\mu = 0$ and $\mu = 1$. Consequently, the solution (4) is asymptotically stable with respect to (3) for $\mu > 1$ and unstable for $1 > \mu > 0$. Equation (21a) can be integrated. The general solution of (3a) is given by

$$r^2(\mu, t) = \frac{\mu r^2(0) \exp(2\mu t)}{r^2(0)(\exp(2\mu t) - 1) + \mu}$$

where

$$r^2(t = 0) = r^2(0).$$

For $\mu < 0$ we find

$$\lim_{t \to \infty} r^2(\mu, t) = 0.$$

For $\mu > 0$ we find

$$\lim_{t \to \infty} r^2(\mu, t) = \mu.$$

Chapter 6

Nonlinear Ordinary Difference Equations

Problem 1. Let

$$u_{t+1} = 2u_t(1 - u_t) \tag{1}$$

where $t = 0, 1, 2, \ldots$ and $u_0 \in [0, 1]$.

(i) Show that the exact solution of the initial-value problem is given by

$$u_t = \frac{1}{2} - \frac{1}{2}(1 - 2u_0)^{2^t} . \tag{2}$$

(ii) Find the fixed points of the equation. Study the stability of the fixed points. Let

$$u_{t+1} = f(u_t), \quad t = 0, 1, 2, \ldots.$$

Then the fixed points are defined as the solutions of the equation

$$f(u^*) = u^* .$$

Solution. (i) If $u_0 \in [0, 1]$, then $u_t \in [0, 1]$ for $t = 1, 2, \ldots$. From (2) it follows that

$$u_{t+1} = \frac{1}{2} - \frac{1}{2}(1 - 2u_0)^{2^{t+1}} \equiv \frac{1}{2} - \frac{1}{2}(1 - 2u_0)^{2 \cdot 2^t} .$$

On the other hand, we have

$$2u_t(1 - u_t) = 2\left[\frac{1}{2} - \frac{1}{2}(1 - 2u_0)^{2^t}\right]\left[1 - \frac{1}{2} + \frac{1}{2}(1 - 2u_0)^{2^t}\right]$$

$$= \frac{1}{2}[1 - (1 - 2u_0)^{2^t}][1 + (1 - 2u_0)^{2^t}]$$

$$= \frac{1}{2}[1 - (1 - 2u_0)^{2 \cdot 2^t}].$$

This proves that (2) is a solution of the initial-value problem of (1).

(ii) The fixed points are determined by the equation

$$u^* = 2u^*(1 - u^*).$$

It follows that $u_1^* = 0$ and $u_2^* = \frac{1}{2}$ are the fixed points. Let $0 < u_0 < 1$. Then from solution (2) we find that

$$u_t \rightarrow \frac{1}{2} \quad \text{as } t \rightarrow \infty.$$

This means that the fixed point u_2^* is stable and the fixed point u_1^* is unstable.

Problem 2.   Let

$$x_{t+1} = \frac{ax_t + b}{cx_t + d} \tag{1}$$

where $c \neq 0$ and

$$D := \begin{vmatrix} a & b \\ c & d \end{vmatrix} \equiv ad - bc \neq 0$$

with $t = 0, 1, 2, \ldots$.

(i) Let

$$x_t = y_t - \frac{d}{c}. \tag{2}$$

Find the difference equation for y_t.

(ii) Let

$$y_t = \frac{w_{t+1}}{w_t}. \tag{3}$$

Find the difference equation for w_t.

Solution. (i) Inserting the transformation (2) into (1) yields

$$y_{t+1} = \frac{a+d}{c} - \frac{D}{c^2 y_t} .$$ (4)

Equation (4) can be written in the form

$$y_{t+1} y_t - \frac{a+d}{c} y_t + \frac{D}{c^2} = 0 .$$ (5)

(ii) The transformation (3) reduces (5) to the linear difference equation

$$w_{t+2} - \frac{a+d}{c} w_{t+1} + \frac{D}{c^2} w_t = 0 .$$

Problem 3. (i) Let r_k satisfy the nonlinear difference equation

$$r_{k+1}(1 + a r_k) = 1$$ (1)

with $k = 0, 1, 2, \ldots$, $r_0 = 0$ and $a \neq 0$. Show that r_k can be expressed as the *continued fraction*

$$r_k = \cfrac{1}{1 + \cfrac{a}{1 + \cfrac{a}{1 + \cfrac{\cdot}{\cdot + \cfrac{\cdot}{\cdot + \cfrac{a}{1}}}}}}$$ (2)

which terminates at the kth stage. Show that (1) can be linearized by

$$r_k = \frac{n_k}{n_{k+1}} .$$ (3)

(ii) Find the solution of the nonlinear difference equation

$$r_{k+1}(b + r_k) = 1$$ (4)

with $r_0 = 0$ and $b \neq 0$.

Solution. (i) From (1) it follows that $r_1 = 1$ and

$$r_k = \frac{1}{1 + a r_{k-1}} .$$

where $k \geq 1$. Since

$$r_{k-1} = \frac{1}{1 + ar_{k-2}}$$

etc., we obtain (2). By making the substitution (3) the nonlinear difference Eq. (1) reduces to the linear difference equation with constant coefficients

$$n_{k+1} - n_k - an_{k-1} = 0, \quad k = 1, 2, \ldots$$

where $n_0 = 0$ and $n_1 = 1$.

(ii) From (4) we obtain

$$r_k = \frac{1}{b + r_{k-1}}.$$

Since

$$r_{k-1} = \frac{1}{b + r_{k-2}}$$

etc., we find that the solution can be written as a continued fraction. Equation (4) can be linearized using ansatz (3). We find

$$n_{k+2} - bn_{k+1} - n_k = 0$$

where $n_0 = 0$ and $n_1 = 1$.

Remark.   *The infinite continued fraction*

$$\cfrac{1}{b + \cfrac{1}{b + \cfrac{1}{b + \cdots}}}$$

converges to

$$\frac{1}{2}(\sqrt{b^2 + 4} - b)$$

when $b > 0$, and to

$$-\frac{1}{2}(\sqrt{b^2 + 4} + b)$$

when $b < 0$. In particular, we have

$$\sqrt{2} = 1 + \cfrac{1}{2 + \cfrac{1}{2 + \cdots}}.$$

Problem 4.   (i) Find the solution of the nonlinear difference equation

$$y_{t+1} = 2y_t^2 - 1 \tag{1}$$

where $t = 0, 1, 2, \ldots$.

(ii) Show that the solution of the initial-value problem of the nonlinear difference equation

$$x_{t+1} = \frac{4x_t(1 - x_t)(1 - k^2 x_t)}{(1 - k^2 x_t^2)^2} \tag{2}$$

where $x_0 \in [0, 1]$ and $0 \le k^2 \le 1$ is given by

$$x_t = \mathrm{sn}^2(2^t \mathrm{sn}^{-1}(\sqrt{x_0}, k), k) \tag{3}$$

where $t = 0, 1, 2, \ldots$ and sn denotes a *Jacobi elliptic function.*

Solution.   (i) By making the substitution

$$y_t = \cos u_t$$

transform (1) into

$$\cos u_{t+1} = \cos 2u_t \tag{4}$$

where we have used that

$$\cos^2 \alpha \equiv \frac{1}{2}(1 + \cos 2\alpha).$$

Hence the solution of (5) is either

$$u_{t+1} = 2u_t + 2m\pi$$

or

$$u_{t+1} = -2u_t + 2n\pi$$

where m and n are arbitrary integers. From the second alternative, we obtain the solution of (1)

$$y_t = \cos\left(2^t \theta + (-1)^t \frac{2n\pi}{3}\right)$$

where θ is an arbitrary constant and n an arbitrary integer. The solution corresponding to the first alternative is contained in this one.

(ii) Let

$$\alpha := 2^t \text{sn}^{-1}(\sqrt{x_0}, k).$$

Then

$$x_{t+1} = \text{sn}^2(2\alpha, k).$$

Consequently, we need the *addition theorem* for Jacobi elliptic functions. In the following the modulus k is omitted. The addition property of the Jacobi elliptic function sn is given by

$$\text{sn}(u \pm v) \equiv \frac{\text{sn}u \, \text{cn}v \, \text{dn}v \pm \text{cn}u \, \text{sn}v \, \text{dn}u}{1 - k^2 \text{sn}^2 u \, \text{sn}^2 v}. \tag{5}$$

From (5) we obtain

$$\text{sn}(2u) \equiv \frac{2\text{sn}u \, \text{cn}u\text{dn}u}{1 - k^2 \text{sn}^4 u}.$$

The addition property of the Jacobi elliptic function cn is given by

$$\text{cn}(u \pm v) \equiv \frac{\text{cn}u \, \text{cn}v \mp \text{sn}u \, \text{sn}v \, \text{dn}u \, \text{dn}v}{1 - k^2 \text{sn}^2 u \, \text{sn}^2 v}. \tag{6}$$

From (6) we obtain

$$\text{cn}(2u) \equiv \frac{\text{cn}^2 u - \text{sn}^2 u \, \text{dn}^2 u}{1 - k^2 \text{sn}^4 u}.$$

The addition property of the Jacobi elliptic function dn is given by

$$\text{dn}(u \pm v) \equiv \frac{\text{dn}u \, \text{dn}v \mp k^2 \text{sn}u \, \text{sn}v \, \text{cn}u \, \text{cn}v}{1 - k^2 \text{sn}^2 u \, \text{sn}^2 v}. \tag{7}$$

From (7) we obtain

$$\text{dn}(2u) \equiv \frac{\text{dn}^2 u - k^2 \text{sn}^2 u \, \text{cn}^2 u}{1 - k^2 \text{sn}^4 u}.$$

Applying these addition theorems we find that (3) is the solution to (2).

Problem 5.   Consider the logistic equation

$$u_{t+1} = 4u_t(1 - u_t)$$

which also can be written as map $f : [0, 1] \to [0, 1]$

$$f(u) = 4u(1 - u).$$

Show that the solution of the algebraic equation

$$f(f(u^*)) = u^* \tag{1}$$

gives the fixed points of f and a periodic orbit of f.

Solution. The fixed points of f are given by

$$u_1^* = 0, \quad u_2^* = \frac{3}{4}.$$

Since $f(f(u)) = 16u(1 - u)(1 - 4u(1 - u))$, we find that the solutions of (1) are

$$u_1^* = 0, \quad u_2^* = \frac{3}{4}, \quad u_3^* = \frac{5 - \sqrt{5}}{8}, \quad u_4^* = \frac{5 + \sqrt{5}}{8}.$$

Thus u_1^* and u_2^* are the fixed points of f. Let

$$u_0 = \frac{5 - \sqrt{5}}{8}.$$

Then

$$u_1 = \frac{5 + \sqrt{5}}{8}, \quad u_2 = \frac{5 - \sqrt{5}}{8} = u_0.$$

Thus we have a periodic orbit with period 2.

Remark. $f(f(u))$ *is the second iterate of* f. *When we consider higher iterates we find other periodic orbits of* f.

Problem 6. Consider a first-order difference equation

$$x_{t+1} = f(x_t), \quad t = 0, 1, \ldots \tag{1}$$

and a second order difference equation

$$x_{t+2} = g(x_t, x_{t+1}), \quad t = 0, 1, \ldots. \tag{2}$$

If

$$g(x, f(x)) = f(f(x)) \tag{3}$$

then (1) is called an *invariant* of (2).

(i) Show that

$$f(x) = 2x^2 - 1$$

is an invariant of

$$g(x, y) = y - 2x^2 + 2y^2 \,.$$

(ii) The *Fibonacci trace map* is given by

$$x_{t+3} = 2x_{t+2}x_{t+1} - x_t \,. \tag{4}$$

Show that the Fibonacci trace map admits the invariant

$$I(x_t, x_{t+1}, x_{t+2}) = x_t^2 + x_{t+1}^2 + x_{t+2}^2 - 2x_t x_{t+1} x_{t+2} - 1 \,. \tag{5}$$

Solution.   (i) The left-hand side of (3) is given by

$$g(x, f(x)) = 2x^2 - 1 - 2x^2 + 2(2x^2 - 1)^2 = -1 + 2(4x^4 - 4x^2 + 1) \,.$$

The right-hand side of (3) is given by

$$f(f(x)) = 2(2x^2 - 1)^2 - 1 = 2(4x^2 - 4x^2 + 1) - 1 \,.$$

Thus we see that f is an invariant of g.

(ii) From (4) we obtain

$$x_{t+2} = 2x_{t+1}x_t - x_{t-1} \,. \tag{6}$$

Inserting (6) into (5) yields

$$x_t^2 + x_{t+1}^2 + (2x_{t+1}x_t - x_{t-1})^2 - 2x_t x_{t+1}(2x_{t+1}x_t - x_{t-1}) - 1$$
$$= x_t^2 + x_{t+1}^2 - 4x_{t-1}x_t x_{t+1} + x_{t-1}^2 + 2x_{t-1}x_t x_{t+1} - 1 \,.$$

Thus we find the expression

$$x_{t-1}^2 + x_t^2 + x_{t+1}^2 - 2x_{t-1}x_t x_{t+1} - 1 \,.$$

Shifting

$$t \rightarrow t + 1 \,,$$

we obtain

$$x_t^2 + x_{t+1}^2 + x_{t+2}^2 - 2x_t x_{t+1} x_{t+2} - 1$$

which is the invariant (5).

Chapter 7

Nonlinear Ordinary Differential Equations

Problem 1.   The nonlinear system of ordinary differential equations

$$\frac{du_1}{dt} = u_1 - u_1 u_2, \tag{1a}$$

$$\frac{du_2}{dt} = -u_2 + u_1 u_2 \tag{1b}$$

is a so-called *Lotka–Volterra model*, where $u_1 > 0$ and $u_2 > 0$.

(i) Give an interpretation of the system assuming that u_1 and u_2 are describing species.

(ii) Find the fixed points (time-independent solutions).

(iii) Find the variational equation. Study the stability of the fixed points.

(iv) Find the first integral of the system.

(v) Describe why the solution cannot be given explicitly.

Solution.   (i) Since the quantities u_1 and u_2 are positive we find the following behaviour: owing to the first term u_1 on the right-hand side of Eq. (1a) u_1 is growing. The second term $-u_1 u_2$ describes the interaction of species 1 with species 2. Owing to the minus sign u_1 is decreasing. For (1b) we have the opposite behaviour. Owing to $-u_2$ we find that u_2 is decreasing and the interacting part $u_1 u_2$ leads to a growing u_2. Consequently, one expects that the quantities u_1 and u_2 are oscillating.

(ii) The equations which determine the *fixed points* are given by

$$u_1^* - u_1^* u_2^* = 0,$$
$$-u_2^* + u_1^* u_2^* = 0.$$

Since $u_1 > 0$ and $u_2 > 0$ we obtain only one fixed point, namely

$$u_1^* = u_2^* = 1. \tag{2}$$

Remark.   *The fixed points are also called* time-independent solutions *or* steady-state solutions *or* equilibrium solutions.

(iii) The variational equation is given by

$$\frac{dv_1}{dt} = v_1 - u_1 v_2 - u_2 v_1, \tag{3a}$$

$$\frac{dv_2}{dt} = -v_2 + u_1 v_2 + u_2 v_1. \tag{3b}$$

Inserting the fixed point solution (2) into (3) yields

$$\frac{dv_1}{dt} = -v_2, \quad \frac{dv_2}{dt} = v_1 \tag{4a}$$

or

$$\begin{pmatrix} \dfrac{dv_1}{dt} \\ \dfrac{dv_2}{dt} \end{pmatrix} = \begin{pmatrix} 0 & -1 \\ 1 & 0 \end{pmatrix} \begin{pmatrix} v_1 \\ v_2 \end{pmatrix}. \tag{4b}$$

Thus in a sufficiently small neighbourhood of the fixed point $\mathbf{u}^* = (1,1)$ we find a periodic solution, since the eigenvalues of the matrix on the right-hand side of (4b) are given by i, $-i$.

(iv) From

$$\frac{du_1}{u_1(1 - u_2)} = \frac{du_2}{u_2(-1 + u_1)},$$

we obtain

$$\frac{(-1 + u_1)du_1}{u_1} = \frac{(1 - u_2)du_2}{u_2}. \tag{5}$$

Integrating (5) leads to the first integral

$$I(\mathbf{u}) = u_1 u_2 e^{-u_1 - u_2}. \tag{6}$$

(v) To find the explicit solution we have to integrate

$$\frac{du_1}{u_1(1 - u_2)} = \frac{du_2}{u_2(-1 + u_1)} = \frac{dt}{1}.$$

Owing to the first integral (6) the constant of motion is given by

$$u_1 u_2 e^{-u_1 - u_2} = c$$

where c is a positive constant. This equation cannot be solved with respect to u_1 or u_2. Since $c > 0$ we find that the system (1) oscillates around the fixed point $\mathbf{u}^* = (1, 1)$.

Problem 2. Let

$$\frac{du_1}{dt} = 2u_1^3, \tag{1a}$$

$$\frac{du_2}{dt} = -(1 + 6u_1^2)u_2. \tag{1b}$$

Show that this autonomous system of ordinary differential equations has negative divergence and a particular solution with *exploding amplitude*.

Remark. *Let $t_c < \infty$. If $u_j(t) \to \infty$ as $t \to t_c$, then $u_j(t)$ is said to have an exploding amplitude.*

Solution. The divergence of system (1) is obviously

$$\frac{\partial}{\partial u_1}(2u_1^3) - \frac{\partial}{\partial u_2}(1 + 6u_1^2)u_2 = 6u_1^2 - 1 - 6u_1^2 = -1.$$

If $u_2(t) = 0$, then (1b) is satisfied and we obtain

$$\frac{du_1}{dt} = 2u_1^3.$$

This equation admits the particular solution

$$u_1(t) = \frac{1}{2}\frac{1}{\sqrt{1 - t}}$$

with

$$u_1(0) = \frac{1}{2}$$

and $0 \leq t < 1$. If $t \to t_c = 1$, then

$$u_1(t) \to \infty.$$

Problem 3. The linear diffusion equation in one-space dimension is given by

$$\frac{\partial u}{\partial t} = \frac{\partial^2 u}{\partial x^2}. \tag{1}$$

Insert the ansatz

$$u(x,t) = \prod_{j=1}^{n}(x - a_j(t)) \tag{2}$$

into (1) and show that the time dependent functions a_j satisfy the nonlinear autonomous system of ordinary differential equations

$$\frac{da_k}{dt} = -2\sum_{k}^{n}{}' \frac{1}{a_k - a_j}$$

where $\sum'$ means that $j \neq k$.

Hint. Use the identity (see Volume I, Chapter 2, Problem 7)

$$\sum_{\substack{j,k=1 \\ j \neq k}}^{n} \frac{1}{x - a_j} \frac{1}{x - a_k} \equiv 2\sum_{\substack{j,k=1 \\ j \neq k}}^{n} \frac{1}{x - a_k} \frac{1}{a_k - a_j}. \tag{3}$$

Solution. Using the result from Problem 7 in Volume I, Chapter 2, we obtain

$$\frac{\partial u}{\partial t} = -u \sum_{j=1}^{n} \frac{1}{x - a_j} \frac{da_j}{dt} \tag{4}$$

and

$$\frac{\partial u}{\partial x} = u \sum_{j=1}^{n} \frac{1}{x - a_j} \tag{5}$$

where u is given by ansatz (2). From (5) it follows that

$$\frac{\partial^2 u}{\partial x^2} = u \left(\sum_{k=1}^{n} \frac{1}{x - a_k} \right) \left(\sum_{j=1}^{n} \frac{1}{x - a_j} \right) - u \sum_{j=1}^{n} \frac{1}{(x - a_j)^2}.$$

It follows that

$$\frac{\partial^2 u}{\partial x^2} = u \sum_{j \neq k}^{n} \frac{1}{x - a_j} \frac{1}{x - a_k}.$$

Using identity (3) we find

$$\frac{\partial^2 u}{\partial x^2} = 2u \sum_{\substack{j \neq k}}^{n} \frac{1}{x - a_k} \frac{1}{a_k - a_j} \,.$$

(6)

Inserting (4) and (6) into the diffusion Eq. (1) gives

$$\sum_{k=1}^{n} \frac{1}{x - a_k} \frac{da_k}{dt} = -2 \sum_{\substack{j \neq k}}^{n} \frac{1}{x - a_k} \frac{1}{a_k - a_j} \,.$$

Consequently,

$$\frac{da_k}{dt} = -2 \sum_{j}^{n}{}' \frac{1}{a_k - a_j} \,.$$

Problem 4. Let

$$\frac{d\mathbf{u}}{dt} = L\mathbf{u} + (\mathbf{a} \cdot \mathbf{u})\mathbf{u}$$

(1)

be a nonlinear autonomous system of n first-order ordinary differential equations, where L is an $n \times n$ matrix with constant coefficients and

$$\mathbf{a} \cdot \mathbf{u} := a_1 u_1 + a_2 u_2 + \cdots + a_n u_n \,.$$

Find the solution of the initial-value problem, where $\mathbf{u}(0) = \mathbf{u}_0$.

Solution. Method 1. We start from the ansatz

$$\mathbf{u}(\mathbf{u}_0, t) = f(t)\mathbf{v}(\mathbf{u}_0, t) \,.$$

(2)

Differentiating (2) gives

$$\frac{d\mathbf{u}}{dt} = \frac{df}{dt}\mathbf{v} + f\frac{d\mathbf{v}}{dt} \,.$$

Inserting this expression into (1) yields

$$\frac{df}{dt}\mathbf{v} + f\frac{d\mathbf{v}}{dt} = Lf\mathbf{v} + f(\mathbf{a} \cdot \mathbf{v})f\mathbf{v} = f(L\mathbf{v}) + f^2(\mathbf{a} \cdot \mathbf{v})\mathbf{v} \,.$$

Therefore

$$\frac{d\mathbf{v}}{dt} = L\mathbf{v}, \quad \mathbf{v}(0) = \mathbf{u}_0$$

(3)

and

$$\frac{df}{dt} = f^2 \mathbf{a} \cdot \mathbf{v}, \quad f(0) = 1 \,.$$

(4)

Remark.   *The method of the solution of system* (1) *is equivalent to the well known method of* variation of constants.

Equation (4) is a *Riccati equation.*
Solving first the system of linear differential Eq. (3) and then the non-linear differential Eq. (4), we obtain

$$\mathbf{u}(\mathbf{u}_0, t) = \frac{\mathbf{v}(\mathbf{u}_0, t)}{1 - \int_0^t \mathbf{a} \cdot \mathbf{v}(\mathbf{u}_0, \tau) d\tau} .$$

Method 2. The transformation

$$\mathbf{u} = \frac{1}{1 - \mathbf{a} \cdot \mathbf{B}} \frac{d\mathbf{B}}{dt}$$

where

$$\mathbf{a} \cdot \mathbf{B} = a_1 B_1 + a_2 B_2 + \cdots + a_n B_n$$

with

$$\frac{d\mathbf{B}(0)}{dt} = \mathbf{u}_0 , \quad \mathbf{B}(0) = 0$$

is the linearization of system (1) to the system of linear differential equations with constant coefficients

$$\frac{d^2 \mathbf{B}}{dt^2} = L \frac{d\mathbf{B}}{dt} .$$

Problem 5.   The mathematical *pendulum* is described by the nonlinear ordinary differential equation

$$\frac{d^2 \theta}{dt^2} + \omega^2 \sin \theta = 0 \tag{1}$$

where $\omega^2 = g/L$ (L length of the pendulum, g acceleration due to gravity) and θ is the angular displacement of the pendulum from its position of equilibrium. Find the solution of the initial-value problem $\theta(t = 0) = \alpha$ and $d\theta(t = 0)/dt = 0$.

Solution.   The constant of motion of Eq. (1) is given by

$$\frac{1}{2} \left(\frac{d\theta}{dt} \right)^2 - \omega^2 \cos \theta = C \tag{2}$$

where C is a constant of integration. The constant of integration is given by the initial condition. We find

$$C = - \left(\frac{g}{L} \cos \alpha \right).$$

We solve (2) for $d\theta/dt$, and thus obtain

$$\frac{d\theta}{dt} = \sqrt{\frac{2g}{L}} \sqrt{\cos \theta - \cos \alpha}. \tag{3}$$

It follows that

$$dt = \sqrt{\frac{L}{2g}} \frac{d\theta}{\sqrt{\cos \theta - \cos \alpha}}. \tag{4}$$

In order to reduce the right-hand member of this equation to standard form, we introduce

$$\cos \theta =: 1 - 2k^2 \sin^2 \phi, \quad k := \sin \frac{\alpha}{2}$$

and use the following equations

$$\cos \theta - \cos \alpha = 2k^2 \cos^2 \phi, \tag{5a}$$

$$\sin \theta = 2k \sin \phi \sqrt{1 - k^2 \sin^2 \phi}, \tag{5b}$$

$$\sin \theta d\theta = 4k^2 \sin \phi \cos \phi d\phi. \tag{5c}$$

When these equations are substituted into (4), we obtain

$$dt = \sqrt{\frac{L}{g}} \frac{d\phi}{\sqrt{1 - k^2 \sin^2 \phi}}. \tag{6}$$

Therefore the time T required for the pendulum to swing from its position of equilibrium at $\theta = 0$ to a displacement of $\theta = \theta_0$ is given by

$$T = \sqrt{\frac{L}{g}} \int_0^{\phi_0} \frac{d\phi}{\sqrt{1 - k^2 \sin^2 \phi}} \tag{7}$$

where ϕ_0 is obtained from

$$\sin^2 \phi_0 = \frac{1 - \cos \theta_0}{2k^2} = \frac{\sin^2 \frac{1}{2}\theta_0}{k^2}.$$

Thus

$$\phi_0 = \arcsin\left(\frac{\sin\frac{1}{2}\theta_0}{k}\right).$$

In terms of elliptic integrals, we can write (7) as

$$T = \sqrt{\frac{L}{g}}F(\phi_0, k).$$

The period of the simple pendulum is defined to be the time required to make a complete oscillation between positions of maximum displacement. To determine this position of maximum displacement, we combine (5) with (3). Consequently

$$\frac{d\theta}{dt} = 2k\sqrt{\frac{g}{L}}\cos\phi.$$

Since the desired value of θ is that for which $d\theta/dt = 0$, we see that this corresponds to $\phi = \frac{1}{2}\pi$. If we let $P(k)$ be the period of the pendulum, we obtain

$$P(k) = 4\sqrt{\frac{L}{g}}\int_0^{\pi/2}\frac{d\phi}{\sqrt{1 - k^2\sin^2\phi}} = 4\sqrt{\frac{L}{g}}K(k)$$

where $K(k)$ is the *complete elliptic integral* of the first kind. When $k = 0$, this reduces to

$$P = 2\pi\sqrt{\frac{L}{g}}.$$

To find the displacement θ as a function of t, we integrate (6). We find

$$t = \sqrt{\frac{L}{g}}\int_0^\phi\frac{d\phi}{\sqrt{1 - k^2\sin^2\phi}}.$$

This equation can be written as

$$\text{sn}\left(t\sqrt{\frac{g}{L}}, k\right) = \sin\phi = \frac{1}{k}\sin\frac{1}{2}\theta,$$

from which we obtain

$$\theta(t) = 2\arcsin\left(k\,\text{sn}\left(t\sqrt{\frac{g}{L}}, k\right)\right)$$

with

$$k = \sin \frac{\alpha}{2}.$$

Problem 6. Consider the nonlinear autonomous system of ordinary differential equations

$$\frac{du_1}{dt} = u_2 + u_1(1 - u_1^2 - u_2^2), \tag{1a}$$

$$\frac{du_2}{dt} = -u_1 + u_2(1 - u_1^2 - u_2^2). \tag{1b}$$

(i) Find the fixed points.

(ii) Show that system (1) admits a periodic solution of the form

$$u_1(t) = \sin(\omega t), \tag{2a}$$

$$u_2(t) = \cos(\omega t). \tag{2b}$$

Determine ω.

(iii) Find the solution for $t \to \infty$.

Solution. (i) The fixed points are determined by

$$u_2^* + u_1^*(1 - u_1^{*2} - u_2^{*2}) = 0,$$

$$-u_1^* + u_2^*(1 - u_1^{*2} - u_2^{*2}) = 0.$$

We find one fixed point, namely

$$u_1^* = u_2^* = 0.$$

(ii) Inserting the periodic solution (2) into system (1) and applying

$$\sin^2 \omega t + \cos^2 \omega t \equiv 1$$

yields

$$\omega = 1.$$

(iii) Introducing the quantity

$$r^2 := u_1^2 + u_2^2$$

we obtain the equation of motion

$$\frac{d}{dt}r^2 = 2r^2(1 - r^2).$$ (3)

Obviously, we find the time-independent solution $r^2 = 0$ (i.e. $u_1 = u_2 = 0$) and the *limit cycle* $r^2 = 1$ as solutions of the original differential equations (1). Let $r^2 \neq 1$ and $r^2 \neq 0$. For $0 < r^2 < 1$, the solution of the above equation is given by

$$r^2(t, r_0^2) = \frac{1}{1 + (1/r_0^2 - 1)e^{-2t}}$$

with $0 < r_0^2 < 1$ and $-\infty < t < \infty$. For $1 < r^2 < \infty$, the solution of (3) is given by

$$r^2(t, r_0^2) = \frac{1}{1 + (1/r_0^2 - 1)e^{-2t}}$$

with $1 < r_0^2 < \infty$ and

$$\frac{1}{2}\ln\left(1 - \frac{1}{r_0^2}\right) < t < \infty.$$

For both cases as $t \to \infty$, we have $r^2(t \to \infty) = 1$. Consequently, the limit cycle is stable. This can also be seen from applying the *Ljapunov theory*. Let $V : \mathbf{R}^2 \to \mathbf{R}$ with

$$V(\mathbf{u}) = \frac{r^2}{2}.$$

Then

$$\frac{dV}{dt} = r\frac{dr}{dt} = r^2(1 - r^2).$$

Thus $dV/dt > 0$ for $0 < r^2 < 1$ and $dV/dt < 0$ for $r^2 > 1$.

Problem 7.   Let

$$\{w, z\} := \frac{w'''}{w'} - \frac{3}{2}\left(\frac{w''}{w'}\right)^2$$

be the *Schwarzian derivative* of w, where $w' \equiv dw/dz$. Let y_1 and y_2 be two linearly independent solutions of the equation

$$y'' + Q(z)y = 0$$ (1)

which are defined and holomorphic in some simply connected domain D in the complex plane.

(i) Show that

$$w(z) = \frac{y_1(z)}{y_2(z)}$$

satisfies the equation

$$\{w, z\} = 2Q(z) \tag{2}$$

at all points of D where $y_2(z) \neq 0$.

(ii) Show that if $w(z)$ is a solution of (2) and is holomorphic in some neighbourhood of a point $z_0 \in D$, then one can find two linearly independent solutions, $u(z)$ and $v(z)$ of (1) defined in D so that

$$w(z) = \frac{u(z)}{v(z)}.$$

If $v(z_0) = 1$ the solutions u and v are uniquely defined.

Solution.   (i) The expression

$$y_1 y_2' - y_2 y_1'$$

is called the *Wronskian*. We may assume that the Wronskian of y_1 and y_2 is identically one. Then

$$w'(z) = (y_2(z))^{-2}$$

so that

$$\frac{w''(z)}{w'(z)} = -2\frac{y_2'(z)}{y_2(z)}$$

and

$$\left(\frac{w''(z)}{w'(z)}\right)' = -2\frac{y_2''(z)}{y_2(z)} + 2\left(\frac{y_2'(z)}{y_2(z)}\right)^2 = 2Q(z) + \frac{1}{2}\left(\frac{w''(z)}{w'(z)}\right)^2$$

from which the first assertion follows.

(ii) Suppose that a solution of Eq. (2) is given by its initial values

$$w(z_0), \quad w'(z_0), \quad w''(z_0)$$

at a point z_0 in D. We may assume that $w'(z_0) \neq 0$ since otherwise $Q(z)$ could not be holomorphic at $z = z_0$. We can now choose two linearly independent solutions, $u(z)$ and $v(z)$, of (1) so that at $z = z_0$ the quotient

$$w_1(z) = \frac{u(z)}{v(z)}$$

has the initial values

$$w(z_0), \quad w'(z_0), \quad w''(z_0)$$

the same as $w(z)$. If $v(z_0) = 1$, the solutions $u(z)$ and $v(z)$ are uniquely determined, and we must have

$$w(z) \equiv w_1(z).$$

Problem 8.   The nonlinear system of ordinary differential equations

$$A\frac{dp}{dt} + (C - B)qr = Mg(y_0\gamma'' - z_0\gamma'), \tag{1a}$$

$$B\frac{dq}{dt} + (A - C)rp = Mg(z_0\gamma - x_0\gamma''), \tag{1b}$$

$$C\frac{dr}{dt} + (B - A)pq = Mg(x_0\gamma' - y_0\gamma), \tag{1c}$$

$$\frac{d\gamma}{dt} = r\gamma' - q\gamma'', \tag{1d}$$

$$\frac{d\gamma'}{dt} = p\gamma'' - r\gamma, \tag{1e}$$

$$\frac{d\gamma''}{dt} = q\gamma - p\gamma' \tag{1f}$$

describes the motion of a heavy rigid body about a fixed point, where M denotes the mass and A, B and C are the principle moments of inertia.

(i) Show that

$$I_1 = Ap^2 + Bq^2 + Cr^2 - 2Mg(x_0\gamma + y_0\gamma' + z_0\gamma''),$$

$$I_2 = Ap\gamma + Bq\gamma' + Cr\gamma'',$$

$$I_3 = \gamma^2 + \gamma'^2 + \gamma''^2$$

are first integrals of system (1).

(ii) Find the conditions on the constants A, B, C, x_0, y_0 and z_0 under which

$$I_4 = A^2 p^2 + B^2 q^2 + C^2 r^2,$$

$$I_5 = r,$$

$$I_6 = x_0 p + y_0 q + z_0 r,$$

$$I_7 = (p^2 - q^2 + c\gamma)^2 + (2pq + c\gamma')^2$$

are first integrals of system (1), where $c = Mgx_0/C$.

Solution.   (i) We recall that I is a *first integral* if $dI/dt = 0$. Straightforward calculation yields

$$\frac{dI_1}{dt} = 2Ap\frac{dp}{dt} + 2Bq\frac{dq}{dt} + 2Cr\frac{dr}{dt} - 2Mg\left(x_0\frac{d\gamma}{dt} + y_0\frac{d\gamma'}{dt} + z_0\frac{d\gamma''}{dt}\right).$$

Thus

$$\frac{dI_1}{dt} = 2p[(B - C)qr + Mg(y_0\gamma'' - z_0\gamma')] + 2q[(C - A)rp$$

$$+ Mg(z_0\gamma - x_0\gamma'')]$$

$$+ 2r[(A - B)pq + Mg(x_0\gamma' - y_0\gamma)] - 2Mg[x_0(r\gamma' - q\gamma'')$$

$$+ y_0(p\gamma'' - r\gamma) + z_0(q\gamma - p\gamma')].$$

Hence we arrive at

$$\frac{dI_1}{dt} = 0.$$

Analogously, for I_2 and I_3 we find

$$\frac{dI_2}{dt} = 0,$$

$$\frac{dI_3}{dt} = 0.$$

(ii) The condition that I_4 is a first integral of system (1), i.e.,

$$\frac{dI_4}{dt} = 0$$

leads to

$$\frac{dI_4}{dt} = 2Mg[\gamma(z_0 Bq - y_0 Cr) + \gamma'(x_0 Cr - z_0 Ap) + \gamma''(y_0 Ap - x_0 Bq)] = 0.$$

Therefore we obtain the condition

$$x_0 = y_0 = z_0 = 0.$$

From (1c) we find that the condition that $I_5 = r$ is a first integral is given by

$$A = B, \quad x_0 = y_0 = 0.$$

The condition that I_6 is a first integral leads to

$$\frac{dI_6}{dt} = x_0 \frac{dp}{dt} + y_0 \frac{dq}{dt} + z_0 \frac{dr}{dt} = 0.$$

Thus

$$\frac{dI_6}{dt} = \frac{x_0 Mg}{A}(y_0\gamma'' - z_0\gamma') + \frac{y_0 Mg}{B}(z_0\gamma - x_0\gamma'') + \frac{z_0 Mg}{C}(x_0\gamma' - y_0\gamma) = 0.$$

Thus we find

$$A = B = C.$$

Taking the time derivative of I_7 we find terms such as

$$4p^3qr\left(\frac{B-C}{A} - \frac{C-A}{B} + 2\frac{C-A}{B}\right)$$

$$+ 4pq^3r\left(\frac{C-B}{A} + \frac{C-A}{B} + 2\frac{B-C}{A}\right).$$

From this we find the condition

$$A = B = 2C. \tag{2}$$

Furthermore we find the term

$$4\frac{Mg}{A}p^3(y_0\gamma'' - z_0\gamma'). \tag{3}$$

From this we conclude that $z_0 = y_0 = 0$. All other terms vanish if the conditions (2) and (3) are satisfied.

Problem 9.   Consider the eigenvalue equation

$$\frac{d^2\psi}{dx^2} = (V(x) - E)\psi(x) \tag{1}$$

where $V(x) = V(-x)$. Let

$$\Phi(x) := x^{-s}\psi(x)$$

where $s = 0$ or $s = 1$ for even or odd states, respectively. The logarithmic derivative

$$f(x) := -\frac{1}{\Phi(x)}\frac{d\Phi(x)}{dx} \tag{2}$$

is regular at the origin for all eigenstates of (1).

(i) Find the differential equation that f satisfies.

(ii) Assume that

$$V(x) = \sum_{j=1}^{K} v_j x^{2j}, \quad v_K > 0. \tag{3}$$

We expand f in a Taylor series around the origin

$$f(x) = \sum_{j=0}^{\infty} f_j x^{2j+1}. \tag{4}$$

Find the condition on the coefficients f_j.

Solution.   (i) Inserting (2) into (1) yields

$$\frac{df(x)}{dx} - f^2(x) + 2s\frac{f(x)}{x} = E - V(x). \tag{5}$$

This is a *Riccati differential equation*.

(ii) Inserting the Taylor expansion (4) and the expansion (3) into (5) we find that the coefficients f_j satisfy the condition

$$f_j = \frac{1}{2j + 2s + 1}\left(\sum_{i=0}^{j-1} f_i f_{j-i-1} + E\delta_{j0} - \sum_{i=1}^{K} v_i \delta_{ij}\right)$$

where δ_{ij} denotes the Kronecker delta. The function f can be approximated by a sequence of rational functions

$$g(x) = \frac{A(x)}{B(x)}$$

where

$$A(x) = \sum_{j=0}^{M} a_j x^{2j+1}, \quad B(x) = \sum_{j=0}^{N} b_j x^{2j}, \quad b_0 = 1.$$

If g is exactly a Padé approximant, then

$$f(x) - g(x) = O(x^{2(M+N)+3}).$$

Problem 10.   Consider the second-order ordinary differential equation

$$\frac{d^2x}{dt^2} + f(x)\frac{dx}{dt} + g(x) = 0.\tag{1}$$

This equation will have a unique periodic solution provided that

$$\oint F(x)dy = 0\tag{2}$$

where

(1) f is even, g is odd, $xg(x) > 0$ for all $x \neq 0$, and $f(0) < 0$;

(2) f and g are continuous and g is Lipschitzian;

(3) $F(x) \to \pm\infty$ as $x \to \pm\infty$, where

$$F(x) := \int_0^x f(s)ds\,;$$

(4) F has one single positive zero $x = a$, for $x \geq a$ the function F increases monotonically with x;

(5) The variable y is defined by $y := dx/dt + F(x)$.

Give a physical interpretation of these conditions.

Solution.   Equation (1) can be written as

$$\frac{d^2x}{dt^2} = -f(x)\frac{dx}{dt} - g(x)$$

and may be thought of as representing a unit mass acted on by a restoring force $-g(x)$ and a damping force $-f(x)dx/dt$. As the unit mass moves a distance dx, the damper does an amount of work dW given by

$$dW = -f(x)\frac{dx}{dt}dx\,.$$

Since $f(0) < 0$, we find that $dW/dx > 0$ when the unit mass passes through the origin with a positive velocity. Thus at this point the damper is putting energy into the system. As x increases, $F(x)$ reaches a minimum, at which point $f(x) = 0$. After this point has been reached f becomes positive and the damper removes energy from the system. Thus, from physical considerations, we realize the possibility of a limit cycle. When the limit cycle is reached, the sum of all of the elements of work done by the damper in one cycle must be zero. Thus, if the system is in a steady-state oscillation

$$W_{cycle} = \int_{cycle} dW = \int_{cycle} \left(-f(x)\frac{dx}{dt}\right) dx = 0.$$ (3)

If we let $dx/dt = v$, then (3), when integrated by parts, becomes

$$(vF(x))_{cycle} - \int_{cycle} F(x)dv = 0.$$ (4)

The first term of (4) is zero because the integration is over a cycle. Thus, the condition for a steady-state oscillation becomes

$$\int_{cycle} F(x)dv = 0.$$ (5)

Equation (5) can be converted into (2) by introducing the variable y. Thus (5) can be written as

$$\int F(x)(dy - f(x)dx) = \int_{cycle} F(x)dy - \int_{cycle} F(x)f(x)dx.$$ (6)

In (6) the second integral is zero because the integration is over a cycle. Hence (6) becomes

$$\oint F(x)dy = 0$$

which is the usual curvilinear integral taken along a trajectory.

Problem 11.   Discuss the solution of the one-dimensional nonlinear differential equation

$$\frac{du}{dt} = -u^{1/3}$$ (1)

where

$$u(t) \geq 0.$$

Solution.   The differential Eq. (1) has a fixed point at

$$u^* = 0.$$

Let

$$f(u) = -u^{1/3}.$$

Then

$$\frac{df}{du} = -\frac{1}{3}u^{-2/3}.$$

Thus

$$\frac{df}{du} \to -\infty \quad \text{at } u \to 0.$$

Thus the *Lipschitz condition* is violated. The fixed point $u^* = 0$ is an attractor with "infinite" local stability. Let

$$u(t = 0) = u_0 > 0.$$

Then the time t_f to reach the fixed point (attractor) is finite, i.e.

$$t_f = -\int_{u_0}^{0} \frac{du}{u^{1/3}} = \frac{3}{2}u_0^{3/2} < \infty.$$

Remark.   *Consider the differential equation*

$$\frac{du}{dt} = f(t, u(t)).$$

Let (t_0, u_0) be a particular pair of values assigned to the real variable (t, u) such that within a rectangular domain D surrounding the point (t_0, u_0) and defined by the inequalities

$$|t - t_0| \leq a, \quad |u - u_0| \leq b.$$

If (t, u) and (t, U) are two points within D, then the Lipschitz condition is

$$|f(t, U) - f(t, u)| < K|(U - u)|$$

where K is a constant.

Problem 12.   Consider the hypothetical chemical reaction mechanism (*Lotka–Volterra model*)

$$A + X \xrightarrow{k_1} 2X, \tag{1a}$$

$$X + Y \xrightarrow{k_2} 2Y, \tag{1b}$$

$$Y \xrightarrow{k_3} B \tag{1c}$$

where X and Y are intermediaries, k_1, k_2, and k_3 are the reaction rate constants, and the concentrations of the reactants A and B are kept constant. Find the kinetic equations.

Solution. Denoting the concentrations of A, B, X, and Y by the same letters for convenience, the law of mass action then gives for (1a) to (1c)

$$\frac{dA}{dt} = k_1 AX \,,$$

$$\frac{dX}{dt} = -k_1 AX + k_1 X^2 - k_2 XY \,,$$

$$\frac{dY}{dt} = -k_2 XY + k_2 Y^2 - k_3 Y \,,$$

$$\frac{dB}{dt} = k_3 Y \,.$$

Since A and B are kept constant the conditions on A and B mean that the system is open and so there must be an exchange of matter with the surroundings. Under this condition the equations reduce to two equations

$$\frac{dX}{dt} = -k_1 AX + k_1 X^2 - k_2 XY \,,$$

$$\frac{dY}{dt} = -k_2 XY + k_2 Y^2 - k_3 Y$$

where A, k_1, k_2 and k_3 are constants.

Problem 13.   Consider the Hamilton function

$$H(p_x, p_y, x, y) = \frac{1}{2}(p_x^2 + p_y^2 + (x^2 y^2)^{1/\alpha}) \tag{1}$$

where $\alpha \in [0, 1]$. Discuss the equation of motion for $x \gg y$.

Solution.   In the limit $\alpha \to 0$ we obtain the hyperbola billard. Increasing the parameter α means gradual softening of the billard walls and when $\alpha = 1$ we recover the $x^2 y^2$ potential. The symmetry group of this family is C_{4v}.

For

$$x \gg y \,,$$

the motion in the x-direction will be much slower than the motion in the y-direction. Thus x may be regarded as a slowly varying parameter. In the *adiabatic approximation*, the motion in the y-direction is described by the Hamilton function

$$H_y = \frac{1}{2}(p_y^2 + (x^2 y^2)^{1/\alpha}) \,.$$

By the *adiabatic theorem* the action integral in the y-direction is given by

$$J_y = \oint p_y dy = \frac{(2H_y)^{(1+\alpha)/2} f(2/\alpha)}{x}$$

where

$$f(s) := \frac{1}{2\pi} \oint \sqrt{1 - |z|^s} dz \,.$$

The quantity J_y is approximately a constant of motion. The full Hamilton function may now be written as

$$H \approx \frac{1}{2} p_x^2 + H_y(x)$$

giving (approximately) the x motion for any J_y. If we transform the (x, p_x) pair to action-angle variables we obtain the following expression for H

$$H \approx \frac{1}{2} \left[\frac{J_x J_y}{f(2/\alpha) f(2/(\alpha + 1))} \right]^{2/(\alpha+2)} \,.$$

Although this expression was derived under the asymmetric assumption $x \gg y$, the Hamilton function (1) is symmetric in x and y. These adiabatic expressions are not valid in the central region.

Problem 14.   (i) Let $f : \mathbf{R} \to \mathbf{R}$ be an analytic function. Solve the functional equation

$$f(x + y) = f(x) + f(y) \,. \tag{1}$$

(ii) Let $f : \mathbf{R} \to \mathbf{R}$ be an analytic function. Solve the functional equation

$$f(x + y) = f(x) f(y) \,. \tag{2}$$

(iii) Let $f : \mathbf{R} \to \mathbf{R}$ be an analytic function. Solve the functional equation

$$f(x + y) + f(x - y) = 2f(x) f(y) \,. \tag{3}$$

Solution.   (i) If we set $y = 0$ in (1) we obtain

$$f(0) = 0 \,. \tag{4}$$

If we differentiate (1) with respect to x we obtain

$$f'(x + y) = f'(x) \,.$$

Thus

$$f'(x) = c, \quad f(x) = cx + b$$

where c and b are constants. Owing to (4) we obtain $b = 0$ and thus

$$f(x) = cx.$$

(ii) Inserting $y = 0$ into (2) yields

$$f(x) = 0 \quad \text{or} \quad f(0) = 1.$$

We find the f with the condition $f(0) = 1$. If we differentiate (2) with respect to x we obtain

$$f'(x + y) = f'(x)f(y).$$

Setting $x = 0$ yields

$$\frac{f'(y)}{f(y)} = f'(0) = c.$$

Thus

$$f(x) = \exp(cx).$$

(iii) Setting $y = 0$ we find

$$f(x) = f(x)f(0) \tag{5}$$

we see that (3) admits the trivial solution. For a non-trival solution of (3) we find, from (5), that

$$f(0) = 1.$$

Setting $x = 0$ we obtain

$$f(-x) = f(x).$$

Taking the second derivative of (3) with respect to x we obtain

$$f''(x + y) + f''(x - y) = 2f''(x)f(y).$$

Taking the second derivative of (3) with respect to y we obtain

$$f''(x + y) + f''(x - y) = 2f(x)f''(y).$$

Thus

$$f''(x)f(y) = f(x)f''(y).$$

It follows that

$$f''(x) = kf(x).$$

For $k = 0$ we obtain $f(x) = 1$, for $k > 0$ we obtain

$$f(x) = \cosh(\sqrt{k}x).$$

For $k < 0$ we obtain

$$f(x) = \cos(\sqrt{-k}x).$$

Problem 15.   Study the singularity structure of the nonlinear ordinary differential equations

$$\frac{d^2x}{dt^2} + \frac{1}{2}x^3 = 0, \tag{1}$$

$$\frac{d^2x}{dt^2} + \lambda\frac{dx}{dt} + \frac{1}{2}x^3 = \epsilon f(t) \tag{2}$$

where $x(t)$ is a real-valued function and $f(t)$ is an analytic function of t.

Solution.   We consider the differential equations in the complex domain. Consider first Eq. (1), i.e.,

$$\frac{d^2w}{dz^2} + \frac{1}{2}w^3 = 0. \tag{3}$$

Inserting the ansatz (locally represented as a Laurent expansion)

$$w(z) = \sum_{j=0}^{\infty} a_j(z - z_0)^{j-1} \tag{4}$$

we find the recursion relation for the expansion coefficients a_j,

$$a_j(j+1)(j-4) = -\frac{1}{2}\sum_k\sum_l a_{j-k-l}a_ka_l, \quad 0 < k+l \le j, \quad 0 \le k,l < j$$

where

$$a_0 = 2i, \quad a_1 = a_2 = a_3 = 0, \quad a_4 = \text{arbitrary}.$$

Owing to the arbitrary pole position z_0 and coefficient a_4 we have a local representation of the general solution to the second-order differential

Eq. (3). Equation (1) can be solved exactly in terms of the Jacobi elliptic functions and the movable singularities in the complex z-plane form a
regular lattice of first-order poles.

Consider now Eq. (2) in the complex plane, i.e.,

$$\frac{d^2w}{dz^2} + \lambda\frac{dw}{dz} + \frac{1}{2}w^3 = \epsilon f(z). \tag{5}$$

The introduction of either the damping term or driving term leads to a
breakdown of the Laurent series (4), since it is not possible to introduce an
arbitrary coefficient at $j = 4$. We have to add logarithmic terms

$$w(z) = \sum_{j=0}^{\infty}\sum_{k=0}^{\infty} a_{jk}(z - z_0)^{j-1}((z - z_0)^4 \ln(z - z_0))^k.$$

Computation of the recursion relation for the a_{jk} yields

$$a_{jk}((j-1)(j-2) + 4k(2j + 4k - 3)) + a_{j-4,k+1}(k+1)(2j + 8k - 3)$$

$$+ a_{j-8,k+2}(k+1)(k+2) + \lambda a_{j-1,k}(j + 4k - 2) + \lambda a_{j-5,k+1}$$

$$= -\frac{1}{2}\sum_{p,q,r,s} a_{j-r,k-s}a_{r-p,s-q}a_{pq} + \epsilon f_{j-3}\delta_{k0}$$

where the summation is for $0 \le p \le r \le j$ and $0 \le q \le s \le k$ and

$$f_j = \frac{1}{j!}\frac{\partial^j f}{\partial t^j}.$$

The values of the first few coefficients are

$$a_{00} = 2i, \quad a_{10} = -\frac{\lambda}{3}, \quad a_{20} = -\frac{i\lambda^2}{18}, \quad a_{30} = -\frac{i\lambda^3}{27} - \frac{\epsilon f_0}{4}.$$

The coefficient a_{40} is arbitrary and therefore a_{01} is given by

$$a_{01} = \frac{4}{135}i\lambda^4 + \frac{1}{5}\epsilon(\lambda f_0 + f_1).$$

The sets of coefficients a_{0k}, $k = 0, 1, 2, \ldots$, satisfy

$$4k(k-1)a_{0k} + ka_{0k} + \frac{1}{2}a_{0k} = -\frac{1}{8}\sum_s\sum_q a_{0,k-s}a_{0,s-q}a_{0q}$$

where the summation is for $0 \leq q \leq s \leq k$. Introducing the generating function

$$\Theta(s) = \sum_{k=0}^{\infty} a_{0k} s^k$$

where s is some independent variable, the following differential equation for Θ is obtained

$$16s^2 \frac{d^2\Theta}{ds^2} + 4s \frac{d\Theta}{ds} + 2\Theta + \frac{1}{2}\Theta^3 = 0. \tag{6}$$

The differential equation can also be obtained by substituting

$$w(z) = \frac{1}{z - z_0} \Theta_0(s)$$

where $s = (z - z_0)^4 \ln(z - z_0)$ into (5). In the limit $z \to z_0$ we find that Θ_0 satisfies (6) provided that there is an ordering in which $|z - z_0| \ll |s|$. Equation (6) can be solved in terms of elliptic functions by making the substitution

$$\Theta_0(s) = s^{1/4} g(s^{1/4})$$

which leads to the equation

$$\frac{d^2 g(y)}{dy^2} + \frac{1}{2} g^3(y) = 0$$

where $y = s^{1/4} = z(\ln z)^{1/4}$.

Chapter 8

Groups

Problem 1. Let

$$A = \begin{pmatrix} 2 & 1 \\ 1 & 2 \end{pmatrix}, \quad I = \begin{pmatrix} 1 & 0 \\ 0 & 1 \end{pmatrix}, \quad C = \begin{pmatrix} 0 & 1 \\ 1 & 0 \end{pmatrix}.$$

(i) Show that

$$[A, I] = 0, \quad [A, C] = 0.$$

(ii) Show that I and C form a group under the matrix multiplication. Find the classes.

(iii) Find the irreducible representations of I and C. Give the *character table*.

(iv) Find the projection matrices from the character table. Use the projection matrices to find the invariant subspaces. Give the eigenvalues and eigenvectors of A.

Solution. (i)

$$[A, I] = AI - IA = A - A = 0,$$

$$[A, C] = AC - CA = \begin{pmatrix} 1 & 2 \\ 2 & 1 \end{pmatrix} - \begin{pmatrix} 1 & 2 \\ 2 & 1 \end{pmatrix} = 0.$$

(ii) We find

$$II = I, \quad IC = C, \quad CI = C, \quad CC = I.$$

Obviously the group properties are satisfied with $C^{-1} = C$. Since

$$III = I, \quad CIC = I, \quad ICI = C, \quad CCC = C,$$

we find the classes $\{I\}$ and $\{C\}$.

(iii) Obviously, $I \to 1$, $C \to 1$ and $I \to 1 C \to -1$ are representations. Since the representations are one-dimensional they are irreducible. The number of irreducible representations is equal to the number of classes. Taking into account the result from (ii) we find that the character table is given by

	$\{I\}$	$\{C\}$
A_1	1	1
A_2	1	-1.

(iv) For the representation A_1, we find the projection matrix

$$\Pi_1 = \frac{1}{2} \sum_{g \in G} \chi_1(g^{-1})g = \frac{1}{2} \sum_{g \in G} \chi_1(g)g = \frac{1}{2}(I + C) = \frac{1}{2} \begin{pmatrix} 1 & 1 \\ 1 & 1 \end{pmatrix}.$$

Analogously

$$\Pi_2 = \frac{1}{2}(I - C) = \frac{1}{2} \begin{pmatrix} 1 & -1 \\ -1 & 1 \end{pmatrix}.$$

Applying the projection matrices to the standard basis in $\mathbf{R}^2$ gives

$$\Pi_1 \begin{pmatrix} 1 \\ 0 \end{pmatrix} = \frac{1}{2} \begin{pmatrix} 1 \\ 1 \end{pmatrix}, \quad \Pi_1 \begin{pmatrix} 0 \\ 1 \end{pmatrix} = \frac{1}{2} \begin{pmatrix} 1 \\ 1 \end{pmatrix},$$

$$\Pi_2 \begin{pmatrix} 1 \\ 0 \end{pmatrix} = \frac{1}{2} \begin{pmatrix} 1 \\ -1 \end{pmatrix}, \quad \Pi_2 \begin{pmatrix} 0 \\ 1 \end{pmatrix} = \frac{1}{2} \begin{pmatrix} -1 \\ 1 \end{pmatrix}.$$

When we normalize the right hand side we obtain the new basis

$$\frac{1}{\sqrt{2}} \begin{pmatrix} 1 \\ 1 \end{pmatrix}, \quad \frac{1}{\sqrt{2}} \begin{pmatrix} 1 \\ -1 \end{pmatrix}. \tag{1}$$

In this new basis the matrix A takes the form

$$\tilde{A} = \begin{pmatrix} 3 & 0 \\ 0 & 1 \end{pmatrix}.$$

It follows that the eigenvalues of the matrix A are given by 3 and 1 and the normalized eigenvectors of A are given by (1).

Problem 2.   (i) Let

$$
U_1 = \begin{pmatrix} 0 & 0 & 0 & 1 \\ 0 & 0 & 1 & 0 \\ 0 & 1 & 0 & 0 \\ 1 & 0 & 0 & 0 \end{pmatrix}, \quad
U_2 = \begin{pmatrix} 0 & 1 & 0 & 0 \\ 1 & 0 & 0 & 0 \\ 0 & 0 & 0 & 1 \\ 0 & 0 & 1 & 0 \end{pmatrix},
$$

$$
U_3 = \begin{pmatrix} 0 & 0 & 1 & 0 \\ 0 & 0 & 0 & 1 \\ 1 & 0 & 0 & 0 \\ 0 & 1 & 0 & 0 \end{pmatrix}.
$$

Show that the set

$$\{I, U_1, U_2, U_3\}$$

forms a group under matrix multiplication. Find the classes. Find the irreducible representation. Give the character table.

(ii) Let

$$
A_1 = \begin{pmatrix} 0 & 1 & 1 & 0 \\ 1 & 0 & 0 & 1 \\ 1 & 0 & 0 & 1 \\ 0 & 1 & 1 & 0 \end{pmatrix}, \quad
A_2 = \begin{pmatrix} 0 & -1 & -1 & 0 \\ -1 & 0 & 0 & -1 \\ -1 & 0 & 0 & -1 \\ 0 & -1 & -1 & 0 \end{pmatrix}
$$

$$
A_3 = \begin{pmatrix} 0 & -1 & 1 & 0 \\ -1 & 0 & 0 & 1 \\ 1 & 0 & 0 & -1 \\ 0 & 1 & -1 & 0 \end{pmatrix}, \quad
A_4 = \begin{pmatrix} 0 & 1 & -1 & 0 \\ 1 & 0 & 0 & -1 \\ -1 & 0 & 0 & 1 \\ 0 & -1 & 1 & 0 \end{pmatrix}.
$$

Show that

$$U_j A_k U_j = A_k \tag{1}$$

for $j = 1, 2, 3$ and $k = 1, 2, 3, 4$. Find the eigenvalues and eigenvectors of A_k.

Solution.   (i) We find

$$IU_1 = U_1, \quad IU_2 = U_2, \quad IU_3 = U_3, \quad U_1 U_2 = U_3$$

$$U_2 U_1 = U_3, \quad U_1 U_3 = U_2, \quad U_3 U_1 = U_2, \quad U_2 U_3 = U_1$$

$$U_3 U_2 = U_1, \quad U_1 U_1 = I, \quad U_2 U_2 = I, \quad U_3 U_3 = I.$$

Thus the group is commutative. Since

$$III = I, \quad U_1IU_1 = I, \quad U_2IU_2 = I, \quad U_3IU_3 = I,$$

$$IU_1I = U_1, \quad U_1U_1U_1 = U_1, \quad U_2U_1U_2 = U_1, \quad U_3U_1U_3 = U_1,$$

$$IU_2I = U_2, \quad U_1U_2U_1 = U_2, \quad U_2U_2U_2 = U_2, \quad U_3U_2U_3 = U_2,$$

$$IU_3I = U_3, \quad U_1U_3U_1 = U_3, \quad U_2U_3U_2 = U_3, \quad U_3U_3U_3 = U_3, \quad (2)$$

we obtain four classes, namely

$$\{\{I\}, \{U_1\}, \{U_2\}, \{U_3\}\}\,.$$

Thus we have four irreducible representations. We find

$$\Gamma_1 : I \mapsto 1, \quad U_1 \mapsto 1, \quad U_2 \mapsto 1, \quad U_3 \mapsto 1,$$

$$\Gamma_2 : I \mapsto 1, \quad U_1 \mapsto -1 \quad U_2 \mapsto -1, \quad U_3 \mapsto 1,$$

$$\Gamma_3 : I \mapsto 1, \quad U_1 \mapsto -1, \quad U_2 \mapsto 1 \quad U_3 \mapsto -1,$$

$$\Gamma_4 : I \mapsto 1, \quad U_1 \mapsto 1 \quad U_2 \mapsto -1 \quad U_3 \mapsto -1\,.$$

Consequently, the character table is given by

	$\{I\}$	$\{U_1\}$	$\{U_2\}$	$\{U_3\}$
Γ_1	1	1	1	1
Γ_2	1	-1	-1	1
Γ_3	1	-1	1	-1
Γ_4	1	1	-1	-1.

(ii) From the representation Γ_1 we find the projection matrix

$$\Pi_1 = \frac{1}{4} \sum_{g \in G} \chi_1(g^{-1})g = \frac{1}{4}(I + U_1 + U_2 + U_3) = \frac{1}{4}\begin{pmatrix} 1 & 1 & 1 & 1 \\ 1 & 1 & 1 & 1 \\ 1 & 1 & 1 & 1 \\ 1 & 1 & 1 & 1 \end{pmatrix}$$

where we have used that $g^{-1} = g$ for all $g \in G$. Analogously

$$\Pi_2 = \frac{1}{4}(I - U_1 - U_2 + U_3)\,,$$

$$\Pi_3 = \frac{1}{4}(I - U_1 + U_2 - U_3)\,,$$

$$\Pi_4 = \frac{1}{4}(I + U_1 - U_2 - U_3)\,.$$

Applying the projection matrix Π_1 to the standard basis in $\mathbf{R}^4$ gives

$$\Pi_1 \begin{pmatrix} 1 \\ 0 \\ 0 \\ 0 \end{pmatrix} = \frac{1}{4} \begin{pmatrix} 1 \\ 1 \\ 1 \\ 1 \end{pmatrix}, \quad \Pi_1 \begin{pmatrix} 0 \\ 1 \\ 0 \\ 0 \end{pmatrix} = \frac{1}{4} \begin{pmatrix} 1 \\ 1 \\ 1 \\ 1 \end{pmatrix},$$

$$\Pi_1 \begin{pmatrix} 0 \\ 0 \\ 1 \\ 0 \end{pmatrix} = \frac{1}{4} \begin{pmatrix} 1 \\ 1 \\ 1 \\ 1 \end{pmatrix}, \quad \Pi_1 \begin{pmatrix} 0 \\ 0 \\ 0 \\ 1 \end{pmatrix} = \frac{1}{4} \begin{pmatrix} 1 \\ 1 \\ 1 \\ 1 \end{pmatrix}.$$

Analogously, we apply the other projection operators to the standard basis. Then the new orthonormal basis is given by

$$\frac{1}{2} \begin{pmatrix} 1 \\ 1 \\ 1 \\ 1 \end{pmatrix}, \quad \frac{1}{2} \begin{pmatrix} 1 \\ -1 \\ 1 \\ -1 \end{pmatrix}, \quad \frac{1}{2} \begin{pmatrix} 1 \\ 1 \\ -1 \\ -1 \end{pmatrix}, \quad \frac{1}{2} \begin{pmatrix} 1 \\ -1 \\ -1 \\ 1 \end{pmatrix}. \tag{3}$$

In this new basis the matrices A_j take the form

$$\begin{pmatrix} 2 & 0 & 0 & 0 \\ 0 & 0 & 0 & 0 \\ 0 & 0 & 0 & 0 \\ 0 & 0 & 0 & -2 \end{pmatrix}, \quad \begin{pmatrix} -2 & 0 & 0 & 0 \\ 0 & 0 & 0 & 0 \\ 0 & 0 & 0 & 0 \\ 0 & 0 & 0 & 2 \end{pmatrix},$$

$$\begin{pmatrix} 0 & 0 & 0 & 0 \\ 0 & 2 & 0 & 0 \\ 0 & 0 & -2 & 0 \\ 0 & 0 & 0 & 0 \end{pmatrix}, \quad \begin{pmatrix} 0 & 0 & 0 & 0 \\ 0 & -2 & 0 & 0 \\ 0 & 0 & 2 & 0 \\ 0 & 0 & 0 & 0 \end{pmatrix}.$$

Consequently, the eigenvalues of the matrices A_k are given by 2, -2, 0, 0 and the normalized eigenvectors are given by (3). Equation (1) follows from $A_1 = U_2 + U_3$, $A_2 = -A_1$, $A_3 = -U_2 + U_3$, $A_4 = -A_3$ and Eq. (2).

Problem 3. Show that the matrices

$$A(\phi, \theta, \psi) = \begin{pmatrix} e^{i(\phi+\psi)/2} \cos \dfrac{\theta}{2} & i e^{i(\phi-\psi)/2} \sin \dfrac{\theta}{2} \\ i e^{i(\psi-\phi)/2} \sin \dfrac{\theta}{2} & e^{-i(\phi+\psi)/2} \cos \dfrac{\theta}{2} \end{pmatrix}$$

form a group under matrix multiplication, where ϕ, θ, ψ are the *Euler angles*. The Euler angles are coordinates on the sphere S^3.

Solution.   Since

$$A_1(\phi_1, \theta_1, \psi_1) A_2(\phi_2, \theta_2, \psi_2) = A(\phi, \theta, \psi)$$

with

$$\cos\theta = \cos\theta_1 \cos\theta_2 - \sin\theta_1 \sin\theta_2 \cos(\phi_2 - \psi_1),$$

$$e^{i\phi} = \frac{e^{i\phi_1}}{\sin\theta}(\sin\theta_1 \cos\theta_2 + \cos\theta_1 \sin\theta_2 \cos(\phi_2 + \psi_1)$$

$$+ i\sin\theta_2 \sin(\phi_2 + \psi_1)),$$

$$e^{i(\phi/\psi)/2} = \frac{e^{i(\phi_1+\psi_1)/2}}{\cos 1/2\theta}\left(\cos\frac{1}{2}\theta_1 \cos\frac{1}{2}\theta_2 e^{i(\phi_2+\psi_1)/2}\right.$$

$$\left. - \sin\frac{1}{2}\theta_1 \sin\frac{1}{2}\theta_2 e^{-i(\phi_2+\psi_2)/2}\right),$$

we find that the matrices $A(\phi, \theta, \psi)$ are closed under matrix multiplication. The neutral element is given by

$$A(\phi = 0, \theta = 0, \psi = 0) = \begin{pmatrix} 1 & 0 \\ 0 & 1 \end{pmatrix}.$$

Since

$$\det A(\phi, \theta, \psi) = \cos^2\left(\frac{\theta}{2}\right) + \sin^2\left(\frac{\theta}{2}\right) = 1,$$

we find that the inverse exists. The matrix A is unitary, i.e.

$$A^{-1}(\phi, \theta, \psi) = A^*(\phi, \theta, \psi) \equiv \bar{A}^T(\phi, \theta, \psi) \tag{1}$$

where A^* denotes the hermitian conjugate matrix of A. Thus the inverse is given by (1). Since for arbitrary $n \times n$ matrices the associative law holds we have proved that the matrices $A(\phi, \theta, \psi)$ form' a group under matrix multiplication.

Problem 4.   (i) The mapping ($z \in \mathbf{C}$)

$$w(z) = \frac{az+b}{cz+d}, \quad ad - bc \neq 0 \tag{1}$$

is called the *fractional linear transformation*, where $a, b, c, d \in \mathbf{R}$. Show that these transformations form a group under the composition of mappings.

(ii) The *Schwarzian derivative* of a function f is defined as

$$D(f)_z := \frac{2f'(z)f'''(z) - 3[f''(z)]^2}{2[f''(z)]^2}$$

where $f'(z) \equiv df/dz$. Calculate $D(w)_z$.

Solution. (i) First we have to show that the transformation is closed under the composition. Let

$$W(w) = \frac{Aw + B}{Cw + D}, \quad AD - BC \neq 0.$$

Inserting (1) yields

$$W(w(z)) = \frac{A(az + b) + B(cz + d)}{C(az + b) + D(cz + d)} = \frac{(Aa + Bc)z + (Ab + Bd)}{(Ca + Dc)z + (Cb + Dd)}. \quad (2)$$

Since

$$ad - bc \neq 0 \quad (3)$$

and

$$AD - BC \neq 0, \quad (4)$$

we find from (2), (3) and (4) that

$$(Aa + Bc)(Cb + Dd) - (Ab + Bd)(Ca + Dc) \equiv (AD - BC)(ad - bc) \neq 0.$$

Thus mapping (1) is closed under composition. For

$$a = 1, \quad b = 0, \quad c = 0, \quad d = 1,$$

we find the unit element

$$w(z) = z.$$

The inverse of mapping (1) is given by

$$z(w) = \frac{-dw + b}{cw - a}$$

since

$$z(w(z)) = \frac{(-d(az + b) + b(cz + d))(cz + d)}{(cz + d)(c(az + b) - a(cz + d))},$$

$$z(w(z)) = \frac{-adz - db + bd + bcz}{bc - ad + acz - acz} = \frac{(-ad + bc)z}{(-ad + bc)} = 1.$$

Finally, it can be proved that the associative law is satisfied.

(ii) We find

$$D(w)_z = 0.$$

This means $w(z)$ is the general solution of the ordinary differential equation

$$D(f)_z = 0$$

since the solution includes three constants of integration.

Problem 5.   Let P be the *parity operator*, i.e.

$$P\mathbf{r} := -\mathbf{r}. \tag{1}$$

Obviously, $P = P^{-1}$. We define

$$\mathbf{O}_P u(\mathbf{r}) := (P^{-1}\mathbf{r}) \equiv u(-\mathbf{r}).$$

Now $\mathbf{r}$ can be expressed in spherical coordinates

$$\mathbf{r} = r(\sin\theta\cos\phi, \sin\theta\sin\phi, \cos\theta)$$

where $0 \le \phi < 2\pi$ and $0 \le \theta < \pi$.

(i) Calculate $P(r, \theta, \phi)$.

(ii) Let

$$Y_{lm}(\theta, \phi) := \frac{(-1)^{l+m}}{2^l l!} \left(\frac{2l + 1}{4\pi} \frac{(l - m)!}{(l + m)!} \right)^{1/2}$$

$$\times (\sin\theta)^m \frac{d^{l+m}}{d(\cos\theta)^{l+m}} (\sin\theta)^{2l} e^{im\phi}$$

where $l = 0, 1, 2, \ldots$ and $m = -l, -l + 1, \ldots, +l$.
Find

$$\mathbf{O}_P Y_{lm}.$$

(iii) Calculate

$$(\mathbf{O}_P \hat{L}_z) Y_{lm}$$

and

$$(\hat{L}_z \mathbf{O}_P) Y_{lm}$$

where

$$\hat{L}_z := -i\hbar \frac{\partial}{\partial \phi} \, .$$

Find $[\hat{L}_z, \mathbf{O}_P] Y_{lm}$.

Remark.   *The functions Y_{lm} are called* spherical harmonics.

Solution.   (i) From (1) we find that

$$P(r, \theta, \phi) = (r, \pi - \theta, \pi + \phi)$$

because

$$\sin(\pi - \theta) \equiv \sin\theta \, ,$$

$$\cos(\pi - \theta) \equiv -\cos\theta \, ,$$

$$\sin(\pi + \phi) \equiv -\sin\phi \, ,$$

$$\cos(\pi + \phi) \equiv -\cos\phi \, .$$

(ii) We find

$$\mathbf{O}_P Y_{lm}(\theta, \phi) = Y_{lm}(\pi - \theta, \pi + \phi) = (-1)^l Y_{lm}(\theta, \phi) \, .$$

(iii) Since

$$\hat{L}_z Y_{lm} = m\hbar Y_{lm} \, ,$$

we find

$$\mathbf{O}_P \hat{L}_z Y_{lm} = \mathbf{O}_P(\hat{L}_z Y_{lm}) = m\hbar \mathbf{O}_P Y_{lm} = m\hbar(-1)^l Y_{lm}$$

and

$$(\hat{L}_z \mathbf{O}_P) Y_{lm} = \hat{L}_z(\mathbf{O}_P Y_{lm}) = (-1)^l \hat{L}_z Y_{lm} = (-1)^l m\hbar Y_{lm} \, .$$

Consequently,

$$[\hat{L}_z, \mathbf{O}_P] Y_{lm} = 0 \, .$$

Problem 6.   Let D_n be the *dihedral group*, the group of rigid motions of an n-gon ($n \leq 3$). It is a noncommutative group of order $2n$. Find the *centre* of D_n, which is defined as

$$C := \{c \in D_n | cx = xc \text{ for all } x \in D_n\}.$$

Solution.   To determine the centre of

$$D_n := \{a, b | a^n = b^2 = 1, \, ba = a^{-1}b\}$$

where a is a rotation by $2\pi/n$ and b is a flip, it suffices to find those elements which commute with the generators a and b. Since $n \geq 3$, we have

$$a^{-1} \neq a.$$

Therefore

$$a^{r+1}b = a(a^r b) = (a^r b)a = a^{r-1}b.$$

If follows that

$$a^2 = 1$$

a contradiction. Thus, no element of the form $a^r b$ is in the centre. Similarly, if for $l \leq s < n$

$$a^s b = ba^s = a^{-s}b,$$

then $a^{2s} = 1$, which is possible only if $2s = n$. Hence, a^s commutes with b if and only if $n = 2s$. Thus, if

$$n = 2s$$

the centre of D_n is $\{1, a^s\}$. If n is odd the centre is $\{1\}$.

Problem 7.   Consider the function $H \in L_2(\mathbf{R})$

$$H(x) := \begin{cases} 1 & 0 \leq x < \dfrac{1}{2}, \\ -1 & \dfrac{1}{2} \leq x \leq 1, \\ 0 & \text{otherwise}. \end{cases}$$

Let

$$H_{mn}(x) := 2^{-m/2} H(2^{-m}x - n)$$

where $m, n \in \mathbf{Z}$ and n is the translation parameter and m the dilation parameter.

(i) Find H_{11}, H_{12}.

(ii) Show that

$$\langle H_{mn}(x), H_{kl}(x) \rangle = \delta_{mk}\delta_{nl}, \quad k, l \in \mathbf{Z}$$

where $\langle \cdot \rangle$ denotes the scalar product in $L_2(\mathbf{R})$.

(iii) Expand the function ($f \in L_2(\mathbf{R})$)

$$f(x) = \exp(-|x|)$$

with respect to H_{mn}.

Remark. *The functions H_{mn} form an orthonormal basis in $L_2(\mathbf{R})$.*

Solution. (i) We have

$$H_{mn}(x) = 2^{-m/2}H(2^{-m}x - n) = \begin{cases} 2^{-m/2} & 0 \le 2^{-m}x - n < \dfrac{1}{2}, \\ -2^{-m/2} & \dfrac{1}{2} \le 2^{-m}x - n \le 1, \\ 0 & \text{otherwise}. \end{cases}$$

Thus

$$H_{mn}(x) = \begin{cases} 2^{-m/2} & 2^m n \le x < 2^m \left(n + \dfrac{1}{2}\right), \\ -2^{-m/2} & 2^m \left(n + \dfrac{1}{2}\right) \le x \le 2^m(n+1), \\ 0 & \text{otherwise}. \end{cases}$$

Thus

$$H_{11}(x) = \begin{cases} \dfrac{1}{\sqrt{2}} & 2 \le x < 3, \\ -\dfrac{1}{\sqrt{2}} & 3 \le x \le 4, \\ 0 & \text{otherwise}, \end{cases} \qquad H_{12}(x) = \begin{cases} \dfrac{1}{\sqrt{2}} & 4 \le x < 5, \\ -\dfrac{1}{\sqrt{2}} & 5 \le x \le 6, \\ ,0 & \text{otherwise}. \end{cases}$$

(ii) We have

$$I_{mnkl} := \langle H_{mn}(x), H_{kl}(x) \rangle = \int_{-\infty}^{\infty} H_{mn}(x)H_{kl}(x)dx.$$

The intervals on which H_{mn} and H_{kl} are non-zero are

$$I_m := (2^m n, 2^m(n+1)), \quad I_{kl} := (2^k l, 2^k(l+1)).$$

We consider the different cases:

Case 1. $m = k$, $n = 1$ with

$$I_{mnkl} = \int_{2^m n}^{2^m(n+1)} 2^{-m} dx = 1.$$

Case 2. $m = k$, $n \neq l$ with

$$I_{mnkl} = 0$$

since $I_{mn} \cap I_{kl} = \emptyset$.

Case 3. $m \neq k$. Suppose without loss of generality that $m < k$. Either $I_{mn} \cap I_{kl} = \emptyset$ ($I_{mnkl} = 0$), or $I_{mn} \subset I_{kl}$ (as shown below). We have the following

$$2^k l \leq 2^m n < 2^k \left(l + \frac{1}{2} \right) \Rightarrow 2^{k-m} l \leq n < 2^{k-m} \left(l + \frac{1}{2} \right)$$

$$\Rightarrow 2^{k-m} l \leq n + 1 \leq 2^{k-m} \left(l + \frac{1}{2} \right) \Rightarrow 2^k l \leq 2^m(n+1) \leq 2^k \left(l + \frac{1}{2} \right).$$

Also

$$2^k \left(l + \frac{1}{2} \right) \leq 2^m n < 2^k(l+1) \Rightarrow 2^k \left(l + \frac{1}{2} \right) \leq 2^m(n+1) \leq 2^k(l+1),$$

$$2^k l < 2^m(n+1) \leq 2^k \left(l + \frac{1}{2} \right) \Rightarrow 2^k l \leq 2^m n \leq 2^k \left(l + \frac{1}{2} \right),$$

$$2^k \left(l + \frac{1}{2} \right) < 2^m(n+1) \leq 2^k(l+1) \Rightarrow 2^k \left(l + \frac{1}{2} \right) \leq 2^m n \leq 2^k(l+1)$$

which gives

$$\langle H_{mn}(x), H_{kl}(x) \rangle = \pm \int_{2^m n}^{2^m(n+1/2)} 2^{-1/2(m+k)} dx$$

$$\mp \int_{2^m(n+1/2)}^{2^m(n+1)} 2^{-1/2(m+k)} dx = 0.$$

Thus $I_{mnkl} = \delta_{mk} \delta_{nl}$.

(iii) The scalar product $\langle f(x), H_{mn}(x) \rangle$ is given by

$$\langle f(x), H_{mn}(x) \rangle = \int_{-\infty}^{\infty} f(x) H_{mn}(x) dx$$

$$= \int_{2^m n}^{2^m (n+1/2)} 2^{-m/2} e^{-|x|} dx - \int_{2^m (n+1/2)}^{2^m (m+1)} 2^{-m/2} e^{-|x|} dx$$

$$= 2^{-m/2} \begin{cases} -e^{-x} \Big|_{2^m n}^{2^m (n+1/2)} + e^{-x} \Big|_{2^m (n+1/2)}^{2^m (n+1)} & n \geq 0 \\[2ex] e^{x} \Big|_{2^m n}^{2^m (n+1/2)} - e^{x} \Big|_{2^m (n+1/2)}^{2^m (n+1)} & n < 0 \end{cases}$$

$$= -\frac{n + \delta_{n,0}}{|n| + \delta_{n,0}} 2^{-m/2} \left(2e^{-2^m |n+1/2|} - e^{-2^m |n|} - e^{-2^m |n+1|}\right).$$

The expansion is given by

$$f(x) = \sum_{m,n \in \mathbf{Z}} \langle f(x), H_{mn}(x) \rangle H_{mn}(x).$$

Chapter 9

Generalized Functions

Problem 1. Let $a > 0$. We define

$$f_a(x) := \begin{cases} \dfrac{1}{2a} & \text{for } |x| \le a \,, \\[2mm] 0 & \text{for } |x| > a \,. \end{cases}$$

(i) Calculate the derivative of f_a in the sense of generalized functions.

(ii) What happens if $a \to +0$?

Solution. (i) The derivative of a function f in the sense of generalized functions is defined as

$$\left(\frac{df(x)}{dx}, \phi(x) \right) := - \left(f(x), \frac{d\phi(x)}{dx} \right)$$

where $\phi \in D(\mathbf{R})$ and $D(\mathbf{R})$ denotes all infinitely-differentiable functions in $\mathbf{R}$ with compact support. In the present case we find

$$\left(\frac{df_a(x)}{dx}, \phi(x) \right) = - \left(f_a(x), \frac{d\phi(x)}{dx} \right) \,.$$

Thus

$$\left(\frac{df_a(x)}{dx}, \phi(x) \right) = - \int_{\mathbf{R}} f_a(x) \frac{d\phi}{dx} dx = - \frac{1}{2a} \int_{-a}^{a} \frac{d\phi}{dx} dx \,.$$

Finally

$$\left(\frac{df_a(x)}{dx}, \phi(x)\right) = \frac{1}{2a}\left(-\phi(a) + \phi(-a)\right)$$

$$= \frac{1}{2a}\left((\delta(x+a), \phi(x)) - (\delta(x-a), \phi(x))\right).$$

We can write

$$\frac{df_a(x)}{dx} = \frac{1}{2a}(\delta(x+a) - \delta(x-a))$$

where δ denotes the *delta function*.

(ii) In the sense of generalized functions the function f_a tends to the delta function for $a \to +0$.

Instead of considering the space $D(\mathbf{R})$ for the test functions we can also consider the space $S(\mathbf{R})$. The space $S(\mathbf{R}^n)$ is the set of all infinitely differentiable functions which decrease as $|\mathbf{x}| \to \infty$, together with all their derivatives, faster than any power of $|\mathbf{x}|^{-1}$. These functions are called test functions.

Problem 2.   Show that

$$\frac{1}{2\pi} \sum_{k=-\infty}^{\infty} e^{ikx} \equiv \sum_{k=-\infty}^{\infty} \delta(x - 2k\pi) \tag{1}$$

in the sense of generalized functions.

Hint.   Expand the 2π-periodic function

$$f(x) = \frac{1}{2} - \frac{x}{2\pi} \tag{2}$$

into a Fourier series. A basis in the Hilbert space $L_2(0, 2\pi)$ is given by

$$\left\{ \phi_k(x) := \frac{1}{\sqrt{2\pi}} \exp(ikx) : k \in \mathbf{Z} \right\}.$$

Solution.   Since we assume that the function f is periodic with respect to 2π we can expand it into a Fourier expansion

$$f(x) = \sum_{k \in \mathbf{Z}} \langle f, \phi_k \rangle \phi_k.$$

Calculating the scalar product $\langle f, \phi_k \rangle$ in the Hilbert space $L_2(0, 2\pi)$ we obtain

$$f(x) = -\frac{i}{2\pi} \sum_{\substack{k=-\infty \\ k \neq 0}}^{\infty} \frac{1}{k} e^{ikx}$$

where

$$\langle f, \phi_0 \rangle = \int_0^{2\pi} \left(\frac{1}{2} - \frac{x}{2\pi} \right) \frac{1}{\sqrt{2\pi}} dx = 0 .$$

In the sense of generalized functions we can differentiate the right-hand side of (2) term by term. We obtain

$$f' = -\frac{1}{2\pi} + \sum_{k=-\infty}^{\infty} \delta(x - 2k\pi) = \frac{1}{2\pi} \sum_{\substack{k=-\infty \\ k \neq 0}}^{\infty} e^{ikx} .$$

Thus we obtain identity (1).

Remark.   *The identity*

$$\delta(f(x)) \equiv \sum_n \frac{1}{|f'(x_n)|} \delta(x - x_n)$$

can also be used to find identity (1). Here the sum runs over all the zeros of f and $f'(x_n) \equiv df(x = x_n)/dx$.

Problem 3.   Calculate the Fourier transform of

$$f(t) = \sin(\Omega t)$$

and

$$g(t) = \cos(\Omega t)$$

in the sense of generalized functions.

Solution.   The Fourier transform for generalized functions is given by

$$(F[f](\omega), \psi(\omega)) := 2\pi(f(t), \phi(t)) \tag{1}$$

where

$$\psi(\omega) := \int_{-\infty}^{+\infty} e^{i\omega t} \phi(t) dt . \tag{2}$$

Inserting f into (1) we obtain

$$(F[f](\omega), \psi(\omega)) = 2\pi \int_{-\infty}^{+\infty} \sin(\Omega t)\phi(t)dt .$$

Since

$$\sin(\Omega t) \equiv \frac{e^{i\Omega t} - e^{-i\Omega t}}{2i} ,$$

we find

$$(F[f](\omega), \psi(\omega)) = \frac{\pi}{i} \int_{-\infty}^{+\infty} e^{i\Omega t}\phi(t)dt - \frac{\pi}{i} \int_{-\infty}^{+\infty} e^{-i\Omega t}\phi(t)dt .$$

From (2) we find

$$\psi(-\omega) = \int_{-\infty}^{+\infty} e^{-i\omega t}\phi(t)dt .$$

Therefore

$$(F[f](\omega), \psi(\omega)) = \frac{\pi}{i}[\psi(\Omega) - \psi(-\Omega)] .$$

Consequently,

$$(F[f](\omega), \psi(\omega)) = \frac{\pi}{i}((\delta(\omega - \Omega), \psi(\omega)) - (\delta(\omega + \Omega), \psi(\omega))) .$$

Thus

$$F[f](\omega) = \frac{\pi}{i}(\delta(\omega - \Omega) - \delta(\omega + \Omega)) . \tag{3}$$

Analogously we find the Fourier transform of g

$$F[g](\omega) = \pi(\delta(\omega + \Omega) + \delta(\omega - \Omega)) . \tag{4}$$

Remark. *The Fourier transform of* $\exp(bx)$ *in the sense of generalized functions is*

$$F[\exp(bx)](\omega) = 2\pi\delta(\omega - ib) . \tag{5}$$

Using

$$\exp(i\Omega t) \equiv \cos(\Omega t) + i\sin(\Omega t) ,$$

$$\exp(-i\Omega t) \equiv \cos(\Omega t) - i\sin(\Omega t)$$

and $b = i\Omega$ *and* $b = -i\Omega$, *we can also find* (3) *and* (4) *from* (5).

Problem 4.   Let $\mathbf{x} = (x_1, x_2, \ldots, x_n)$. Show that

$$E(\mathbf{x}, t) := \frac{1}{(4\pi a^2 t)^{n/2}} \exp\left(-\frac{|\mathbf{x}|^2}{4a^2 t}\right) \to \delta(\mathbf{x}), \quad t \to +0$$

in the sense of generalized functions, where $n = 1, 2, \ldots$.

Solution.   Let ϕ be a test function. The space $S(\mathbf{R}^n)$ is the set of all infinitely differentiable functions which decrease as $|\mathbf{x}| \to \infty$, together with all their derivatives, faster than any power of $|\mathbf{x}|^{-1}$. These functions are called test functions. Then we have the estimates

$$\left| \int_{\mathbf{R}^n} E(\mathbf{x}, t)[\phi(\mathbf{x}) - \phi(\mathbf{0})]d\mathbf{x} \right| \leq \frac{K}{(4\pi a^2 t)^{n/2}} \int_{\mathbf{R}^n} \exp\left(-\frac{|\mathbf{x}|^2}{4a^2 t}\right) |\mathbf{x}| d\mathbf{x}$$

$$= \frac{K\sigma_n}{(4\pi a^2 t)^{n/2}} \int_0^\infty \exp\left(-\frac{r^2}{4a^2 t}\right) r^n dr$$

$$= K'\sqrt{t} \int_0^\infty \exp(-u^2) u^n du$$

$$= C\sqrt{t}$$

where

$$\sigma_n := \frac{2\pi^{n/2}}{\Gamma(n/2)}$$

is the area of the surface of a unit sphere in $\mathbf{R}^n$ and

$$\Gamma(z) := \int_0^\infty t^{z-1} e^{-t} dt, \quad \Re z > 0$$

where Γ denotes the *Gamma function*. For $t > 0$ we have

$$\int_{\mathbf{R}^n} E(\mathbf{x}, t)d\mathbf{x} = \frac{1}{(2a\sqrt{\pi t})^n} \int_{\mathbf{R}^n} \exp\left(-\frac{|\mathbf{x}|^2}{4a^2 t}\right) d\mathbf{x}$$

$$= \prod_{j=1}^n \frac{1}{\sqrt{\pi}} \int_{-\infty}^\infty \exp(-\xi_j^2) d\xi_j^2$$

$$= 1.$$

Thus we obtain

$$\begin{aligned}
(E(\mathbf{x},t),\phi) &= \int_{\mathbf{R}^n} E(\mathbf{x},t)\phi(\mathbf{x})dx \\
&= \phi(0)\int_{\mathbf{R}^n} E(\mathbf{x},t)dx + \int_{\mathbf{R}^n} E(\mathbf{x},t)[\phi(\mathbf{x}) - \phi(0)]dx \\
&\to \phi(0) \\
&= (\delta,\phi)\,.
\end{aligned}$$

Problem 5.   Find a particular solution to the differential equation

$$\frac{d^2u}{dt^2} + a\frac{du}{dt} + bu = \delta(t) \quad a,b \in \mathbf{R} \tag{1}$$

in the sense of generalized functions.

Solution.   Consider the ansatz

$$u(t) = \theta(t)f(t) \tag{2}$$

where

$$\theta(t) := \begin{cases} 1 & t \geq 0\,, \\ 0 & \text{otherwise}\,, \end{cases} \tag{3}$$

and f is a differentiable function. Let ϕ be a test function $\phi \in S(\mathbf{R})$. The space $S(\mathbf{R})$ is the set of all infinitely-differentiable functions which decrease as $|x| \to \infty$, together with all their derivatives, faster than any power of $|x|^{-1}$. These functions are called test functions. Then from (1) we obtain

$$\left(\frac{d^2u}{dt^2},\phi\right) + a\left(\frac{du}{dt},\phi\right) + b(u,\phi) = (\delta(t),\phi) = \phi(0)\,. \tag{4}$$

Applying differentiation in the sense of generalized functions we obtain from (4)

$$\left(u,\frac{d^2\phi}{dt^2}\right) - a\left(u,\frac{d\phi}{dt}\right) + b(u,\phi) = \phi(0)\,. \tag{5}$$

Inserting the ansatz (2) into (5) it follows that

$$\int_{-\infty}^{\infty} \theta(t)f(t)\frac{d^2\phi}{dt^2}dt - a\int_{-\infty}^{+\infty} \theta(t)f(t)\frac{d\phi}{dt}dt + b\int_{-\infty}^{\infty} \theta(t)f(t)\phi(t)dt = \phi(0)\,.$$

Applying the property of the step-function (3) leads to

$$\int_0^\infty f(t)\frac{d^2\phi}{dt^2}dt - a\int_0^\infty f(t)\frac{d\phi}{dt}dt + b\int_0^\infty f(t)\phi(t)dt = \phi(0). \qquad (6)$$

Applying integration by parts and taking into account the boundary conditions for ϕ

$$\phi(|\infty|) = 0, \quad \phi'(|\infty|) = 0,$$

we find

$$\int_0^\infty f(t)\frac{d\phi}{dt}dt = -f(0)\phi(0) - \int_0^\infty \frac{df}{dt}\phi dt \qquad (7)$$

and

$$\int_0^\infty f(t)\frac{d^2\phi}{dt^2}dt = -f(0)\frac{d\phi(0)}{dt} - \int_0^\infty \frac{df}{dt}\frac{d\phi}{dt}dt$$

$$= -f(0)\frac{d\phi(0)}{dt} + \frac{df(0)}{dt}\phi(0) + \int_0^\infty \frac{d^2f}{dt^2}\phi dt. \qquad (8)$$

Inserting (7) and (8) into (6) we find

$$\int_0^\infty \left(\frac{d^2f}{dt^2} + a\frac{df}{dt} + bf\right)\phi(t)dt - f(0)\frac{d\phi(0)}{dt} + \frac{df(0)}{dt}\phi(0) + af(0)\phi(0) = \phi(0). \qquad (9)$$

To satisfy Eq. (9) we have to fulfill the following conditions

$$\frac{d^2f}{dt^2} + a\frac{df}{dt} + bf = 0,$$

$$f(0) = 0,$$

$$\frac{df(0)}{dt} + af(0) = 1.$$

Problem 6.   Use the identity

$$\sum_{k=-\infty}^\infty \delta(x - 2\pi k) = \frac{1}{2\pi}\sum_{k=-\infty}^\infty e^{ikx} \qquad (1)$$

to show that

$$2\pi\sum_{k=-\infty}^\infty \phi(2\pi k) = \sum_{k=-\infty}^\infty F[\phi](k) \qquad (2)$$

where F denotes the Fourier transform and ϕ is a test function.

Remark. *Equation (2) is called Poisson's summation formula.*

Solution. We rewrite (1) in the form

$$2\pi \sum_{k=-\infty}^{\infty} \delta(x - 2\pi k) = \sum_{k=-\infty}^{\infty} F[\delta(x - k)] .$$

Applying this equation to $\phi \in S(\mathbf{R})$, we obtain

$$2\pi \left(\sum_{k=-\infty}^{\infty} \delta(x - 2\pi k), \phi \right) = 2\pi \sum_{k=-\infty}^{\infty} \phi(2\pi k) = \left(\sum_{k=-\infty}^{\infty} F[\delta(x - k)], \phi \right) .$$

Finally we arrive at

$$2\pi \left(\sum_{k=-\infty}^{\infty} \delta(x - 2\pi k), \phi \right) = \sum_{k=-\infty}^{\infty} (\delta(x - k), F[\phi]) = \sum_{k=-\infty}^{\infty} F[\phi](k) .$$

This proves Poisson's summation formula.

Problem 7. The *Hermite polynomials* are defined by

$$H_n(x) := (-1)^n e^{x^2} \frac{d^n}{dx^n} e^{-x^2} , \qquad n = 0, 1, 2, \ldots \tag{1}$$

where $H_0(x) = 1$, $H_1(x) = 2x$, $H_2(x) = 4x^2 - 2$ etc. A basis in the Hilbert space $L_2(\mathbf{R})$ is given by

$$B := \left\{ \frac{(-1)^n}{2^{n/2}\sqrt{n!}\sqrt[4]{\pi}} e^{x^2/2} \frac{d^n}{dx^n} e^{-x^2} : n = 0, 1, 2, \ldots \right\} . \tag{2}$$

Use the completeness relation to find an expansion of a function $f \in L_2(\mathbf{R})$.

Solution. Using the basis (2) and the Hermite polynomials (1) the completeness relation is given by

$$\sum_{n=0}^{\infty} \frac{1}{2^n n! \sqrt{\pi}} e^{-(x^2 + x'^2)/2} H_n(x) H_n(x') = \delta(x - x') , \qquad x, x' \in \mathbf{R} . \tag{3}$$

Using completeness relation (3) and the properties of the delta function we obtain

$$f(x) = \sum_{n=0}^{\infty} \frac{1}{2^n n! \sqrt{\pi}} H_n(x) e^{-x^2/2} \int_{-\infty}^{\infty} f(x') H_n(x') e^{-x'^2/2} dx' .$$

Problem 8.   The probability amplitude for the three-dimensional transition of a particle with mass m from point $\mathbf{r}_0$ at the time t_0 to the point $\mathbf{r}_N$ at time t_N is given by the kernel

$$K(\mathbf{r}_N, t_N; \mathbf{r}_0, t_0) = \lim_{N \to \infty} \prod_{n=1}^{N} \left(\frac{m}{2\pi i\hbar(t_n - t_{n-1})} \right)^{3/2}$$

$$\times \int \cdots \int d^3\mathbf{r}_1 \cdots d^3\mathbf{r}_{N-1} \exp\left(\left(\frac{i}{\hbar} \right) S(\mathbf{r}_N, \mathbf{r}_0) \right). \quad (1)$$

In a compact form one writes

$$K(\mathbf{r}_N, t_N; \mathbf{r}_0, t_0) = \int_0^N \exp\left(\left(\frac{i}{\hbar} \right) S(\mathbf{r}_N, \mathbf{r}_0) \right) D\mathbf{r}(t).$$

The integrals extend over all paths in three-dimensional coordinate space and S is the *action* of the corresponding classical system. Thus

$$S(\mathbf{r}_N, \mathbf{r}_0) = \int_{t_0}^{t_N} L(\dot{\mathbf{r}}, \mathbf{r}, t)dt \quad (2)$$

where $L(\dot{\mathbf{r}}, \mathbf{r}, t)$ represents the *Lagrangian function*. We can also write (2) as

$$S(\mathbf{r}_N, \mathbf{r}_0) = \sum_{n=1}^{N} (t_n - t_{n-1}) L\left(\frac{\mathbf{r}_n - \mathbf{r}_{n-1}}{t_n - t_{n-1}}, \mathbf{r}_n, t_n \right).$$

The wave function ψ satisfies the integral equation

$$\psi(\mathbf{r}, t) = \int K(\mathbf{r}, t; \mathbf{r}_0, t_0) \psi(\mathbf{r}_0, t_0) d^3\mathbf{r}_0.$$

The momentum amplitude is given by the Fourier transform of the wave function

$$\phi(\mathbf{p}, t) = \int \exp\left(-\left(\frac{i}{\hbar} \right) \mathbf{p} \cdot \mathbf{r} \right) \psi(\mathbf{r}, t) d^3\mathbf{r}.$$

With $\mathbf{p} = \hbar\mathbf{k}$ we obtain

$$\phi(\mathbf{k}, t) = \int \exp(-2\pi i\mathbf{k} \cdot \mathbf{r}) \psi(\mathbf{r}, t) d^3\mathbf{r}.$$

The momentum amplitude satisfies a similar integral equation as the wave function

$$\phi(\mathbf{k}, t) = \int \kappa(\mathbf{k}, t; \mathbf{k}_0, t_0)\phi(\mathbf{k}_0, t_0)d^3\mathbf{k}_0 .$$

The momentum kernel κ represents the probability amplitude of the transition from the wave vector $\mathbf{k}_0$ at the time t_0 to the wave vector $\mathbf{k}$ at the time t. We have

$$\kappa(\mathbf{k}, t; \mathbf{k}_0, t_0) = \int \int \exp(-2\pi i \mathbf{k} \cdot \mathbf{r}) K(\mathbf{r}, t; \mathbf{r}_0, t_0) \exp(2\pi \mathbf{k}_0 \cdot \mathbf{r}_0) d^3\mathbf{r}_0 d^3\mathbf{r} .$$

$$(3)$$

Describe the motion of a particle in a local potential $V(\mathbf{r}, t)$. The Lagrangian function is given by

$$L(\dot{\mathbf{r}}, \mathbf{r}, t) = \frac{1}{2}m\dot{\mathbf{r}}^2 - V(\mathbf{r}, t)$$

where $V(\mathbf{r}, t)$ represents the potential energy.

Solution. The action S becomes

$$S(\mathbf{r}_N, \mathbf{r}_0) = \sum_{n=1}^{N}(t_n - t_{n-1})\left(\frac{1}{2}m\left(\frac{\mathbf{r}_n - \mathbf{r}_{n-1}}{t_n - t_{n-1}}\right)^2 - V(\mathbf{r}_n, t_n)\right) . \qquad (4)$$

Inserting (4) into (1) yields

$$K(\mathbf{r}_N, t_N; \mathbf{r}_0, t_0) = \lim_{N \to \infty} \prod_{n=1}^{N}\left(\frac{m}{ih(t_n - t_{n-1})}\right)^{3/2} \int \cdots \int d^3\mathbf{r}_1 \cdots d^3\mathbf{r}_{N-1}$$

$$\times \exp\left(\left(\frac{i\pi m}{h}\right)\sum_{n=1}^{N}\frac{(\mathbf{r}_n - \mathbf{r}_{n-1})^2}{(t_n - t_{n-1})}\right)$$

$$\times \exp\left(-\left(\frac{2\pi i}{h}\right)\sum_{n=1}^{N}(t_n - t_{n-1})V(\mathbf{r}_n, t_n)\right) . \qquad (5)$$

Using the identity

$$\left(\frac{m}{iht}\right)^{3/2}\exp(i\pi m\mathbf{r}^2/ht) = \int d^3\mathbf{k}\exp(2\pi i(\mathbf{k} \cdot \mathbf{r} - ht\mathbf{k}^2/2m))$$

where the integration extends over the three-dimensional momentum space, (5) can be transformed into

$$K(\mathbf{r}_N, t_N; \mathbf{r}_0, t_0)$$

$$= \lim_{N \to \infty} \int \cdots \int d^3\mathbf{r}_1 \cdots d^3\mathbf{r}_{N-1} d^3\mathbf{k}_0 \cdots d^3\mathbf{k}_{N-1}$$

$$\times \exp\left(2\pi i \sum_{n=1}^{N} \mathbf{k}_{n-1} \cdot (\mathbf{r}_n - \mathbf{r}_{n-1}) - \left(\frac{h}{2m}\right)(t_n - t_{n-1})\mathbf{k}_{n-1}^2\right)$$

$$\times \exp\left(-\frac{2\pi i}{h} \sum_{n=1}^{N} (t_n - t_{n-1})V(\mathbf{r}_n, t_n)\right).$$

From (3) we obtain

$$\kappa(\mathbf{k}_N, t_N; \mathbf{k}_0', t_0)$$

$$= \lim_{N \to \infty} \int \cdots \int d^3\mathbf{r}_0 \cdots d^3\mathbf{r}_N d^3\mathbf{k}_0 \cdots d^3\mathbf{k}_{N-1}$$

$$\times \exp(-2\pi i(\mathbf{k}_0 - \mathbf{k}_0') \cdot \mathbf{r}_0 \exp\left(-2\pi i \sum_{n=1}^{N}(\mathbf{k}_n - \mathbf{k}_{n-1}) \cdot \mathbf{r}_n\right)$$

$$\times \exp\left(-\left(\frac{2\pi i}{h}\right) \sum_{n=1}^{N}(t_n - t_{n-1})V(\mathbf{r}_n, t_n)\right)$$

$$\times \exp\left(-\left(\frac{\pi i h}{m}\right) \sum_{n=1}^{N}(t_n - t_{n-1})\mathbf{k}_{n-1}^2\right).$$

The integration over $\mathbf{r}_0$ can be carried out giving

$$\int d^3\mathbf{r}_0 \exp(-2\pi i(\mathbf{k}_0 - \mathbf{k}_0') \cdot \mathbf{r}_0) = \delta(\mathbf{k}_0 - \mathbf{k}'_0).$$

Using the property of the delta function we can also calculate the integral over $\mathbf{k}_0$. We obtain

$$\kappa(\mathbf{k}_N, t_N; \mathbf{k}_0, t_0) = \lim_{N \to \infty} \int \cdots \int d^3\mathbf{r}_1 \cdots d^3\mathbf{r}_N d^3\mathbf{k}_1 \cdots \mathbf{k}_{N-1}$$

$$\times \exp\left(-2\pi i \sum_{n=1}^{N}(\mathbf{k}_n - \mathbf{k}_{n-1}) \cdot \mathbf{r}_n\right)$$

$$\times \exp\left(-\left(\frac{2\pi i}{h}\right)\sum_{n=1}^{N}(t_n - t_{n-1})V(\mathbf{r}_n, t_n)\right)$$

$$\times \exp\left(-\left(\frac{i\pi h}{m}\right)\sum_{n=1}^{N}(t_n - t_{n-1})\mathbf{k}_{n-1}^2\right).$$

Defining

$$F(\mathbf{k}_n - \mathbf{k}_{n-1}) := \int d^3\mathbf{r}_n \exp(-2\pi i(\mathbf{k}_n - \mathbf{k}_{n-1}) \cdot \mathbf{r}_n)$$

$$\times \exp\left(-\left(\frac{2\pi i}{h}\right)(t_n - t_{n-1})V(\mathbf{r}_n, t_n)\right),$$

we can write

$$\kappa(\mathbf{k}_N, t_N; \mathbf{k}_0, t_0) = \lim_{N\to\infty}\int\cdots\int d^3\mathbf{k}_1\cdots d^3\mathbf{k}_{N-1}\prod_{n=1}^{N}F(\mathbf{k}_n - \mathbf{k}_{n-1})$$

$$\times \exp\left(-\left(\frac{i\pi h}{m}\right)(t_n - t_{n-1})\mathbf{k}_{n-1}^2\right).$$

The physical interpretation of this formula is that the transmission of the wave, with wave vector $\mathbf{k}_{n-1}$ through the nth slice can be described by the propagation factor $\exp(\ )$ multiplied by the probability amplitude $F(\mathbf{k}_n - \mathbf{k}_{n-1})$ for the scattering into the new wave vector $\mathbf{k}_n$. Integration over all possible wave vectors and time slices gives the expression for the momentum kernel.

Chapter 10

Linear Partial Differential Equations

Problem 1. Show that

$$u(x,t) = \sum_{n=1}^{\infty} c_n e^{-n^2 t} \sin(nx)$$

satisfies the one-dimensional linear heat equation

$$\frac{\partial u}{\partial t} = \frac{\partial^2 u}{\partial x^2} \tag{1}$$

together with the initial and boundary conditions

$$
\begin{aligned}
u(x,0) &= f(x), & 0 < x < \pi, \\
u(0,t) &= 0, & t > 0, \\
u(\pi,t) &= 0, & t > 0.
\end{aligned}
$$

Solution. We start with the *separation ansatz* $u(x,t) = X(x)T(t)$. Substituting the separation ansatz into (1) yields

$$\frac{1}{X(x)} \frac{d^2 X}{dx^2} = \frac{1}{T(t)} \frac{dT}{dt} = -\lambda$$

where λ is an arbitrary real constant. The eigenvalue problem

$$-\frac{d^2 X}{dx^2} = \lambda X(x), \quad X(0) = X(\pi) = 0$$

gives an infinite countable number of solutions

$$X_n(x) = C \sin(nx)$$

112

with the eigenvalues $\lambda_n = n^2$ where $n = 1, 2, \ldots, \infty$. Therefore, for the function T it follows that

$$\frac{dT}{dt} + n^2 T = 0, \quad t > 0$$

with the solution $T(t) = Ce^{-n^2 t}$. Since $(2/\pi)^{1/2} \sin(nx)$ with $n = 1, 2, \ldots, \infty$ is a basis in the Hilbert space $L_2(0, \pi)$ and $f \in L_2(0, \pi)$ we find

$$f(x) = \sqrt{\frac{2}{\pi}} \sum_{n=1}^{\infty} d_n \sin(nx)$$

with

$$d_n := \left(f(x), \sqrt{\frac{2}{\pi}} \sin(nx) \right) = \sqrt{\frac{2}{\pi}} \int_0^\pi f(x) \sin(nx) dx$$

where $(\,,)$ denotes the scalar product in the Hilbert space $L_2(0, \pi)$. Since

$$u(x, 0) = \sum_{n=1}^{\infty} c_n \sin(nx) = f(x),$$

we obtain

$$c_n = \frac{2}{\pi} \int_0^\pi f(x) \sin(nx) dx.$$

Problem 2.   Show that

$$E(\mathbf{x}) = -\frac{e^{ik|\mathbf{x}|}}{4\pi|\mathbf{x}|}$$

satisfies the partial differential equation

$$\Delta E + k^2 E = \delta$$

in the sense of generalized functions, where

$$\mathbf{x} = (x_1, x_2, x_3)$$

and

$$|\mathbf{x}| \equiv r = \sqrt{x_1^2 + x_2^2 + x_3^2}$$

and δ is the delta function.

Solution.　First we have to show that

$$\Delta\frac{1}{|\mathbf{x}|} = -4\pi\delta(\mathbf{x})$$

in the sense of generalized functions. Let ϕ be a test function. Without loss of generality we can assume that ϕ is spherically symmetric. From the definition of the derivative of a generalized function and using spherical coordinates we obtain

$$\left(\Delta\frac{1}{r}, \phi(r)\right) := \left(\frac{1}{r}, \Delta\phi(r)\right) = \int_{\mathbf{R}^3}\frac{\Delta\phi}{r}r^2\sin\theta\, dr d\theta d\phi. \tag{1}$$

Since ϕ depends only on r, the angle part of Δ can be omitted. We find that the Laplace operator takes the form

$$\Delta := \frac{1}{r^2}\frac{\partial}{\partial r}\left(r^2\frac{\partial}{\partial r}\right). \tag{2}$$

Inserting (2) into (1) gives

$$\left(\frac{1}{r}, \Delta\phi\right) = 4\pi\int_{r=0}^{\infty}\frac{1}{r}\left(2r\frac{d}{dr}\phi + r^2\frac{d^2}{dr^2}\phi\right)dr.$$

Thus

$$\left(\frac{1}{r}, \Delta\phi\right) = -4\pi\phi(0) = -4\pi(\delta, \phi)$$

where we have used integration by parts and that $\phi(r) \to 0$ as $r \to \infty$. We can write

$$\Delta\frac{1}{r} = -4\pi\delta.$$

Since for $r > 0$ we have

$$\frac{\partial}{\partial x_j}\frac{1}{|\mathbf{x}|} = -\frac{x_j}{|\mathbf{x}|^3},$$

$$\frac{\partial}{\partial x_j}e^{ik|\mathbf{x}|} = ik\frac{x_j}{|\mathbf{x}|}e^{ik|\mathbf{x}|},$$

$$\Delta e^{ik|\mathbf{x}|} = \left(\frac{2ik}{|\mathbf{x}|} - k^2\right)e^{ik|\mathbf{x}|}.$$

Using (3) we obtain

$$(\Delta + k^2)\frac{1}{|\mathbf{x}|}e^{ik|\mathbf{x}|} = -4\pi\delta(r).$$

Problem 3. Consider the linear partial differential equation

$$\frac{\partial u}{\partial t} + c_0 \frac{\partial u}{\partial x} + \beta \frac{\partial^3 u}{\partial x^3} = 0 \tag{1}$$

where β and c_0 are positive constants.

(i) Find the dispersion relation.

(ii) Let

$$x'(x,t) = x - c_0 t, \quad t'(x,t) = t,$$

$$u'(x'(x,t), t'(x,t)) = u(x,t).$$

Find the partial differential equation for $u'(x', t')$.

(iii) Show that the solution to the initial-value problem of the partial differential equation

$$\frac{\partial u}{\partial t} + \beta \frac{\partial^3 u}{\partial x^3} = 0 \tag{2}$$

is given by

$$u(x,t) = (\pi)^{-1/2}(3\beta t)^{-1/3} \int_{-\infty}^{\infty} \mathrm{Ai}\left(\frac{x-s}{(3\beta t)^{1/3}}\right) u(s, t = 0) ds \tag{3}$$

where

$$\mathrm{Ai}(s) := \frac{1}{\sqrt{\pi}} \int_0^{\infty} \cos\left(\frac{v^3}{3} + vs\right) dv$$

is the *Airy function*.

Solution. (i) Inserting the ansatz

$$u(x,t) = u_0 \exp(i(kx - \omega(k)t))$$

into (1) leads to the *dispersion relation* (i.e. the relation between the frequency ω and the wave vector k)

$$\omega(k) = c_0 k - \beta k^3.$$

(ii) Straightforward application of the chain rule yields

$$\frac{\partial u'}{\partial t'} + \beta \frac{\partial^3 u'}{\partial x'^3} = 0.$$

(iii) The Airy function has the asymptotic representation

$$
\mathrm{Ai}(s) = \begin{cases} \dfrac{1}{2} s^{-1/4} \exp\left(-\dfrac{2}{3} s^{3/2}\right) & \text{for } s \to \infty, \\[2ex] |s|^{-1/4} \cos\left(\dfrac{2}{3}|s|^{3/2} - \dfrac{\pi}{4}\right) & \text{for } s \to -\infty. \end{cases}
$$

If we assume that the initial distribution $u(x, t = 0)$ tends to zero sufficiently fast as $x \to \pm\infty$ we find by differentiating that (3) is the solution of the initial-value problem of (2).

Remark.   *The Airy functions* $\mathrm{Ai}(z)$ *and* $\mathrm{Bi}(z)$ *can be expressed in terms of Bessel functions. Vice versa the Bessel functions can be expressed in terms of the Airy functions* $\mathrm{Ai}(z)$ *and* $\mathrm{Bi}(z)$.

Problem 4.   Consider a wave function $u(x, y)$ which is confined within a regular triangle. The regular triangle is bound by the three lines

$$
x = -A, \quad x = \sqrt{3}y + 2A, \quad x = \sqrt{3}y + 2A
$$

where $A > 0$. The eigenvalue problem is given by

$$
\Delta u(x, y) \equiv \frac{\partial^2 u}{\partial x^2} + \frac{\partial^2 u}{\partial y^2} = -\frac{mE}{\hbar^2} u(x, y) \tag{1}
$$

where at the boundary

$$
u|_B = 0. \tag{2}
$$

(i) Find the symmetry group of the system.

(ii) Find the eigenvalues and eigenfunctions for an invariant subspace.

Solution.   (i) Diffraction in a corner of a polygon does not occur whenever the angle of the corner is π divided by an integer. In our case we have $\pi/3$. Therefore it is possible to express the wave function in terms of elementary functions, i.e. sin and cos. The triangle is invariant under the following six symmetry operations:

> identity, labelled I,
>
> rotation by $\dfrac{2}{3}\pi$ around the origin, labelled C_3,

rotation by $\dfrac{4}{3}\pi$ around the origin, labelled C_3^2,

reflection in the line $x = \left(\dfrac{1}{\sqrt{3}}\right) y$, labelled σ,

reflection in the x axis, labelled σ,

reflection in the line $x = \left(\dfrac{1}{\sqrt{3}}\right) y$, labelled σ''.

These operations form the group C_{3v}. The character table is given by

$$
\begin{array}{cccc}
 & \{I\} & \{2C_3\} & \{3\sigma\} \\
A1 & 1 & 1 & 1 \\
A2 & 1 & 1 & -1 \\
E & 2 & -1 & 0\,.
\end{array}
\tag{3}
$$

(ii) To apply group theory we first have to find the projection operators. From the character table (3) we find that the projection operators are given by

$$\Pi_{A1} = I + C_3 + C_3^2 + \sigma + \sigma' + \sigma'' \,,$$

$$\Pi_{A2} = I + C_3 + C_3^2 - \sigma - \sigma' - \sigma'' \,,$$

$$\Pi_E = 2I - C_3 - C_3^2 \,.$$

We construct the eigenfunctions as linear combinations of

$$f(x,y) := \sin(P(x + x_0))\sin(Qy)\,,$$

$$g(x,y) := \sin(P(x + x_0))\cos(Qy)\,.$$

Inserting these functions into (1) yields

$$E = \frac{\hbar^2}{m}(P^2 + Q^2)\,.$$

Now we have to apply the projection operators to f and g and then impose the boundary condition (2). We find $\Pi_{A1}f = 0$ and the eigenfunctions of the symmetry type $A1$ are given by

$$\phi(x,y) := \frac{1}{2}\Pi_{A1}g(x,y)\,.$$

Thus

$$\phi(x,y) = \sin(P(x+x_0))\cos(Qy))$$

$$+ \sin\left(P\left(-\frac{1}{2}x - \frac{\sqrt{3}}{2}y + x_0\right)\right)\cos\left(Q\left(-\frac{1}{2}y + \frac{\sqrt{3}}{2}x\right)\right)$$

$$+ \sin\left(P\left(-\frac{1}{2}x + \frac{\sqrt{3}}{2}y + x_0\right)\right)\cos\left(Q\left(-\frac{1}{2}y - \frac{\sqrt{3}}{2}x\right)\right).$$

Analogously, we find $\Pi_{A2}g = 0$ and the eigenfunctions of the symmetry type $A2$ are given by

$$\zeta(x,y) := \frac{1}{2}\Pi_{A2}f(x,y)$$

$$= \sin(P(x+x_0))\sin(Qy)$$

$$+ \sin\left(P\left(-\frac{1}{2}x - \frac{\sqrt{3}}{2}y + x_0\right)\right)\sin\left(Q\left(-\frac{1}{2}y + \frac{\sqrt{3}}{2}x\right)\right)$$

$$+ \sin\left(P\left(-\frac{1}{2}x - \frac{\sqrt{3}}{2}y + x_0\right)\right)\sin\left(Q\left(-\frac{1}{2}y + \frac{\sqrt{3}}{2}x\right)\right).$$

Since the functions of the symmetry class $A1$ are symmetric under C_3 it is sufficient to require that $\phi(x,y) = 0$ for $x = -A$ and any y. Then the functions automatically satisfy the boundary conditions along the two other sides of the triangle. Thus we obtain

$$P = \frac{k\pi}{3A},$$

$$Q = \frac{n\pi}{\sqrt{3}A}$$

where k and n are integers which are either both even or both odd. We can restrict the values of k and n as follows: k and n are positive integers with $k \le n$ and are either both even or both odd. For the symmetry class $A2$ it is sufficient to require that

$$\zeta(x,y) = 0 \quad \text{for } x = -A$$

and any y. The function ζ is identically zero for $k = n$. Therefore we can restrict the values of k as follows: k and n are positive integers with $k > n$ and are either both even or both odd. Consequently, we have an infinite countable number of eigenfunctions of the symmetry class $A1$ and $A2$. Inserting these eigenfunctions into the eigenvalue equation (2) we find the eigenvalues

$$E_{n,k} = \frac{\pi^2 \hbar^2}{mA^2} \left(\frac{k^2}{9} + \frac{n^2}{3} \right).$$

Problem 5. The *telegraph equation* is given by

$$\frac{\partial^2 w}{\partial t^2} + a \frac{\partial w}{\partial t} + bw = c^2 \frac{\partial^2 w}{\partial x^2}.$$ (1)

Show that this equation can be transformed into the canonical form

$$\frac{\partial^2 u}{\partial \eta \partial \xi} + ku = 0$$ (2)

where $k = (a^2 - 4b^2)/(16c^2)$ by applying the transformation

$$\eta(x,t) = x - ct, \quad \xi(x,t) = x + ct,$$ (3a)

$$w(x(\eta,\xi), t(\eta,\xi)) = u(\eta,\xi) \exp\left(-\frac{at(\eta,\xi)}{2} \right).$$ (3b)

Solution. From (3a) and (3b) we find by applying the chain rule

$$\frac{\partial w}{\partial \eta} = \frac{\partial w}{\partial x}\frac{\partial x}{\partial \eta} + \frac{\partial w}{\partial t}\frac{\partial t}{\partial \eta} = \frac{\partial u}{\partial \eta} \exp\left(-\frac{at(\eta,\xi)}{2} \right)$$

$$- \frac{a}{2} u(\eta,\xi) \exp\left(-\frac{at(\eta,\xi)}{2} \right) \frac{\partial t}{\partial \eta}.$$ (4a)

Using (3a) we obtain

$$\frac{\partial w}{\partial x}\frac{1}{2} - \frac{\partial w}{\partial t}\frac{1}{2c} = \frac{\partial u}{\partial \eta} \exp\left(-\frac{at(\eta,\xi)}{2} \right)$$

$$+ \frac{a}{2}\frac{1}{2c} u(\eta,\xi) \exp\left(-\frac{at(\eta,\xi)}{2} \right).$$ (4b)

Analogously when we take the derivative of w with respect to ξ we find

$$\frac{\partial w}{\partial x}\frac{1}{2} + \frac{\partial w}{\partial t}\frac{1}{2c} = \frac{\partial u}{\partial \xi} \exp\left(-\frac{at(\eta,\xi)}{2} \right) - \frac{a}{2}\frac{1}{2c} \exp\left(-\frac{at(\eta,\xi)}{2} \right).$$ (5)

Taking the second derivative of (4b) with respect to η and the second derivative of (5) with respect to ξ and inserting these expressions and $\partial w/\partial t$ into (1) we arrive at (2).

Problem 6.   Consider the eigenvalue equation

$$-\frac{d^2\psi}{dx^2} + V(x)\psi(x) = E\psi(x). \tag{1}$$

Let the potential V be given by

$$V(x) = a^2 x^6 - 3ax^2 \tag{2}$$

where $a > 0$.

(i) Show that for this potential the ground-state wave function ψ_0 is

$$\psi_0(x) = \exp\left(-\frac{a}{4}x^4\right). \tag{3}$$

(ii) What happens when we analytically continue the eigenvalue $E_0(a)$ from positive to negative values of the parameter a?

Solution.   (i) Inserting (3) into (1) yields

$$E(a) = 0.$$

The eigenfunction corresponds to the lowest-lying energy eigenvalue because it has no nodes.

(ii) When we analytically continue the function $E_0(a)$ from positive to negative values, the result is still $E_0(a) = 0$. However, this conclusion is wrong. The spectrum of the potential (2) for $a < 0$ is strictly positive. When we continue the wave function ψ_0 to a negative value of a we find a nonnormalizable solution of the eigenvalue equation (1), which indicates that a simple replacement of a by $-a$ is not allowed.

Problem 7.   Consider the *Dirac–Hamilton operator*

$$\hat{H} = mc^2\beta + c\alpha \cdot \mathbf{p} \equiv mc^2\beta + c(\alpha_1 p_1 + \alpha_2 p_2 + \alpha_3 p_3)$$

where m is the rest mass, c is the speed of light and

$$\beta := \begin{pmatrix} 1 & 0 & 0 & 0 \\ 0 & 1 & 0 & 0 \\ 0 & 0 & -1 & 0 \\ 0 & 0 & 0 & -1 \end{pmatrix},$$

$$
\alpha_1 := \begin{pmatrix} 0 & 0 & 0 & 1 \\ 0 & 0 & 1 & 0 \\ 0 & 1 & 0 & 0 \\ 1 & 0 & 0 & 0 \end{pmatrix}, \quad
\alpha_2 := \begin{pmatrix} 0 & 0 & 0 & -i \\ 0 & 0 & i & 0 \\ 0 & -i & 0 & 0 \\ i & 0 & 0 & 0 \end{pmatrix},
$$

$$
\alpha_3 := \begin{pmatrix} 0 & 0 & 1 & 0 \\ 0 & 0 & 0 & -1 \\ 1 & 0 & 0 & 0 \\ 0 & -1 & 0 & 0 \end{pmatrix},
$$

$$
p_1 := -i\hbar \frac{\partial}{\partial x_1}, \quad p_2 := -i\hbar \frac{\partial}{\partial x_2}, \quad p_3 := -i\hbar \frac{\partial}{\partial x_3}.
$$

Thus the Dirac–Hamilton operator takes the form

$$
\hat{H} = c \begin{pmatrix}
mc & 0 & -\hbar\dfrac{\partial}{\partial x_3} & -i\hbar\dfrac{\partial}{\partial x_1} - \hbar\dfrac{\partial}{\partial x_2} \\
0 & mc & -i\hbar\dfrac{\partial}{\partial x_1} + \hbar\dfrac{\partial}{\partial x_2} & i\hbar\dfrac{\partial}{\partial x_3} \\
-i\hbar\dfrac{\partial}{\partial x_3} & -i\hbar\dfrac{\partial}{\partial x_1} - \hbar\dfrac{\partial}{\partial x_2} & -mc & 0 \\
-i\hbar\dfrac{\partial}{\partial x_1} + \hbar\dfrac{\partial}{\partial x_2} & i\hbar\dfrac{\partial}{\partial x_3} & 0 & -mc
\end{pmatrix}.
$$

Let I_4 be the 4×4 unit matrix. Use the *Heisenberg equation of motion* to find the time evolution of β, α_j and $I_4 p_j$, where $j = 1, 2, 3$.

Solution. The Heisenberg equation of motion is given by

$$
i\hbar \frac{d\hat{A}}{dt} = [\hat{A}, \hat{H}](t).
$$

We have

$$
[\beta, \hat{H}] = 2c \begin{pmatrix}
0 & 0 & -i\hbar\dfrac{\partial}{\partial x_3} & -i\hbar\dfrac{\partial}{\partial x_1} - \hbar\dfrac{\partial}{\partial x_2} \\
0 & 0 & -i\hbar\dfrac{\partial}{\partial x_1} + \hbar\dfrac{\partial}{\partial x_2} & i\hbar\dfrac{\partial}{\partial x_3} \\
i\hbar\dfrac{\partial}{\partial x_3} & i\hbar\dfrac{\partial}{\partial x_1} + \hbar\dfrac{\partial}{\partial x_2} & 0 & 0 \\
i\hbar\dfrac{\partial}{\partial x_1} - \hbar\dfrac{\partial}{\partial x_2} & -\hbar\dfrac{\partial}{\partial x_3} & 0 & 0
\end{pmatrix}.
$$

Thus we find

$$
i\hbar \frac{d\beta}{dt} = 2c\beta\alpha_1(p_1 - ip_2)(t) + 2c\alpha_3 p_3)(t).
$$

Since

$$[\alpha_1, \hat{H}] = -2\hat{H}\alpha_1 + 2cI_4p_1,$$

we find

$$i\hbar\frac{d\alpha_1}{dt} = -(2\hat{H}\alpha_1 + 2cI_4p_1)(t).$$

Analogously,

$$i\hbar\frac{d\alpha_2}{dt} = -(2\hat{H}\alpha_2 + 2cI_4p_2)(t),$$

$$i\hbar\frac{d\alpha_3}{dt} = -(2\hat{H}\alpha_3 + 2cI_4p_3)(t).$$

Furthermore we obviously have

$$[I_4p_j, \hat{H}] = 0, \quad j = 1, 2, 3.$$

Thus

$$i\hbar\frac{d}{dt}I_4p_j = 0, \quad j = 1, 2, 3.$$

From these equations we obtain

$$i\hbar\frac{d^2\alpha}{dt^2} = 2\frac{d\alpha}{dt}\hat{H} = -2\hat{H}\frac{d\alpha}{dt}.$$

This equation can be integrated and we find

$$\frac{d\alpha}{dt} = \frac{d\alpha(0)}{dt}\exp\left(-\frac{2i\hat{H}t}{\hbar}\right).$$

Remark.   *We also have the identities*

$$\{\hat{H}, \dot{\alpha}\}_+ = 0,$$

$$[\hat{H}, \dot{\alpha}] = 2\hat{H}\alpha.$$

Problem 8.   Consider the *linear diffusion equation*

$$\frac{\partial u}{\partial t} = \frac{\partial^2 u}{\partial x^2}$$

with

$$u(x, 0) = f(x), \quad x \in [0, \infty),$$
$$u(0, t) = U, \quad t \geq 0.$$

The exact solution is given by

$$u(x,t) = U \operatorname{erfc}\left(\frac{1}{2}\frac{x}{t^{1/2}}\right) + \frac{1}{2(\pi t)^{1/2}} \int_0^\infty f(x_1)\left(\exp\left(-\frac{(x-x_1)^2}{4t}\right)\right.$$

$$\left. - \exp\left(-\frac{(x+x_1)^2}{4t}\right)\right) dx_1 . \tag{1}$$

Find the asymptotic expansion of (1) as $t \to \infty$, $\eta = O(1)$ with $\frac{1}{2}\eta x/t^{1/2}$.

Solution.   The solution $u(x,t)$ can be written as

$$u(x,t) \sim U \operatorname{erfc}\left(\frac{1}{2}\frac{x}{t^{1/2}}\right) + \frac{1}{\pi^{1/2}}$$

$$\times \left(\frac{I_2(0)\eta \exp(-\eta^2)}{t} + \frac{I_4(0)(\eta^3 - \frac{3}{2}\eta)\exp(-\eta^2)}{t^2} + \cdots\right). \tag{2}$$

The coefficients being proportional to *Hermite polynomials*. If we assume that the integrals converge, then the numbers $I_2(0)$, $I_4(0), \ldots$, are given by

$$I_2(x) = \int_x^\infty dx_1 \int_{x_1}^\infty f(x_2)dx_2 , \quad I_4(x) = \int_x^\infty dx_{x_1} \int_{x_1}^\infty I_2(x_2)dx_2, \ldots$$

at $x = 0$. The expansion (2) depends on the initial condition only via these coefficients.

Problem 9.   If u is a differentiable function of x, y, and z which satisfies the partial differential equation

$$(y - z)\frac{\partial u}{\partial x} + (z - x)\frac{\partial u}{\partial y} + (x - y)\frac{\partial u}{\partial z} = 0$$

show that u contains x, y and z only in combinations $x + y + z$ and $x^2 + y^2 + z^2$.

Solution.   The auxiliary equations are

$$\frac{dx}{y - z} = \frac{dy}{z - x} = \frac{dz}{x - y} = \frac{du}{0}$$

and they are equivalent to the three relations

$$du = 0, \quad dx + dy + dz = 0, \quad xdx + ydy + zdz = 0 .$$

Thus the integrals are

$$u = c_1 , \quad x + y + z = c_2 , \quad x^2 + y^2 + z^2 = c_3$$

where c_1, c_2, c_3 are constants. Thus the general solution is given by

$$u(x, y, z) = f(x + y + z, x^2 + y^2 + z^2)$$

where f is an arbitrary differentiable function.

Chapter 11

Nonlinear Partial Differential Equations

Problem 1. The nonlinear partial differential equation

$$\frac{\partial \mathbf{S}}{\partial t} = \mathbf{S} \times \frac{\partial^2 \mathbf{S}}{\partial x^2} \tag{1}$$

where (constraint)

$$S_1^2 + S_2^2 + S_3^2 = 1 \tag{2}$$

is called the *Heisenberg ferromagnet equation* in one-space dimension. Here $\times$ denotes the vector product and

$$\mathbf{S} = \begin{pmatrix} S_1 \\ S_2 \\ S_3 \end{pmatrix}.$$

(i) Write Eq. (1) in components.

(ii) Let

$$Q := 1 + u^2 + v^2$$

and set

$$S_1 := \frac{2u}{Q}, \quad S_2 := \frac{2v}{Q}, \quad S_3 := \frac{-1 + u^2 + v^2}{Q}. \tag{3}$$

Show that condition (2) is satisfied identically. Find the time evolution of u and v.

Remark. *The transformation (3) is called stereographic projection.*

125

Solution.   (i) Since

$$\mathbf{a} \times \mathbf{b} := \begin{pmatrix} a_2 b_3 - a_3 b_2 \\ a_3 b_1 - a_1 b_3 \\ a_1 b_2 - a_2 b_1 \end{pmatrix},$$

we have

$$\frac{\partial S_1}{\partial t} = S_2 \frac{\partial^2 S_3}{\partial x^2} - S_3 \frac{\partial^2 S_2}{\partial x^2}, \tag{4a}$$

$$\frac{\partial S_2}{\partial t} = S_3 \frac{\partial^2 S_1}{\partial x^2} - S_1 \frac{\partial^2 S_3}{\partial x^2}, \tag{4b}$$

$$\frac{\partial S_3}{\partial t} = S_1 \frac{\partial^2 S_2}{\partial x^2} - S_2 \frac{\partial^2 S_1}{\partial x^2}. \tag{4c}$$

(ii) From (3) we find

$$\frac{\partial S_1}{\partial t} = \frac{2}{Q} \frac{\partial u}{\partial t} - \frac{2u}{Q^2} \frac{\partial Q}{\partial t} = \frac{1}{Q^2} \left(2Q \frac{\partial u}{\partial t} - 2u \frac{\partial Q}{\partial t} \right), \tag{5a}$$

$$\frac{\partial S_2}{\partial t} = \frac{2}{Q} \frac{\partial v}{\partial t} - \frac{2v}{Q^2} \frac{\partial Q}{\partial t} = \frac{1}{Q^2} \left(2Q \frac{\partial v}{\partial t} - 2v \frac{\partial Q}{\partial t} \right). \tag{5b}$$

Since

$$S_3 = \frac{-1 + u^2 + v^2}{Q} \equiv \frac{Q-2}{Q} = 1 - \frac{2}{Q},$$

we have

$$\frac{\partial S_3}{\partial t} = \frac{2}{Q^2} \frac{\partial Q}{\partial t}. \tag{5c}$$

Now

$$\frac{\partial^2 S_1}{\partial x^2} = \frac{2}{Q} \frac{\partial^2 u}{\partial x^2} - \frac{4}{Q^2} \frac{\partial u}{\partial x} \frac{\partial Q}{\partial x} - \frac{2u}{Q^2} \frac{\partial^2 Q}{\partial x^2} + \frac{4u}{Q^3} \left(\frac{\partial Q}{\partial x} \right)^2,$$

$$\frac{\partial^2 S_2}{\partial x^2} = \frac{2}{Q} \frac{\partial^2 v}{\partial x^2} - \frac{4}{Q^2} \frac{\partial v}{\partial x} \frac{\partial Q}{\partial x} - \frac{2v}{Q^2} \frac{\partial^2 Q}{\partial x^2} + \frac{4v}{Q^3} \left(\frac{\partial Q}{\partial x} \right)^2$$

and

$$\frac{\partial^2 S_3}{\partial x^2} = \frac{2}{Q^2} \frac{\partial^2 Q}{\partial x^2} - \frac{4}{Q^3} \left(\frac{\partial Q}{\partial x} \right)^2. \tag{6}$$

Inserting (5) through (6) into (4c) yields

$$\frac{2}{Q^2}\frac{\partial Q}{\partial t} = \frac{2u}{Q}\left(\frac{2}{Q}\frac{\partial^2 v}{\partial x^2} - \frac{4}{Q^2}\frac{\partial v}{\partial x}\frac{\partial Q}{\partial x} - \frac{2v}{Q^2}\frac{\partial^2 Q}{\partial x^2} + \frac{4v}{Q^3}\left(\frac{\partial Q}{\partial x}\right)^2\right)$$

$$-\frac{2v}{Q}\left(\frac{2}{Q}\frac{\partial^2 u}{\partial x^2} - \frac{4}{Q^2}\frac{\partial u}{\partial x}\frac{\partial Q}{\partial x} - \frac{2u}{Q^2}\frac{\partial^2 Q}{\partial x^2} + \frac{4u}{Q^3}\left(\frac{\partial Q}{\partial x}\right)^2\right)$$

or

$$\frac{\partial Q}{\partial t} = u\left(2\frac{\partial^2 v}{\partial x^2} - \frac{4}{Q}\frac{\partial v}{\partial x}\frac{\partial Q}{\partial x}\right) - v\left(2\frac{\partial^2 u}{\partial x^2} - \frac{4}{Q}\frac{\partial u}{\partial x}\frac{\partial Q}{\partial x}\right). \tag{7}$$

Inserting (5) through (6) and (7) into (4a) and (4b) yields

$$Q\frac{\partial v}{\partial t} + Q\frac{\partial^2 u}{\partial x^2} - 2\left(\left(\frac{\partial u}{\partial x}\right)^2 - \left(\frac{\partial v}{\partial x}\right)^2\right)u - 4v\frac{\partial u}{\partial x}\frac{\partial v}{\partial x} = 0,$$

$$-Q\frac{\partial u}{\partial t} + Q\frac{\partial^2 v}{\partial x^2} + 2\left(\left(\frac{\partial u}{\partial x}\right)^2 - \left(\frac{\partial v}{\partial x}\right)^2\right)v - 4u\frac{\partial u}{\partial x}\frac{\partial v}{\partial x} = 0.$$

Problem 2.   (i) Solve the partial differential equation

$$F\left(u, \frac{\partial u}{\partial t}, \frac{\partial u}{\partial x}\right) \equiv \frac{\partial u}{\partial t} + u\frac{\partial u}{\partial x} = 0 \tag{1}$$

with the initial condition

$$u(x, t = 0) = x. \tag{2}$$

(ii) Show that the nonlinear partial differential equation

$$\frac{\partial u}{\partial t} + (\alpha + \beta u)\frac{\partial u}{\partial x} = 0 \tag{3}$$

admits the general solution

$$u(x, t) = f(x - (\alpha + \beta u(x, t))t) \tag{4}$$

to the initial-value problem, where

$$u(x, t = 0) = f(x)$$

and $\alpha, \beta \in \mathbf{R}$.

(iii) Simplify case (ii) to case (i).

Solution.   (i) From (1) we obtain the surface

$$F(u, p, q) \equiv p + uq = 0 \,.\tag{5}$$

Then we obtain the autonomous system of first-order ordinary differential equations

$$\frac{dt}{ds} = \frac{\partial F}{\partial p} = 1 \,,\tag{6a}$$

$$\frac{dx}{ds} = \frac{\partial F}{\partial q} = u \,,\tag{6b}$$

$$\frac{du}{ds} = p\frac{\partial F}{\partial p} + q\frac{\partial F}{\partial q} = p + uq = 0\tag{6c}$$

where we have used (5) in (6c). These equations determine the *characteristic strip*. The solution of the initial-value problem of system (6) is given by

$$x(s) = u_0 s + x_0 \,,$$
$$t(s) = s + t_0 \,,$$
$$u(s) = u_0 \,.$$

To impose the initial condition (2) we have to set

$$x_0(\alpha) = \alpha \,,$$
$$t_0(\alpha) = 0 \,,$$
$$u_0(\alpha) = \alpha \,.$$

Thus we obtain

$$x(s, \alpha) = \alpha s + \alpha \,,\tag{7a}$$

$$t(s, \alpha) = s \,,\tag{7b}$$

$$u(s, \alpha) = \alpha \,.\tag{7c}$$

Since

$$D := \frac{\partial(t, x)}{\partial(s, \alpha)} = (1 + s) \,,$$

we can solve (7a) and (7b) with respect to s and α if $D \neq 0$. We find

$$s(x,t) = t,$$

$$\alpha(x,t) = \frac{x}{1+t}.$$

Inserting s and α into (7c) gives the solution

$$u(x,t) = \frac{x}{1+t} \tag{8}$$

of the initial-value problem.

(ii) Since

$$\frac{\partial u}{\partial t} = \left(-\alpha - \beta u - \beta t \frac{\partial u}{\partial t} \right) f',$$

$$\frac{\partial u}{\partial x} = \left(1 - \beta t \frac{\partial u}{\partial x} \right) f'$$

where f' is the derivative of f with respect to the argument we obtain

$$\left(1 - \beta t \frac{\partial u}{\partial x} \right) \frac{\partial u}{\partial t} = \left(-\alpha - \beta u - \beta t \frac{\partial u}{\partial t} \right) \frac{\partial u}{\partial x}$$

or

$$\frac{\partial u}{\partial t} = (-\alpha - \beta u) \frac{\partial u}{\partial x} \tag{9}$$

which is (3).

(iii) Let us now simplify (9) to case (i). Let $\alpha = 0$ and $\beta = 1$. Then we obtain from (4) that

$$u(x,t) = f(x - u(x,t)t).$$

Since

$$u(x, t = 0) = f(x) = x,$$

it follows that

$$u(x,t) = f(x - u(x,t)t) = x - u(x,t)t. \tag{10}$$

From (10) we easily see that solution (8) follows.

Problem 3.   (i) Show that the nonlinear partial differential equation

$$\frac{\partial^2 u}{\partial t^2} - \frac{\partial^2 u}{\partial x^2} = \mu^2 y - \lambda u^3 \tag{1}$$

cab be derived from the *Lagrange density*

$$\mathcal{L} = \frac{1}{2}\left(\left(\frac{\partial u}{\partial t}\right)^2 - \left(\frac{\partial u}{\partial x}\right)^2\right) + \frac{1}{2}\mu^2 u^2 - \frac{1}{4}\lambda u^4 \tag{2}$$

where $\mu^2 > 0$ and $\lambda > 0$. From the Lagrange density $\mathcal{L}$ it follows that we can introduce the Hamilton density and therefore the *energy functional*

$$E(u) = \int_{-\infty}^{+\infty} dx \left(\frac{1}{2}\left(\frac{\partial u}{\partial t}\right)^2 + \frac{1}{2}\left(\frac{\partial u}{\partial x}\right)^2 - \frac{1}{2}\mu^2 u^2 + \frac{1}{4}\lambda u^4\right).$$

(ii) Find the space-time independent solutions of (1).

(iii) Calculate the energy difference between the different solutions of (i).

(iv) Show that (1) admits the time-independent solution

$$u_{\text{kink}}(x) = \pm \frac{\mu}{\sqrt{\lambda}} \tanh\left(\frac{\mu}{\sqrt{2}}(x - x_0)\right) \tag{3}$$

and calculate the energy difference.

Remark.   *Owing to its shape, the solution* (3) *is called the kink solution.*

Solution.   (i) The *Euler–Lagrange equation* is given by

$$\frac{\partial \mathcal{L}}{\partial u} - \frac{\partial}{\partial x}\left(\frac{\partial \mathcal{L}}{\partial(\partial u/\partial x)}\right) - \frac{\partial}{\partial t}\left(\frac{\partial \mathcal{L}}{\partial(\partial u/\partial t)}\right) = 0. \tag{4}$$

Inserting the Lagrange density (2) into (4) gives (1).

(ii) The space-time independent solutions are determined by

$$\mu^2 u - \lambda u^3 = 0.$$

We obtain

$$u = 0, \quad u = \pm\frac{\mu}{\sqrt{\lambda}}.$$

(iii) Straightforward calculation shows that

$$E(u = 0) - E\left(u = \pm\frac{\mu}{\sqrt{\lambda}}\right) = \lim_{L \to \infty}\int_{-L}^{+L} dx \frac{1}{4}\frac{\mu^4}{\lambda} = +\infty$$

which implies that the difference between the energy densities is positive and moreover that $u = \pm\mu/\sqrt{\lambda}$ are the minima ($u = 0$ is a maximum). The set of vacua is degenerate.

(iv) If u is independent of t we obtain from (1)

$$\frac{d^2u}{dx^2} + \mu^2 u - \lambda u^3 = 0.$$

Consequently

$$\frac{1}{2}\left(\frac{du}{dx}\right)^2 + \mu^2 \frac{u^2}{2} - \frac{\lambda u^4}{4} = \text{const}.$$

It follows that

$$x - x_0 = \int_0^u \frac{du'}{\sqrt{(\lambda u'^4/2) - \mu^2 u'^2 + k}}, \quad x_0, k = \text{const}$$

which represents an elliptic integral. Therefore $u(x)$ is periodic on the complex x-plane. Hence $E[u]$ is of infinite energy, unless the two zeroes of the square root coalesce. This happens for $k = \mu^4/2\lambda$. Then we obtain (3). Straightforward calculation gives

$$E(u_{\text{kink}}) - E(u_{\text{vac}}) = \frac{2\sqrt{2}}{3}\frac{\mu^3}{\lambda} < +\infty$$

where $u_{\text{vac}} = \pm\mu/\sqrt{\lambda}$.

Remark.   *A space-time dependent solution can be found from* (3) *by applying the Lorentz transformation.*

Problem 4.   The Navier–Stokes equation of a viscous incompressible fluid is given by

$$\left(\frac{\partial}{\partial t} + \mathbf{u} \cdot \nabla\right)\mathbf{u} = -\frac{1}{\rho}\nabla p + \nu\Delta\mathbf{u}. \tag{1}$$

We assume that the fluid is incompressible, i.e.

$$\nabla \cdot \mathbf{u} = \mathbf{0}. \tag{2}$$

(i) Express the Navier–Stokes equation in cylindrical coordinates r, ϕ, z.

(ii) Consider a flow between concentric rotating cylinders. The cylinders are infinitely long. Let r_1, r_2 and Ω_1, Ω_2 denote the radii and angular velocities of the inner and outer cylinders, respectively. We denote the velocity components in the increasing r, ϕ and z directions by u, v and w. The boundary conditions are

$$u = w = 0 \quad \text{at } r = r_1 \text{ and } r = r_2,$$

$$v(r_1, \phi, z, t) = r_1 \Omega_1,$$

$$v(r_2, \phi, z, t) = r_2 \Omega_2.$$

These conditions are called *no-slip conditions*. Find an exact solution of the form

$$u = w = 0, \tag{3a}$$

$$v = V(r). \tag{3b}$$

Solution.   The polar coordinates are given by

$$x(\phi, r) = r \cos \phi,$$

$$y(\phi, r) = r \sin \phi.$$

Applying the chain rule we find that systems (1) and (2) take the form

$$\left(\frac{\partial}{\partial t} + \mathbf{u} \cdot \nabla \right) u - \frac{v^2}{r} = -\frac{\partial}{\partial r} \frac{p}{\rho} + \nu \left(\Delta - \frac{1}{r^2} \right) u - \frac{2\nu}{r^2} \frac{\partial}{\partial \phi} v, \tag{4a}$$

$$\left(\frac{\partial}{\partial t} + \mathbf{u} \cdot \nabla \right) v - \frac{uv^2}{r} = -\frac{1}{r} \frac{\partial}{\partial \phi} \frac{p}{\rho} + \nu \left(\Delta - \frac{1}{r^2} \right) v - \frac{2\nu}{r^2} \frac{\partial}{\partial \phi} u, \tag{4b}$$

$$\left(\frac{\partial}{\partial t} + \mathbf{u} \cdot \nabla \right) w = -\frac{\partial}{\partial z} \frac{p}{\rho} + \nu \Delta w, \tag{4c}$$

$$\left(\frac{\partial}{\partial r} + \frac{1}{r} \right) u = -\frac{1}{r} \frac{\partial}{\partial \phi} v - \frac{\partial}{\partial z} w \tag{4d}$$

where

$$\mathbf{u} \cdot \nabla := u \frac{\partial}{\partial r} + \frac{1}{r} v \frac{\partial}{\partial \phi} + w \frac{\partial}{\partial z},$$

$$\Delta := \frac{\partial^2}{\partial r^2} + \frac{1}{r} \frac{\partial}{\partial r} + \frac{1}{r^2} \frac{\partial^2}{\partial \phi^2} + \frac{\partial^2}{\partial z^2}.$$

Inserting (3a) and (3b) into (4a) yields

$$-\frac{V^2(r)}{r} = -\frac{1}{\rho} \frac{\partial p}{\partial r}.$$

Thus we can assume that the pressure p depends only on r. Equation (4c) is satisfied identically. From (4b) we find

$$\nu \left(\frac{\partial^2}{\partial r^2} + \frac{1}{r}\frac{\partial}{\partial r} - \frac{1}{r^2} \right) V(r) = 0.$$

Therefore the solution is given by

$$V(r) = Ar + \frac{B}{r}$$

where A and B are the constants of integration. Imposing the boundary conditions gives

$$A = \frac{\Omega_2 r_2^2 - \Omega_1 r_1^2}{r_2^2 - r_1^2}, \quad B = \frac{r_1^2 r_2^2 (\Omega_1 - \Omega_2)}{r_2^2 - r_1^2}.$$

Problem 5.   Consider the nonlinear diffusion equation

$$\frac{\partial u}{\partial t} = D\frac{\partial}{\partial x}\left(\frac{1}{u}\frac{\partial u}{\partial x} \right)$$

where $u(x,t)$ is the density and x,t are the space, time coordinates and $x \in [0,1]$. D is a positive constant. We set at the boundaries

$$u(0,t) = u(1,t) = u_0 \geq 0.$$

(i) Show the problem is well-posed.

(ii) Introduce the new independent and dependent variables

$$\tau(t,x) = \frac{D}{u_0}t, \quad y(t,x) = x, \tag{1a}$$

$$v(\tau(t,x), y(t,x)) = \ln\left(\frac{u(x,t)}{u_0} \right). \tag{1b}$$

Find the partial differential equation for these variables.

(iii) How do the boundary conditions change?

Solution.   (i) Since

$$\frac{d}{dt}\int_0^1 u(x,t)dx = D\left(\frac{1}{u}\frac{\partial u}{\partial x} \right)\Big|_0^1 \tag{2}$$

the flux (i.e. the right-hand side of (2) will be finite and the problem is well posed.

(ii) From (1b) we find

$$u(x,t) = u_0 \exp(v(\tau(t,x), y(t,x))).$$ (3)

Using (3) and (1b) we find, by applying the chain rule,

$$\frac{\partial^2 v}{\partial y^2} = \exp(v)\frac{\partial v}{\partial \tau} = \frac{\partial}{\partial \tau}\exp(v).$$

(iii) The boundary conditions change to

$$v(0,t) = v(1,t) = 0.$$

The quantity v is nonnegative and the new time scale differs from the old by a factor of u_0/D.

Problem 6.   Consider the complex *Ginzburg–Landau equation*

$$\frac{\partial A}{\partial t} = (1 + ic_1)\frac{\partial^2 A}{\partial x^2} + A - (-1 + ic_2)|A|^2 A.$$ (1a)

Find a solution of the form

$$A(x,t) = R(x,t)\exp(i\Theta(x,t))$$ (1b)

where R and Θ are real-valued functions.

Solution.   Inserting (2) into (1) we obtain two real equations

$$\epsilon^2\frac{\partial R}{\partial t} + \epsilon R\frac{\partial \Theta}{\partial t} = (1 + \epsilon^2)\left(\frac{\partial^2 R}{\partial x^2} - \left(\frac{\partial \Theta}{\partial x}\right)^2 R\right) + \epsilon^2 R + (\beta + \epsilon^2)R^3,$$

$$-\frac{1}{2}\epsilon\frac{\partial R^2}{\partial t} + \epsilon^2 R^2\frac{\partial \Theta}{\partial t} = (1 + \epsilon^2)\frac{\partial}{\partial x}\left(R^2\frac{\partial \Theta}{\partial x}\right) - \epsilon(1 + (1 - \beta)R^2)R^2$$

where $c_2 = -\beta c_1$ and $\epsilon = 1/c_1$. For sufficiently large values of c_1 an expansion in ϵ becomes meaningful

$$R = R_0 + \epsilon^2 R_2 + \cdots,$$

$$\Theta = \frac{1}{\epsilon}(\Theta_{-1} + \epsilon^2\Theta_1 + \cdots).$$

For the order ϵ^{-2} we find

$$R_0\left(\frac{\partial \Theta_{-1}}{\partial x}\right)^2 = 0.$$

For the order ϵ^{-1} we find

$$\frac{\partial}{\partial x}\left(R_0^2 \frac{\partial \Theta_{-1}}{\partial x}\right) = 0.$$

Excluding $R_0 = 0$ we obtain $\partial \Theta_{-1}/\partial x = 0$, so that Θ_{-1} only depends on t. Setting

$$\gamma(t) := \frac{\partial \Theta_{-1}}{\partial t},$$

we obtain for the ϵ^0,

$$0 = \frac{\partial^2 R_0}{\partial x^2} - \gamma(t)R_0 + \beta R_0^3. \tag{3a}$$

For ϵ^1 we find

$$-\frac{\partial}{\partial x}\left(R_0^2 \frac{\partial \Theta_1}{\partial x}\right) = \frac{1}{2}\frac{\partial R_0^2}{\partial t} - (1 + \gamma(t))R_0^2 - (1 - \beta)R_0^4. \tag{3b}$$

For $\beta, \gamma > 0$, we obtain from (3a) the spatially periodic solution

$$R_0(x,t) = \left(\frac{2\gamma(t)}{(2 - m(t))\beta}\right)^{1/2} \mathrm{dn}\left(\left(\frac{\gamma(t)}{2 - m(t)}\right)^{1/2} x, m(t)\right). \tag{4}$$

Here $\mathrm{dn}(u, m)$ is a Jacobian elliptic function that varies between $(1 - m)$ and 1 with period $2K(m)$. The parameter m is between 0 and 1, and $K(m)$ is the complete elliptic integral of the first kind. For $m \to 1$ the period of the function dn goes to infinity and (4) degenerates into the pulse

$$\left(\frac{2\gamma}{\beta}\right)^{1/2} \mathrm{sech}(\gamma^{1/2}x) \tag{5}$$

while for $m \to 0$ one has small, harmonic oscillations.

Problem 7. In plasma physics the differential equations for the density of the ion-fluid n and its velocity u, in dimensionless form, is given by

$$\frac{\partial n}{\partial t} + \frac{\partial (nu)}{\partial x} = 0 \qquad \text{equation of continuity}$$

$$\frac{\partial u}{\partial t} + u\frac{\partial u}{\partial x} = E \qquad \text{equation of motion}$$

$$\frac{\partial n_e}{\partial x} = -n_e E \qquad \text{balance of pressure and electric force}$$

$$\frac{\partial E}{\partial x} = n - n_e \qquad \text{the Poisson equation}$$

where n_e is the electron density, E the electric field, and the inertia term is neglected because of the small mass of electrons.

(i) Eliminate n and E to find the differential equations for u and n_e.

(ii) Introduce

$$\xi = \epsilon^{1/2}(x - t), \quad \eta = \epsilon^{3/2}x$$

and find a solution of the form

$$u = \epsilon u^{(1)} + \epsilon^2 u^{(2)} + \cdots, \quad n_e = 1 + \epsilon n_e^{(1)} + \epsilon^2 n_e^{(2)} + \cdots. \qquad (1a)$$

Solution.   (i) Eliminating n and E yields

$$\frac{\partial u}{\partial t} + u\frac{\partial u}{\partial x} + \frac{1}{n_e}\frac{\partial n_2}{\partial x} = 0, \qquad (2a)$$

$$\frac{\partial n_e}{\partial t} + \frac{\partial (n_e u)}{\partial x} + \frac{\partial P}{\partial x} = 0, \qquad (2b)$$

where

$$P := -\left(\frac{\partial}{\partial t} + u\frac{\partial}{\partial x}\right)\left(\frac{1}{n_e}\frac{\partial n_e}{\partial x}\right).$$

(ii) Inserting expansion (1) into (2) we obtain

$$u^{(1)} = n_e^{(1)},$$

$$\frac{\partial u^{(1)}}{\partial \eta} + u^{(1)}\frac{\partial u^{(1)}}{\partial \xi} + \frac{\partial^3 u^{(1)}}{\partial \xi^3} = 0,$$

$$\frac{\partial n_e^{(1)}}{\partial \eta} + n_e^{(1)}\frac{\partial n_e^{(1)}}{\partial \xi} + \frac{\partial^3 n_e^{(1)}}{\partial \xi^3} = 0.$$

We see that u, the ion-fluid velocity, and n_e, the electron density, obey the same *Korteweg–de Vries equation* and move with the same phase $u^{(1)} = n_e^{(1)}$. The nonlinear term

$$u^{(1)} \frac{\partial u^{(1)}}{\partial \xi}$$

comes from the interaction of the ions with electrons affecting ions themselves, and

$$n_e^{(1)} \frac{\partial n_e^{(1)}}{\partial \xi}$$

expresses similar effects on electrons through the interaction with ions. We have

$$n^{(1)}(\eta, \xi) = - \int^{\xi} E^{(1)}(\eta, s)ds = u^{(1)}(\eta, \xi) = n_e^{(1)}(\eta, \xi).$$

Problem 8.   The spherically symmetric SU(2) Yang–Mills equations can be written in the form

$$\frac{\partial \phi_1}{\partial t} - \frac{\partial \phi_2}{\partial r} = -A_0 \phi_2 - A_1 \phi_1, \tag{1a}$$

$$\frac{\partial \phi_2}{\partial t} - \frac{\partial \phi_1}{\partial r} = -A_1 \phi_2 + A_0 \phi_1, \tag{1b}$$

$$r^2 \left(\frac{\partial A_1}{\partial t} - \frac{\partial A_0}{\partial r} \right) = 1 - (\phi_1^2 + \phi_2^2) \tag{1c}$$

where r is the spatial radius vector and t the time. Introduce the new variables

$$z(r, t) = \phi_1(r, t) + i\phi_2(r, t), \tag{2a}$$

$$z(r, t) = R(r, t) \exp(i\theta(r, t)). \tag{2b}$$

Find the differential equation for R.

Solution.   Using (2a) we obtain from (1a) and (1b) that

$$\frac{\partial z}{\partial t} + i \frac{\partial z}{\partial r} = (iA_0 - A_1)z.$$

Using (2b) we obtain after separating the imaginary and real parts

$$A_0 = \frac{\partial \theta}{\partial r} + \frac{\partial \ln R}{\partial r}, \quad A_1 = \frac{\partial \theta}{\partial r} + \frac{\partial \ln R}{\partial t}.$$

Substitution of this equation into (1c) yields

$$r^2 \left(\frac{\partial^2}{\partial r^2} + \frac{\partial^2}{\partial t^2} \right) \ln R = R^2 - 1. \tag{3}$$

Thus θ does not arise in (3). Changing the variables

$$R(r, t) = r \exp(g(r, t)),$$

we find the *Liouville equation*

$$\left(\frac{\partial^2}{\partial r^2} + \frac{\partial^2}{\partial t^2} \right) g = \exp(2g) \tag{4}$$

which has the general solution in terms of two harmonic functions $a(r, t)$ and $b(r, t)$ related by the *Cauchy–Riemann condition*

$$\exp(2g) = 4 \left(\left(\frac{\partial a}{\partial r} \right)^2 + \left(\frac{\partial a}{\partial t} \right)^2 \right) \frac{1}{(1 - a^2 - b^2)^2},$$

$$\left(\frac{\partial^2}{\partial r^2} + \frac{\partial^2}{\partial t^2} \right) a = 0, \quad \left(\frac{\partial^2}{\partial r^2} + \frac{\partial^2}{\partial t^2} \right) b = 0,$$

$$\frac{\partial a}{\partial r} = \frac{\partial b}{\partial t}, \quad \frac{\partial a}{\partial t} = \frac{\partial b}{\partial t}.$$

Equation (4) was obtained with an arbitrary function $\theta(r, t)$.

Chapter 12

Group Theoretical Reductions

Problem 1. The motion of an incompressible constant-property fluid in two-space dimension is described by the *stream function equation*

$$\nabla^2 \frac{\partial \psi}{\partial t} + \frac{\partial \psi}{\partial y} \nabla^2 \frac{\partial \psi}{\partial x} - \frac{\partial \psi}{\partial x} \nabla^2 \frac{\partial \psi}{\partial y} = \nu \nabla^4 \psi \qquad (1)$$

where $\nabla^2 := \partial^2/\partial x^2 + \partial^2/\partial y^2$ and ν is a positive constant. The *stream function* ψ is defined by

$$u = \frac{\partial \psi}{\partial y}, \qquad v = -\frac{\partial \psi}{\partial x}$$

in order to satisfy the continuity equation identically, i.e. $\partial u/\partial x + \partial v/\partial y = 0$.

(i) Show that (1) is invariant under

$$x'(x, y, t, \epsilon) = x \cos(\epsilon t) + y \sin(\epsilon t),$$
$$y'(x, y, t, \epsilon) = -x \sin(\epsilon t) + y \cos(\epsilon t),$$
$$t'(x, y, t, \epsilon) = t, \qquad (2)$$
$$\psi'(x'(x, y, t), y'(x, y, t), t'(x, y, t), \epsilon) = \psi(x, y, t) + \frac{1}{2}\epsilon(x^2 + y^2)$$

where ϵ is a real parameter.

(ii) Find the infinitesimal generator of this transformation.

Solution.   (i) Applying the chain rule we find

$$\frac{\partial \psi'}{\partial x'} \cos(\epsilon t) - \frac{\partial \psi'}{\partial y'} \sin(\epsilon t) = \frac{\partial \psi}{\partial x} + \epsilon x, \tag{3a}$$

$$\frac{\partial \psi'}{\partial x'} \sin(\epsilon t) + \frac{\partial \psi'}{\partial y'} \cos(\epsilon t) = \frac{\partial \psi}{\partial y} + \epsilon y \tag{3b}$$

and

$$\frac{\partial \psi'}{\partial x'}(-\epsilon x \sin(\epsilon t) + \epsilon y \cos(\epsilon t))$$

$$+ \frac{\partial \psi'}{\partial y'}(-\epsilon x \cos(\epsilon t) - \epsilon y \sin(\epsilon t)) + \frac{\partial \psi'}{\partial t'} = \frac{\partial \psi}{\partial t}. \tag{3c}$$

From (3a) and (3b) it follows that

$$\frac{\partial^2 \psi'}{\partial x'^2} \cos^2(\epsilon t) - 2\frac{\partial^2 \psi'}{\partial x' \partial y'} \cos(\epsilon t)\sin(\epsilon t) + \frac{\partial^2 \psi'}{\partial y'^2}\sin^2(\epsilon t) = \frac{\partial^2 \psi}{\partial x^2} + \epsilon,$$

$$\frac{\partial^2 \psi'}{\partial x'^2} \sin^2(\epsilon t) + 2\frac{\partial^2 \psi'}{\partial x' \partial y'} \cos(\epsilon t)\sin(\epsilon t) + \frac{\partial^2 \psi'}{\partial y'^2}\cos^2(\epsilon t) = \frac{\partial^2 \psi}{\partial y^2} + \epsilon.$$

Therefore

$$\frac{\partial^2 \psi'}{\partial x'^2} + \frac{\partial^2 \psi'}{\partial y'^2} = \frac{\partial^2 \psi}{\partial x^2} + \frac{\partial^2 \psi}{\partial y^2} + 2\epsilon.$$

Thus it is obvious that

$$\nu \nabla'^4 \psi' = \nu \nabla^4 \psi.$$

From (3c) we find that

$$\nabla'^2 \frac{\partial \psi'}{\partial t'} = \nabla^2 \frac{\partial \psi}{\partial t} + \epsilon x \nabla^2 \frac{\partial \psi}{\partial y} - \epsilon y \nabla^2 \frac{\partial \psi}{\partial x}.$$

From (3a) and (3b) we also find that

$$\frac{\partial \psi'}{\partial y'} \nabla'^2 \frac{\partial \psi'}{\partial x'} - \frac{\partial \psi'}{\partial x'} \nabla'^2 \frac{\partial \psi'}{\partial y'}$$

$$= \frac{\partial \psi}{\partial y} \nabla^2 \frac{\partial \psi}{\partial x} - \frac{\partial \psi}{\partial x} \nabla^2 \frac{\partial \psi}{\partial y} - \epsilon x \nabla^2 \frac{\partial \psi}{\partial y} + \epsilon y \nabla^2 \frac{\partial \psi}{\partial x}.$$

Thus we arrive at

$$\nabla'^2 \frac{\partial \psi'}{\partial t'} + \frac{\partial \psi'}{\partial y'} \nabla'^2 \frac{\partial \psi'}{\partial x'} - \frac{\partial \psi'}{\partial x'} \nabla'^2 \frac{\partial \psi'}{\partial y'} - \nu \nabla'^4 \psi' = 0.$$

(ii) From (2) we obtain the mapping

$$x'(x, y, t, \epsilon) = x \cos(\epsilon t) + y \sin(\epsilon t),$$

$$y'(x, y, t, \epsilon) = -x \sin(\epsilon t) + y \cos(\epsilon t),$$

$$t'(x, y, t, \epsilon) = t,$$

$$\psi'(x, y, t, \psi, \epsilon) = \psi + \frac{1}{2}\epsilon(x^2 + y^2).$$

It follows that

$$\left.\frac{dx'}{d\epsilon}\right|_{\epsilon=0} = yt, \quad \left.\frac{dy'}{d\epsilon}\right|_{\epsilon=0} = -xt, \quad \left.\frac{dt'}{d\epsilon}\right|_{\epsilon=0} = 0, \quad \left.\frac{d\psi'}{d\epsilon}\right|_{\epsilon=0} = \frac{1}{2}(x^2 + y^2).$$

Thus the infinitesimal generator is given by

$$U = yt\frac{\partial}{\partial x} - xt\frac{\partial}{\partial y} + \frac{1}{2}(x^2 + y^2)\frac{\partial}{\partial \psi}.$$

We find the transformation (2) from the generator U when we apply the exponential map, i.e.,

$$\begin{pmatrix} x'(x, y, t, \epsilon) \\ y'(x, y, t, \epsilon) \\ t'(x, y, t, \epsilon) \\ \psi'(x'(x, y, t), y'(x, y, t), t'(x, y, t), \epsilon) \end{pmatrix} = e^{\epsilon U} \left.\begin{pmatrix} x \\ y \\ t \\ \psi \end{pmatrix}\right|_{\psi \to \psi(x, y, t)}$$

Problem 2. (i) Show that the linear one-dimensional diffusion equation

$$\frac{\partial u}{\partial t} = \frac{\partial^2 u}{\partial x^2} \tag{1}$$

is invariant under the transformation

$$t'(x, t, \epsilon) = t, \quad x'(x, t, \epsilon) = t\epsilon + x,$$

$$u'(x'(x, t), t'(x, t), \epsilon) = u(x, t)e^{-1/2(1/2t\epsilon^2 + x\epsilon)} \tag{2}$$

where ϵ is a real parameter.

(ii) Find the infinitesimal generator of this transformation.

(iii) Show that the transformation given by (2) can be derived from the infinitesimal generator.

(iv) Find a *similarity ansatz* and *similarity solution* from transformation (2).

Solution.   (i) From (2) we find, by applying the chain rule, that

$$\frac{\partial u'}{\partial t} = \frac{\partial u' \partial x'}{\partial x' \partial t} + \frac{\partial u' \partial t'}{\partial t' \partial t} = \frac{\partial u}{\partial t} e^{-1/2(1/2t\epsilon^2 + x\epsilon)} - \frac{1}{4}\epsilon^2 u e^{-1/2(1/2t\epsilon^2 + x\epsilon)},$$

$$\frac{\partial u'}{\partial x} = \frac{\partial u' \partial x'}{\partial x' \partial x} + \frac{\partial u' \partial t'}{\partial t' \partial x} = \frac{\partial u}{\partial x} e^{-1/2(1/2t\epsilon^2 + x\epsilon)} - \frac{1}{2}\epsilon u e^{-1/2(1/2t\epsilon^2 + x\epsilon)}.$$

Since $\partial x'/\partial t = \epsilon$, $\partial x'/\partial x = 1$, $\partial t'/\partial t = 1$ and $\partial t'/\partial x = 0$, we find

$$\frac{\partial u'}{\partial x'}\epsilon + \frac{\partial u'}{\partial t'} = \frac{\partial u}{\partial t} e^{-1/2(1/2t\epsilon^2 + x\epsilon)} - \frac{1}{4}\epsilon^2 u e^{-1/2(1/2t\epsilon^2 + x\epsilon)}, \tag{3a}$$

$$\frac{\partial u'}{\partial x'} = \frac{\partial u}{\partial x} e^{-1/2(1/2t\epsilon^2 + x\epsilon)} - \frac{1}{2}\epsilon u e^{-1/2(1/2t\epsilon^2 + x\epsilon)}. \tag{3b}$$

From (3b) we obtain

$$\frac{\partial^2 u'}{\partial x'^2} = \frac{\partial^2 u}{\partial x^2} e^{-1/2(1/2t\epsilon^2 + x\epsilon)} - \epsilon\frac{\partial u}{\partial x} e^{-1/2(1/2t\epsilon^2 + x\epsilon)} + \frac{1}{4}\epsilon^2 u e^{-1/2(1/2t\epsilon^2 + x\epsilon)}.$$

Therefore we find from (1) that

$$\frac{\partial u'}{\partial t'} = \frac{\partial^2 u'}{\partial x'^2}.$$

(ii) From transformation (2), we obtain the mapping

$$t'(x,t,\epsilon) = t, \quad x'(x,t,\epsilon) = t\epsilon + x, \quad u'(x,t,u,\epsilon) = ue^{-1/2(1/2t\epsilon^2 + x\epsilon)}. \tag{4}$$

From (4) we find

$$\left.\frac{dt'}{d\epsilon}\right|_{\epsilon=0} = 0, \quad \left.\frac{dx'}{d\epsilon}\right|_{\epsilon=0} = t, \quad \left.\frac{du'}{d\epsilon}\right|_{\epsilon=0} = -\frac{1}{2}xu.$$

Therefore the infinitesimal generator is given by

$$G = t\frac{\partial}{\partial x} - \frac{1}{2}xu\frac{\partial}{\partial u}.$$

(iii) The autonomous system associated with the symmetry generator G is

$$\frac{dt}{d\epsilon} = 0, \tag{5a}$$

$$\frac{dx}{d\epsilon} = t, \tag{5b}$$

$$\frac{du}{d\epsilon} = -\frac{1}{2}xu. \tag{5c}$$

Solving Eq. (5a) we find $t = t_0$. Inserting this solution into (5b) and integrating, we find $x = t_0\epsilon + x_0$. Inserting this solution into (5c) and integrating, we arrive at

$$u = u_0 e^{-1/2(1/2t_0\epsilon^2 + x_0\epsilon)}$$

which is the general solution to the system (10). When we set

$$t \to t', \quad t_0 \to t, \quad x \to x', \quad x_0 \to x, \quad u \to u', \quad u_0 \to u$$

we obtain the mapping (4). To find transformation (2) we have to calculate

$$\begin{pmatrix} x'(x,t,\epsilon) \\ t'(x,t,\epsilon) \\ u'(x'(x,t),t'(x,t),\epsilon) \end{pmatrix} = e^{\epsilon G} \begin{pmatrix} x \\ t \\ u \end{pmatrix} \Bigg|_{u \to u(x,t)}$$

where

$$e^{\epsilon G} x = x + \epsilon \left(t\frac{\partial}{\partial x} - \frac{1}{2}xu\frac{\partial}{\partial u} \right) x + \frac{\epsilon^2}{2!} \left(t\frac{\partial}{\partial x} - \frac{1}{2}xu\frac{\partial}{\partial u} \right)^2 x + \cdots = x + \epsilon t\,,$$

$$e^{\epsilon G} t = t\,,$$

$$e^{\epsilon G} u = u - \frac{\epsilon}{2}xu + \frac{\epsilon^2}{2!} \left(-1/2tu + \frac{1}{4}x^2 u \right) + \cdots = u e^{-1/2(1/2t\epsilon^2 + x\epsilon)}\,.$$

Thus we find the transformation group (2).

(iv) To find a similarity ansatz we write the mapping (4) as

$$t(x_0, t_0, \epsilon) = t_0\,, \tag{6a}$$

$$x(x_0, t_0, \epsilon) = t_0\epsilon + x_0\,, \tag{6b}$$

$$u(x_0, t_0, u_0, \epsilon) = u_0 e^{-1/2(1/2t_0\epsilon^2 + x_0\epsilon)}\,. \tag{6c}$$

We set $t_0 = s$ and $x_0 = 0$, where s is the *similarity variable*. Then from (6a) and (6b) we find $s = t$, $\epsilon = x/s$. Therefore (6c) gives the similarity ansatz

$$u(x,t) = f(s)e^{-1/4x^2/s}\,. \tag{7}$$

Inserting the similarity ansatz (7) into (1) leads to the ordinary differential equation

$$\frac{df}{ds} + \frac{1}{2s}f = 0\,. \tag{8}$$

The general solution of (8) is given by $f(s) = C/\sqrt{s}$, where C is the constant of integration. Since $s = t$ is the similarity variable, we find from (7) and (8) an exact solution

$$u(x,t) = \frac{C}{\sqrt{t}} e^{-\frac{1}{4}(x^2/t)}$$

of (1).

Remark.   *Transformation (4) is called the Galilean transformation.*

Chapter 13

Bäcklund Transformations

Problem 1. The one-dimensional *sine-Gordon equation* in light-cone coordinates ξ, η is given by

$$\frac{\partial^2 u}{\partial \xi \partial \eta} = \sin u . \tag{1}$$

Show that the transformation

$$\xi'(\xi, \eta) = \xi , \tag{2a}$$

$$\eta'(\xi, \eta) = \eta , \tag{2b}$$

$$\frac{\partial u'(\xi'(\xi,\eta), \eta'(\xi,\eta))}{\partial \xi'} = \frac{\partial u}{\partial \xi} - 2\lambda \sin\left(\frac{u(\xi,\eta) + u'(\xi'(\xi,\eta), \eta'(\xi,\eta))}{2}\right) , \tag{2c}$$

$$\frac{\partial u'(\xi'(\xi,\eta), \eta'(\xi,\eta))}{\partial \eta'} = -\frac{\partial u}{\partial \eta} + \frac{2}{\lambda} \sin\left(\frac{u(\xi,\eta) - u'(\xi'(\xi,\eta), \eta'(\xi,\eta))}{2}\right) \tag{2d}$$

defines an *auto-Bäcklund transformation*, where λ is a nonzero real parameter. This means one has to show that

$$\frac{\partial^2 u'}{\partial \xi' \partial \eta'} = \sin u' \tag{3}$$

follows from (1) and (2). Vice versa one has to show that (1) follows from (3) and (2).

Solution.   Differentiating (2c) with respect to η, taking into account (2a) and (2b) yields

$$\frac{\partial^2 u'}{\partial \eta' \partial \xi'} = \frac{\partial^2 u}{\partial \eta \partial \xi} - \lambda \cos\left(\frac{u+u'}{2}\right)\left(\frac{\partial u}{\partial \eta} + \frac{\partial u'}{\partial \eta'}\right). \tag{4}$$

Differentiating (2d) with respect to ξ, taking into account (2a) and (2b) yields

$$\frac{\partial^2 u'}{\partial \xi' \partial \eta'} = -\frac{\partial^2 u}{\partial \xi \partial \eta} + \frac{1}{\lambda} \cos\left(\frac{u-u'}{2}\right)\left(\frac{\partial u}{\partial \xi} - \frac{\partial u'}{\partial \xi'}\right). \tag{5}$$

For the sake of simplicity we have omitted the arguments. Adding (4) and (5) gives

$$2\frac{\partial^2 u'}{\partial \eta' \partial \xi'} = -\lambda \cos\left(\frac{u+u'}{2}\right)\left(\frac{\partial u}{\partial \eta} + \frac{\partial u'}{\partial \eta'}\right)$$
$$+ \frac{1}{\lambda} \cos\left(\frac{u-u'}{2}\right)\left(\frac{\partial u}{\partial \xi} - \frac{\partial u'}{\partial \xi'}\right). \tag{6}$$

Inserting $\partial u/\partial \xi$ from (2c) and $\partial u/\partial \eta$ from (2d) into (6) gives

$$2\frac{\partial^2 u'}{\partial \eta' \partial \xi'} = -\lambda \cos\left(\frac{u+u'}{2}\right)\frac{2}{\lambda}\sin\left(\frac{u-u'}{2}\right)$$
$$+ \frac{1}{\lambda} \cos\left(\frac{u-u'}{2}\right)2\lambda \sin\left(\frac{u+u'}{2}\right).$$

Using the identity

$$\sin \alpha \cos \beta \equiv \frac{1}{2}\sin(\alpha+\beta) + \frac{1}{2}\sin(\alpha-\beta),$$

we obtain

$$\frac{\partial^2 u'}{\partial \eta' \partial \xi'} = -\frac{1}{2}\sin u + \frac{1}{2}\sin u' + \frac{1}{2}\sin u + \frac{1}{2}\sin u'.$$

Consequently

$$\frac{\partial^2 u'}{\partial \eta' \partial \xi'} = \sin u'.$$

In the same manner we can show that (1) follows from (3).

Problem 2. In Problem 1 we showed that the one-dimensional sine-Gordon equation in light-cone coordinates ξ, η

$$\frac{\partial^2 u}{\partial \xi \partial \eta} = \sin u \tag{1}$$

admits an auto-Bäcklund transformation. Obviously

$$u(\xi, \eta) = 0 \tag{2}$$

is a trivial solution of (1). Apply the auto-Bäcklund transformation given in Problem 1 to find a nontrivial solution of (1).

Remark. *The solution* (2) *is also called the vacuum solution of* (1).

Solution. Inserting (2) into the auto-Bäcklund transformation yields

$$\frac{\partial u'(\xi', \eta')}{\partial \xi'} = -2\lambda \sin \left(\frac{u'(\xi', \eta')}{2} \right) ,$$

$$\frac{\partial u'(\xi', \eta')}{\partial \eta'} = \frac{2}{\lambda} \sin \left(\frac{-u'(\xi', \eta')}{2} \right) .$$

Since

$$\int \frac{du}{\sin(\frac{u}{2})} = 2 \ln \left(\tan \left(\frac{u}{4} \right) \right) ,$$

we obtain

$$u'(\xi', \eta') = 4 \tan^{-1} \exp(\lambda \xi' + \lambda^{-1} \eta' + C)$$

where C is a constant of integration.

Problem 3. Consider the two nonlinear ordinary differential equations

$$\frac{d^2 u}{dt^2} = \sin u , \qquad \frac{d^2 v}{dt^2} = \sinh v \tag{1}$$

where u and v are real-valued functions. Show that

$$\frac{du}{dt} - i\frac{dv}{dt} = 2e^{i\lambda} \sin \left(\frac{1}{2}(u + iv) \right) \tag{2}$$

defines a Bäcklund transformation, where λ is a real parameter.

Solution. Taking the time derivative of (2) gives

$$\frac{d^2 u}{dt^2} - i\frac{d^2 v}{dt^2} = e^{i\lambda} \cos \left(\frac{1}{2}(u + iv) \right) \left(\frac{du}{dt} + i\frac{dv}{dt} \right) . \tag{3}$$

The complex conjugate of (2) is

$$\frac{du}{dt} + i\frac{dv}{dt} = 2e^{-i\lambda}\sin\left(\frac{1}{2}(u - iv)\right) \tag{4}$$

where we have used $\overline{\sin z} = \sin \bar{z}$. Inserting (4) into (3) yields

$$\frac{d^2u}{dt^2} - i\frac{d^2v}{dt^2} = 2\cos\left(\frac{1}{2}(u + iv)\right)\sin\left(\frac{1}{2}(u - iv)\right).$$

Using the identity

$$\cos\left(\frac{1}{2}(u + iv)\right)\sin\left(\frac{1}{2}(u - iv)\right) \equiv \frac{1}{2}\sin u + \frac{1}{2}\sin(-iv)$$

$$\equiv \frac{1}{2}\sin u - \frac{1}{2}i\sinh v,$$

we find $d^2u/dt^2 - id^2v/dt^2 = \sin u - i\sinh v$. Therefore (1) follows.

Problem 4.   The nonlinear partial differential equation

$$\frac{\partial u}{\partial t} + u\frac{\partial u}{\partial x} = D\frac{\partial^2 u}{\partial x^2} \tag{1}$$

is called *Burgers equation*. Let

$$u(x, t) = -2D\frac{\partial}{\partial x}\ln v(x, t). \tag{2}$$

Show that v satisfies the linear diffusion equation $\partial v/\partial t = D\partial^2 v/\partial x^2$.

Solution.   From (2) we find

$$u = -2D\frac{1}{v}\frac{\partial v}{\partial x}.$$

It follows that

$$\frac{\partial u}{\partial t} = 2D\frac{1}{v^2}\frac{\partial v}{\partial t}\frac{\partial v}{\partial x} - 2D\frac{1}{v}\frac{\partial^2 v}{\partial t\partial x}, \quad \frac{\partial u}{\partial x} = 2D\frac{1}{v^2}\left(\frac{\partial v}{\partial x}\right)^2 - 2D\frac{1}{v}\frac{\partial^2 v}{\partial x^2}, \tag{3}$$

$$\frac{\partial^2 u}{\partial x^2} = -4D\frac{1}{v^3}\left(\frac{\partial v}{\partial x}\right)^3 + 6D\frac{1}{v^2}\frac{\partial v}{\partial x}\frac{\partial^2 v}{\partial x^2} - 2D\frac{1}{v}\frac{\partial^3 v}{\partial x^3}. \tag{4}$$

Inserting (3) and (4) into (1) yields

$$\frac{1}{v}\frac{\partial v}{\partial t}\frac{\partial v}{\partial x} - \frac{\partial^2 v}{\partial t\partial x} = \frac{D}{v}\frac{\partial v}{\partial x}\frac{\partial^2 v}{\partial x^2} - D\frac{\partial^3 v}{\partial x^3}$$

or

$$\frac{\partial v}{\partial x}\left(\frac{\partial v}{\partial t} - D\frac{\partial^2 v}{\partial x^2}\right) = v\frac{\partial}{\partial x}\left(\frac{\partial v}{\partial t} - D\frac{\partial^2 v}{\partial x^2}\right). \tag{5}$$

Equation (5) is satisfied if v satisfies the linear diffusion equation.

Problem 5. The *Korteweg–de Vries equation* is given by

$$\frac{\partial u}{\partial t} + 6u\frac{\partial u}{\partial x} + \frac{\partial^3 u}{\partial x^3} = 0 \tag{1}$$

and the *modified Korteweg–de Vries equation* takes the form

$$\frac{\partial v}{\partial t} - 6v^2\frac{\partial v}{\partial x} + \frac{\partial^3 v}{\partial x^3} = 0. \tag{2}$$

Show that the Korteweg–de Vries equation (1) and the modified Korteweg–de Vries equation (2) are related by the Bäcklund transformation

$$\frac{\partial v}{\partial x} = u + v^2, \tag{3a}$$

$$\frac{\partial v}{\partial t} = -\frac{\partial^2 u}{\partial x^2} - 2\left(v\frac{\partial u}{\partial x} + u\frac{\partial v}{\partial x}\right). \tag{3b}$$

Solution. First we show that the Korteweg–de Vries equation follows from system (3). Taking the derivative of (3a) with respect to t yields

$$\frac{\partial^2 v}{\partial x \partial t} = \frac{\partial u}{\partial t} + 2v\frac{\partial v}{\partial t}.$$

Inserting (3b) gives

$$\frac{\partial^2 v}{\partial x \partial t} = \frac{\partial u}{\partial t} - 2v\frac{\partial^2 u}{\partial x^2} - 4v^2\frac{\partial u}{\partial x} - 4uv\frac{\partial v}{\partial x}$$

or

$$\frac{\partial^2 v}{\partial x \partial t} = \frac{\partial u}{\partial t} - 2v\frac{\partial^2 u}{\partial x^2} - 4v^2\frac{\partial u}{\partial x} - 4uv(u + v^2). \tag{4}$$

Taking the derivative of (3b) with respect to x yields

$$\frac{\partial^2 v}{\partial x \partial t} = -\frac{\partial^3 u}{\partial x^3} - 2\left(2\frac{\partial u}{\partial x}\frac{\partial v}{\partial x} + v\frac{\partial^2 u}{\partial x^2} + u\frac{\partial^2 v}{\partial x^2}\right). \tag{5}$$

Taking the derivative of (3a) with respect to x yields

$$\frac{\partial^2 v}{\partial x^2} = \frac{\partial u}{\partial x} + 2v\frac{\partial v}{\partial x} = \frac{\partial u}{\partial x} + 2v(u + v^2). \tag{6}$$

Inserting (3a) and (6) into (5) leads to

$$\frac{\partial^2 v}{\partial x \partial t} = -\frac{\partial^3 u}{\partial x^3} - 2\left(2\frac{\partial u}{\partial x}(u+v^2) + v\frac{\partial^2 u}{\partial x^2} + u\left(\frac{\partial u}{\partial x} + 2uv + 2v^3\right)\right). \quad (7)$$

Subtracting (7) from (4) gives the Korteweg–de Vries equation. Now we have to show that the modified Korteweg–de Vries equation follows from system (3). Differentiating (3a) with respect to x yields

$$\frac{\partial^2 v}{\partial x^2} = \frac{\partial u}{\partial x} + 2v\frac{\partial v}{\partial x} = \frac{\partial u}{\partial x} + 2v(u+v^2),$$

$$\frac{\partial^3 v}{\partial x^3} = \frac{\partial^2 u}{\partial x^2} + 2u\frac{\partial v}{\partial x} + 2v\frac{\partial u}{\partial x} + 6v^2\frac{\partial v}{\partial x}$$

$$= \frac{\partial^2 u}{\partial x^2} + 2v\frac{\partial u}{\partial x} + (2u + 6v^2)(u+v^2). \quad (8)$$

Adding (8) and (3b) gives

$$\frac{\partial v}{\partial t} + \frac{\partial^3 v}{\partial x^3} = -2u\frac{\partial v}{\partial x} + (2u + 6v^2)(u+v^2). \quad (9)$$

From (3) we obtain $u = \partial v/\partial x - v^2$. Inserting this equation into (9) leads to the modified Korteweg–de Vries equation (2).

Remark.   *Equation (3a) is called the Miura transformation.*

Problem 6.   Consider the partial differential equations

$$\frac{\partial^2 u}{\partial x \partial y} = \exp(u), \qquad \frac{\partial^2 v}{\partial x \partial y} = 0. \quad (1)$$

The first equation is called the *Liouville equation*. Show that

$$\frac{\partial v}{\partial x} = \frac{\partial u}{\partial x} + \lambda \exp\left(\frac{1}{2}(v+u)\right), \quad (2a)$$

$$\frac{\partial v}{\partial y} = -\frac{\partial u}{\partial y} - \frac{2}{\lambda}\exp\left(\frac{1}{2}(u-v)\right) \quad (2b)$$

provides a Bäcklund transformation of these two equations, where λ is a nonzero parameter.

Solution.   Differentiating (2a) with respect to y yields

$$\frac{\partial^2 v}{\partial x \partial y} = \frac{\partial^2 u}{\partial x \partial y} + \lambda \exp\left(\frac{1}{2}(u+v)\right)\frac{1}{2}\left(\frac{\partial v}{\partial y} + \frac{\partial u}{\partial y}\right). \quad (3)$$

Inserting (2b) into (3) yields

$$\frac{\partial^2 v}{\partial x \partial y} = \frac{\partial^2 u}{\partial x \partial y} - \exp\left(\frac{1}{2}(u+v)\right)\exp\left(\frac{1}{2}(u-v)\right) = \frac{\partial^2 u}{\partial x \partial y} - \exp(u). \quad (4)$$

Differentiating (2b) with respect to x leads to

$$\frac{\partial^2 v}{\partial x \partial y} = -\frac{\partial^2 u}{\partial x \partial y} + \frac{1}{\lambda}\exp\left(\frac{1}{2}(u-v)\right)\left(\frac{\partial v}{\partial x} - \frac{\partial u}{\partial x}\right). \quad (5)$$

Inserting (2a) into (5) gives

$$\frac{\partial^2 v}{\partial x \partial y} = -\frac{\partial^2 u}{\partial x \partial y} + \exp\left(\frac{1}{2}(u-v)\right)\exp\left(\frac{1}{2}(u+v)\right) = -\frac{\partial^2 u}{\partial x \partial y} + \exp(u).$$

$$(6)$$

Adding (4) and (6) leads to (1b). Subtracting (4) from (6) we find (1a).

Chapter 14

Soliton Equations

Problem 1. The Korteweg–de Vries equation is given by

$$\frac{\partial u}{\partial t} - 6u\frac{\partial u}{\partial x} + \frac{\partial^3 u}{\partial x^3} = 0. \tag{1}$$

Find a solution of the form

$$u(x, t) = -2f(x - ct). \tag{2}$$

Solution. From ansatz (2) we find

$$\frac{\partial u}{\partial t} = 2cf', \tag{3a}$$

$$\frac{\partial u}{\partial x} = -2f', \tag{3b}$$

$$\frac{\partial^3 u}{\partial x^3} = -2f''' \tag{3c}$$

where f' denotes differentiation with respect to the argument $s := x - ct$. Inserting (3) into (1) yields

$$2cf' - 24ff' - 2f''' = 0. \tag{4}$$

Integrating (4) once gives

$$cf - 6f^2 - f'' + \frac{a}{2} = 0.$$

This equation can be integrated again and we find

$$(f')^2 = -4f^3 + cf^2 + af + b \tag{5}$$

where a and b are the constants of integration. Equation (5) can written as

$$\left(\frac{df}{dx}\right)^2 = -4(f - \alpha_1)(f - \alpha_2)(f - \alpha_3)$$

with real roots α_1, α_2 and α_3 and $\alpha_1 < \alpha_2 < \alpha_3$. Let

$$f - \alpha_3 =: -g.$$

Then we obtain

$$\left(\frac{dg}{dx}\right)^2 = 4g(\alpha_3 - \alpha_2 - g)(\alpha_3 - \alpha_1 - g).$$

Setting

$$g = (\alpha_3 - \alpha_2)v^2,$$

we arrive at

$$\left(\frac{dv}{dx}\right)^2 = (\alpha_3 - \alpha_1)(1 - v^2)(1 - k^2 v^2) \tag{6}$$

where $k^2 = (\alpha_3 - \alpha_2)/(\alpha_3 - \alpha_1)$. The equation $(dw/dz)^2 = (1 - w^2)(1 - k^2 w^2)$ is the equation that defines the *Jacobi elliptic function* sn(z, k). The constant k is the modulus of the elliptic function. The solution of (6) is therefore

$$v(x) = \text{sn}(\sqrt{\alpha_3 - \alpha_1}\, x, k).$$

Consequently

$$f(x - ct) = \alpha_3 - (\alpha_3 - \alpha_2)\,\text{sn}^2(\sqrt{\alpha_3 - \alpha_1}(x - ct), k).$$

Problem 2.   The Korteweg–de Vries equation is given by

$$\frac{\partial u}{\partial t} + u\frac{\partial u}{\partial x} + \beta\frac{\partial^3 u}{\partial x^3} = 0. \tag{1}$$

A *conservation law* is given by

$$\frac{\partial Q}{\partial t} + \frac{\partial P}{\partial x} = 0.$$

(i) Show that (1) can be written as a conservation law.

(ii) Show that

$$
\frac{\partial}{\partial t}\left(\frac{u^2}{2}\right) + \frac{\partial}{\partial x}\left(\frac{u^3}{3} + \beta\left(u\frac{\partial^2 u}{\partial x^2} - \frac{1}{2}\left(\frac{\partial u}{\partial x}\right)^2\right)\right) = 0, \qquad \text{(2a)}
$$

$$
\frac{\partial}{\partial t}\left(\frac{u^3}{3} - \beta\frac{\partial^2 u}{\partial x^2}\right) + \frac{\partial}{\partial x}\left(\frac{u^4}{4} + \beta\left(u^2\frac{\partial^2 u}{\partial x^2} + 2\frac{\partial u}{\partial t}\frac{\partial u}{\partial x}\right) + \beta^2\left(\frac{\partial^2 u}{\partial x^2}\right)^2\right) = 0
$$

$$
\text{(2b)}
$$

are conservation laws of the Korteweg–de Vries equation (1).

Solution.   (i) Since $\partial u^2/\partial x \equiv 2u\partial u/\partial x$ we find that the Korteweg–de Vries equation (1) can be written as

$$
\frac{\partial u}{\partial t} + \frac{\partial}{\partial x}\left(\frac{u^2}{2} + \beta\frac{\partial^2 u}{\partial x^2}\right) = 0 \,.
$$

(ii) Multiplying (1) by u and using the identity

$$
\frac{\partial}{\partial x}\left(u\frac{\partial^2 u}{\partial x^2} - \frac{1}{2}\left(\frac{\partial u}{\partial x}\right)^2\right) \equiv u\frac{\partial^3 u}{\partial x^3}
$$

we obtain (2a). Multiplying (1) by u^2 we obtain (2b).

Problem 3.   Consider the *Schrödinger eigenvalue equation*

$$
\frac{d^2 y}{dx^2} + (k^2 - u(x))y = 0
$$

where k is an arbitrary real parameter.

(i) Find a solution of the form

$$
y(x) = e^{ikx}(2k + if(x)),
$$

i.e., determine f and u.

(ii) Find the condition that $u(ax + bt)$ satisfies the Korteweg–de Vries equation

$$
\frac{\partial u}{\partial t} - 6u\frac{\partial u}{\partial x} + \frac{\partial^3 u}{\partial x^3} = 0 \,. \qquad \text{(1)}
$$

Solution.   (i) Since

$$
\frac{d^2 y}{dx^2} = -k^2 e^{ikx}(2k + if(x)) - 2ke^{ikx}\frac{df}{dx} + ie^{ikx}\frac{d^2 f}{dx^2} \,,
$$

we obtain

$$i\frac{d^2 f}{dx^2} - iu(x)f(x) + k\left(-2\frac{df}{dx} - 2u(x)\right) = 0.$$

Separating terms according to powers of k, we arrive at

$$\frac{d^2 f}{dx^2} = fu, \tag{2a}$$

$$\frac{df}{dx} = -u. \tag{2b}$$

Elimination of u and integration yields

$$\frac{df}{dx} + \frac{1}{2}f^2 = 2c^2$$

where $2c^2$ is the constant of integration. The substitution

$$f = \frac{2}{w}\frac{dw}{dx} \tag{3}$$

leads to the linear equation

$$\frac{d^2 w}{dx^2} = c^2 w$$

with the general solution

$$w(x) = Ae^{cx} + Be^{-cx}. \tag{4}$$

From (3) and (2b) we obtain

$$u(x) = -2\frac{d^2}{dx^2}\ln w(x). \tag{5}$$

Inserting the solution (4) into (5) leads to

$$u(x) = -2c^2 \operatorname{sech}^2(cx - \phi)$$

where $\phi = \frac{1}{2}\ln(B/A)$ and $\operatorname{sech}\alpha := 1/\cosh\alpha$.

(ii) Inserting the ansatz

$$u(x,t) = -2c^2 \operatorname{sech}^2(cx - \phi(t))$$

into the Korteweg–de Vries equation (1) we find

$$\frac{d\phi}{dt} = 4c^3.$$

Therefore

$$u(x,t) = -2c^2 \operatorname{sech}^2(cx - 4c^3t).$$

Problem 4. The function

$$u(x,t) = -12\frac{4\cosh(2x - 8t) + \cosh(4x - 64t) + 3}{(3\cosh(x - 28t) + \cosh(3x - 36t))^2}$$

is a so-called *two-soliton solution* of the Korteweg–de Vries equation

$$\frac{\partial u}{\partial t} - 6u\frac{\partial u}{\partial x} + \frac{\partial^3 u}{\partial x^3} = 0.$$

(i) Find the solution for $t = 0$.

(ii) Find the solution for $|2x - 32t| \gg 0$ and $|x - 4t| < 1$.

(iii) Find the solution for $|x - 4t| \gg 0$ and $|2x - 32t| < 1$.

Solution. For $t = 0$ the function (1) simplifies to

$$u(x,t = 0) = -12\frac{4\cosh(2x) + \cosh(4x) + 3}{(3\cosh(x) + \cosh(3x))^2} = -6\operatorname{sech}^2 x$$

where we have used the identity .

$$\cosh(2x) \equiv \sinh^2 x + \cosh^2 x.$$

etc. Thus the two solitons completely overlap.

(ii) For $|2x - 32t| \gg 1$ and $|x - 4t| < 1$, we find

$$u(x,t) = -2\operatorname{sech}^2\left(x - 4t \mp \tanh^{-1}\frac{1}{2}\right).$$

(iii) For $|x - 4t| \gg 1$ and $|2x - 32t| < 1$, we find

$$u(x,t) = -8\operatorname{sech}^2\left(2x - 32t \mp \tanh^{-1}\frac{1}{2}\right).$$

Problem 5. Consider the Korteweg–de Vries–Burgers equation

$$\frac{\partial u}{\partial t} + u\frac{\partial u}{\partial x} + b\frac{\partial^3 u}{\partial x^3} - a\frac{\partial^2 u}{\partial x^2} = 0 \tag{1}$$

where u is a conserved quantity

$$\frac{d}{dt}\int_{-\infty}^{\infty} u(x,t)dx = 0,$$

i.e., the area under u is conserved for all t. Find solutions of the form

$$u(x,t) = U(\xi(x,t)), \quad \xi(x,t) := c(x-vt)$$

where

$$U(\xi) = S(Y(\xi)) = \sum_{n=0}^{N} a_n Y(\xi)^n, \quad Y(\xi) = \tanh(\xi(x,t)) = \tanh(c(x-vt)).$$

$$(2)$$

Solution. Inserting (2) into (1) yields

$$-cvU(\xi) + \frac{1}{2}cU(\xi)^2 + bc^3\frac{d^2U}{d\xi^2} - ac^2\frac{dU(\xi)}{d\xi} = C.$$

Requiring that

$$U(\xi) \to 0, \quad \frac{dU(\xi)}{d\xi} \to 0, \quad \frac{d^2U(\xi)}{d\xi^2} \to 0 \text{ as } \xi \to \infty, \tag{3}$$

we find that the integration constant C must be set to zero. Equation (3) can be expressed in the variable Y as

$$-vS(Y) + \frac{1}{2}S(Y)^2 + bc^2(1-Y^2)\left(-2Y\frac{dS(Y)}{dY} + (1-Y^2)\frac{d^2S(Y)}{dY^2}\right)$$

$$-ac(1-Y^2)\frac{dS(Y)}{dY} = 0. \tag{4}$$

Substitution of the expansion (2) into (4) and balancing the highest degree in Y, yields $N = 2$. The boundary condition

$$U(\xi) \to 0$$

for $\xi \to +\infty$ or $\xi \to -\infty$ implies that $S(Y) \to 0$ for $Y \to +1$ or $Y \to -1$. Without loss of generality we only consider the limit $Y \to 1$. Two possible solutions arise in this case:

$$S(Y) = F(Y) = b_0(1-Y)(1+b_1Y)$$

or

$$S(Y) = G(Y) = d_0(1-Y)^2.$$

In the first case $S(Y)$ decays as $\exp(2\xi)$, whereas in the second case, $S(Y)$ decays as $\exp(-4\xi)$ as $\xi \to +\infty$. We first take $F(Y) = (1-Y)b_0(1+b_1Y)$.

Upon substituting this ansatz into (4) and subsequent cancellation of a common factor $(1 - Y)$, we take the limit $Y \to 1$, which gives

$$v = 4bc^2 + 2ac.$$

The remaining constants b_0 and b_1 can be found through simple algebra. We obtain

$$c = \frac{a}{10b}, \quad v = 24bc^2.$$

Thus

$$F(Y) = 36bc^2(1 - Y)\left(1 + \frac{1}{3}Y\right).$$

The second possible solution $G(Y) = d_0(1-Y)^2$ also leads to a real solution.

Chapter 15

Lax Pairs for Partial Differential Equations

Problem 1. Let L and M be two linear operators. Assume that

$$Lv = \lambda v\,, \tag{1}$$

$$\frac{\partial v}{\partial t} = Mv \tag{2}$$

where L is the linear operator of the *spectral problem* (1), M is the linear operator of an associated *time-evolution equation* (2) and λ is a parameter.

(i) Show that

$$L_t = [M, L](t) \tag{3}$$

where

$$[L, M] := LM - ML$$

is the commutator.

Remark. *Equation (3) is called the Lax representation. L and M are called a Lax pair.*

Equation (3) contains a nonlinear evolution equation if L and M are correctly chosen.

(ii) Let

$$L := \frac{\partial^2}{\partial x^2} + u(x, t)$$

and

$$M := 4\frac{\partial^3}{\partial x^3} + 6u(x,t)\frac{\partial}{\partial x} + 3\frac{\partial u}{\partial x}.$$

Calculate the evolution equation for u.

Solution.   (i) From (1) we obtain

$$\frac{\partial}{\partial t}(Lv) = \frac{\partial}{\partial t}(\lambda v).$$

It follows that

$$L_t v + L\frac{\partial v}{\partial t} = \lambda\frac{\partial v}{\partial t} \tag{4}$$

where we assume that λ does not depend on t. Inserting (1) and (2) into (4) gives

$$L_t v + LMv = \lambda Mv. \tag{5}$$

From (1) we obtain $MLv = \lambda Mv$. Inserting this expression into (5) yields

$$L_t v = -LMv + MLv.$$

Consequently,

$$L_t v = [M, L]v.$$

Since the smooth function v is arbitrary we obtain (3).

(ii) Since

$$L_t v = \frac{\partial}{\partial t}(Lv) - L\left(\frac{\partial v}{\partial t}\right),$$

we obtain

$$L_t v = \frac{\partial}{\partial t}\left(\left(\frac{\partial^2}{\partial x^2} + u\right)v\right) - \left(\frac{\partial^2}{\partial x^2} + u\right)\frac{\partial v}{\partial t} = \frac{\partial u}{\partial t}v.$$

Now

$$[M, L]v = (ML)v - (LM)v = M(Lv) - L(Mv).$$

Thus

$$[M, L]v = \left(4\frac{\partial^3}{\partial x^3} + 6u\frac{\partial}{\partial x} + 3\frac{\partial u}{\partial x}\right)\left(\frac{\partial^2 v}{\partial x^2} + uv\right)$$

$$- \left(\frac{\partial^2}{\partial x^2} + u\right)\left(4\frac{\partial^3 v}{\partial x^3} + 6u\frac{\partial v}{\partial x} + 3\frac{\partial u}{\partial x}v\right)$$

Finally

$$[M, L]v = 6u\frac{\partial u}{\partial x}v + \frac{\partial^3 u}{\partial x^3}v = \left(6u\frac{\partial u}{\partial x} + \frac{\partial^3 u}{\partial x^3}\right)v.$$

Since v is arbitrary, (3) yields

$$\frac{\partial u}{\partial t} = 6u\frac{\partial u}{\partial x} + \frac{\partial^3 u}{\partial x^3}.$$

Problem 2.   Let

$$L := \frac{\partial^2}{\partial x^2} + bu(x, y, t) + \frac{\partial}{\partial y}$$

and

$$T := -4\frac{\partial^3}{\partial x^3} - 6bu(x, y, t)\frac{\partial}{\partial x} - 3b\frac{\partial u}{\partial x} - 3b\partial_x^{-1}\frac{\partial u}{\partial y} + \frac{\partial}{\partial t}$$

where $b \in \mathbf{R}$ and

$$\partial_x^{-1}f := \int_{-\infty}^{x} f(s)ds.$$

Calculate the nonlinear evolution equation which follows from the condition

$$[T, L]v = 0 \tag{1}$$

where v is a smooth function of x, y, t and $[\,,\,]$ denotes the commutator.

Solution.   Since

$$[T, L]v = (TL)v - (LT)v = T(Lv) - L(Tv),$$

we have

$$[T, L]v = \left(-4\frac{\partial^3}{\partial x^3} - 6bu\frac{\partial}{\partial x} - 3b\frac{\partial u}{\partial x} - 3b\partial_x^{-1}\frac{\partial u}{\partial y} + \frac{\partial}{\partial t}\right)\left(\frac{\partial^2 v}{\partial x^2} + buv + \frac{\partial v}{\partial y}\right)$$

$$- \left(\frac{\partial^2}{\partial x^2} + bu + \frac{\partial}{\partial y}\right)\left(-4\frac{\partial^3 v}{\partial x^3} - 6bu\frac{\partial v}{\partial x} - 3b\frac{\partial u}{\partial x}v\right)$$

$$- 3b\partial_x^{-1}\left(\frac{\partial u}{\partial y}\right)v + \frac{\partial v}{\partial t}\bigg).$$

Thus we arrive at

$$[T, L]v = b\frac{\partial u}{\partial t}v - b\frac{\partial^3 u}{\partial x^3}v - 6b^2u\frac{\partial u}{\partial x}v - 3b\left(\partial_x^{-1}\frac{\partial^2 u}{\partial y^2}\right)v.$$

Since v is arbitrary, condition (1) yields

$$\frac{\partial u}{\partial t} = \frac{\partial^3 u}{\partial x^3} + 6bu\frac{\partial u}{\partial x} + 3\partial_x^{-1}\frac{\partial^2 u}{\partial y^2} \,. \tag{2}$$

To find the local form we differentiate (2) with respect to x and find

$$\frac{\partial^2 u}{\partial t \partial x} = \frac{\partial^4 u}{\partial x^4} + 6b\left(\frac{\partial u}{\partial x}\right)^2 + 6bu\frac{\partial^2 u}{\partial x^2} + 3\frac{\partial^2 u}{\partial y^2} \,. \tag{3}$$

For our calculation we used the property that

$$\frac{\partial}{\partial x}\left(\partial_x^{-1} f\right) = f(x) \,.$$

Equation (3) is called the *Kadomtsev–Petviashvili equation.*

Problem 3.   Consider the system of partial differential equations

$$\frac{\partial \psi_1}{\partial x} = \lambda \psi_1 + q\psi_2 \,, \tag{1a}$$

$$\frac{\partial \psi_2}{\partial x} = r\psi_1 - \lambda \psi_2 \,, \tag{1b}$$

$$\frac{\partial \psi_1}{\partial t} = A\psi_1 + B\psi_2 \,, \tag{1c}$$

$$\frac{\partial \psi_2}{\partial t} = C\psi_1 - A\psi_2 \tag{1d}$$

where λ is a real parameter and q and r are smooth functions of the independent variables x and t. The coefficients A, B and C are one-parameter (λ) families of x, t and q and r with their derivatives.

(i) Find the equation which follows from the *compatibility condition* (also called *integrability condition*)

$$\frac{\partial^2 \psi_1}{\partial x \partial t} = \frac{\partial^2 \psi_1}{\partial t \partial x} \,, \qquad \frac{\partial^2 \psi_2}{\partial x \partial t} = \frac{\partial^2 \psi_2}{\partial t \partial x} \,.$$

(ii) Let w be a complex-valued function of x and t. Assume that

$$A := 2i\lambda^2 + i|w|^2 \,, \quad B := i\frac{\partial w}{\partial x} + 2i\lambda w \,, \quad C := i\frac{\partial \bar{w}}{\partial x} - 2i\lambda \bar{w} \,,$$

$$q := w \,, \quad r := -\bar{q} \equiv -\bar{w} \,. \tag{2}$$

Find the partial differential equation of w.

Solution. Taking the derivative of (1a) and (1b) with respect to t yields

$$\frac{\partial^2 \psi_1}{\partial t \partial x} = \lambda \frac{\partial \psi_1}{\partial t} + \frac{\partial q}{\partial t} \psi_2 + q \frac{\partial \psi_2}{\partial t}$$

$$= (\lambda A + qC)\psi_1 + \left(\lambda B - qA + \frac{\partial q}{\partial t}\right)\psi_2, \tag{3a}$$

$$\frac{\partial^2 \psi_2}{\partial t \partial x} = -\lambda \frac{\partial \psi_2}{\partial t} + \frac{\partial r}{\partial t} \psi_1 + r \frac{\partial \psi_1}{\partial t}$$

$$= \left(-\lambda C + \frac{\partial r}{\partial t} + rA\right)\psi_1 + (\lambda A + rB)\psi_2 \tag{3b}$$

where we have used (1a) through (1d). Taking the derivative of (1c) and (1d) with respect to x yields

$$\frac{\partial^2 \psi_1}{\partial x \partial t} = \frac{\partial A}{\partial x}\psi_1 + A\frac{\partial \psi_1}{\partial x} + \frac{\partial B}{\partial x}\psi_2 + B\frac{\partial \psi_2}{\partial x}$$

$$= \left(\frac{\partial A}{\partial x} + \lambda A + rB\right)\psi_1 + \left(Aq + \frac{\partial B}{\partial x} - \lambda B\right)\psi_2 \tag{3c}$$

$$\frac{\partial^2 \psi_2}{\partial x \partial t} = \frac{\partial C}{\partial x}\psi_1 + C\frac{\partial \psi_1}{\partial x} - \frac{\partial A}{\partial x}\psi_2 - A\frac{\partial \psi_2}{\partial x}$$

$$= \left(\frac{\partial C}{\partial x} + \lambda C - rA\right)\psi_1 + \left(qC - \frac{\partial A}{\partial x} + \lambda A\right)\psi_2 \tag{3d}$$

where we have used (1a) through (1d). From (3a) and (3c) we obtain

$$\frac{\partial A}{\partial x} = qC - rB, \tag{4a}$$

$$\frac{\partial B}{\partial x} = 2\lambda B + \frac{\partial q}{\partial t} - 2Aq. \tag{4b}$$

From (3b) and (3d) we find

$$\frac{\partial C}{\partial x} = -2\lambda C + \frac{\partial r}{\partial t} + 2rA, \tag{5a}$$

$$\frac{\partial A}{\partial x} = qC - rB. \tag{5b}$$

We see that (4a) and (4b) are the same.

(ii) Inserting (2) into (4b) or (5a) we obtain the partial differential equation

$$i\frac{\partial w}{\partial t} + \frac{\partial^2 w}{\partial x^2} + 2|w|^2 w = 0. \tag{6}$$

Equation (3a) is satisfied identically. We used that $w = w_1 + iw_2$, where w_1 and w_2 are real-valued functions and $|w|^2 = w_1^2 + w_2^2$.

Remark.   *System (1) is called the AKNS system. Equation (6) is the so-called one-dimensional* cubic Schrödinger equation.

Problem 4.   Consider a system consisting of a linear equation for an eigenfunction and an evolution equation

$$L(x, \partial)\psi(x, \lambda) = \lambda\psi(x, \lambda), \tag{1}$$

$$\frac{\partial\psi(x, \lambda)}{\partial x_n} = B_n(x, \partial)\psi(x, \lambda). \tag{2}$$

(i) Show that

$$\frac{\partial L}{\partial x_n} = [B_n, L] \equiv B_n L - LB_n$$

and

$$\frac{\partial B_m}{\partial x_n} - \frac{\partial B_n}{\partial x_m} = [B_n, B_m]. \tag{3}$$

(ii) Let

$$L := \partial + u_2(x)\partial^{-1} + u_3(x)\partial^{-2} + u_4(x)\partial^{-3} + \cdots$$

where $x = (x_1, x_2, x_3, \ldots)$, $\partial \equiv \partial/\partial x_1$ and

$$\partial^{-1}f(x) = \int^{x_1} f(s, x_2, x_3, \ldots)ds.$$

Define $B_n(x, \partial)$ as the differential part of $(L(x, \partial))^n$. Show that

$$B_1 = \partial,$$

$$B_2 = \partial^2 + 2u_2,$$

$$B_3 = \partial^3 + 3u_2\partial + 3u_3 + 3\frac{\partial u_2}{\partial x_1},$$

$$B_4 = \partial^4 + 4u_2\partial^2 + \left(4u_3 + 6\frac{\partial u_2}{\partial x_1}\right)\partial + 4u_4 + 6\frac{\partial u_3}{\partial x_1} + 4\frac{\partial^2 u_2}{\partial x_1^2} + 6u_2^2.$$

(iii) Find the equations of motion from (3) for $n = 2$ and $m = 3$.

Solution.   (i) From (1) we obtain

$$\frac{\partial}{\partial x_n}(L\psi) = \frac{\partial L}{\partial x_n}\psi + L\frac{\partial \psi}{\partial x_n} = \frac{\partial}{\partial x_n}(\lambda\psi) = \lambda\frac{\partial \psi}{\partial x_n}. \tag{4}$$

Inserting (2) into (4) yields

$$\frac{\partial L}{\partial x_n}\psi + LB_n\psi = \lambda B_n\psi = B_n\lambda\psi$$

or

$$\frac{\partial L}{\partial x_n}\psi = -LB_n\psi + B_nL\psi = [B_n, L]\psi. \tag{5}$$

(ii) From (5) we obtain

$$\frac{\partial}{\partial x_n}(L\psi) - L\frac{\partial \psi}{\partial x_n} = B_nL\psi - LB_n\psi \tag{6}$$

and

$$\frac{\partial}{\partial x_m}(L\psi) - L\frac{\partial \psi}{\partial x_m} = B_mL\psi - LB_m\psi. \tag{7}$$

Taking the derivatives of (6) with respect to x_m and of (7) with respect to x_n gives

$$\frac{\partial^2}{\partial x_m \partial x_n}(L\psi) - \frac{\partial}{\partial x_m}\left(L\frac{\partial \psi}{\partial x_n}\right) = \frac{\partial}{\partial x_m}(B_nL\psi) - \frac{\partial}{\partial x_m}(LB_n\psi), \tag{8}$$

$$\frac{\partial^2}{\partial x_n \partial x_m}(L\psi) - \frac{\partial}{\partial x_n}\left(L\frac{\partial \psi}{\partial x_m}\right) = \frac{\partial}{\partial x_n}(B_mL\psi) - \frac{\partial}{\partial x_n}(LB_m\psi). \tag{9}$$

Subtracting (9) from (8) we obtain

$$-\frac{\partial}{\partial x_m}\left(L\frac{\partial \psi}{\partial x_n}\right) + \frac{\partial}{\partial x_n}\left(L\frac{\partial \psi}{\partial x_m}\right)$$

$$= \frac{\partial}{\partial x_m}(B_nL\psi) - \frac{\partial}{\partial x_m}(LB_n\psi) + \frac{\partial}{\partial x_n}(LB_m\psi) - \frac{\partial}{\partial x_m}(LB_n\psi). \tag{10}$$

Inserting

$$\frac{\partial \psi}{\partial x_n} = B_n\psi, \qquad \frac{\partial \psi}{\partial x_m} = B_m\psi \tag{11}$$

into (10) and taking into account (1) gives

$$\frac{\partial}{\partial x_m}(B_nL\psi) - \frac{\partial}{\partial x_n}(LB_m\psi) = 0. \tag{12}$$

From (12) we obtain

$$\frac{\partial B_n}{\partial x_m}\psi + B_n\frac{\partial\psi}{\partial x_m} - \frac{\partial B_m}{\partial x_n}\psi - B_m\frac{\partial\psi}{\partial x_n} = 0. \tag{13}$$

Inserting (11) into (13) gives

$$\frac{\partial B_n}{\partial x_m}\psi + B_nB_m\psi - \frac{\partial B_m}{\partial x_n}\psi - B_mB_n\psi = 0.$$

Since ψ is arbitrary, Eq. (4) follows.

(ii) From L it is obvious that $B_1 = \partial$. Let f be a smooth function. Then

$$L^2f = L(Lf)$$

$$= L(\partial f + u_2\partial^{-1}f + u_3\partial^{-2}f + u_4\partial^{-3}f + u_5\partial^{-4}f + \cdots)$$

$$= \partial^2 f + u_2 f + (\partial u_2)\partial^{-1}f + u_3(\partial^{-1}f) + (\partial u_3)\partial^{-2}f + u_4\partial^{-2}f$$

$$+ (\partial u_4)\partial^{-2}f + \cdots$$

where

$$\partial^{-2}f := \int^{x_1}\left(\int^{s_1} f(s_2)ds_2\right)ds_1.$$

Since f is an arbitrary smooth function we have

$$B_2 = \partial^2 + u_2.$$

In the same manner we find B_3 and B_4.

(iii) Since

$$\frac{\partial B_3}{\partial x_2}\psi = \frac{\partial}{\partial x_2}(\partial^3\psi) + 3\frac{\partial u_2}{\partial x_2}(\partial\psi) + 3u_2\partial\left(\frac{\partial\psi}{\partial x_2}\right) + 3\frac{\partial u_3}{\partial x_2}\psi + 3u_3\frac{\partial\psi}{\partial x_2}$$

$$+ 3\frac{\partial^2 u_2}{\partial x_1\partial x_2}\psi + 3\frac{\partial u_2}{\partial x_1}\frac{\partial\psi}{\partial x_2}$$

and

$$-\frac{\partial B_2}{\partial x_3}\psi = -\partial^2\frac{\partial\psi}{\partial x_3} - 2\frac{\partial u_2}{\partial x_3}\partial\psi - 2u_2\frac{\partial\psi}{\partial x_3},$$

we find

$$\frac{\partial}{\partial x_1}\left(\frac{\partial u_2}{\partial x_3} - \frac{1}{4}\frac{\partial^3 u_2}{\partial x_1^3} - 3u_2\frac{\partial u_2}{\partial x_1}\right) - \frac{3}{4}\frac{\partial^2 u_2}{\partial x_2^2} = 0. \tag{14}$$

Equation (14) is called the *Kadomtsev–Petviashvili equation*.

Problem 5. The nonlinear partial differential equation

$$\frac{\partial^2 u}{\partial x \partial y} + \alpha \frac{\partial u}{\partial x} + \beta \frac{\partial u}{\partial y} + \gamma \frac{\partial u}{\partial x} \frac{\partial u}{\partial y} = 0 \tag{1}$$

is called *Thomas equation*, where α, β and γ are real constants with $\gamma \neq 0$. Show that

$$\frac{\partial \phi}{\partial x} = -\frac{\partial u}{\partial x} - \left(2\beta + \gamma \frac{\partial u}{\partial x} \right) \phi, \tag{2a}$$

$$\frac{\partial \phi}{\partial y} = \frac{\partial u}{\partial y} - \left(2\alpha + \gamma \frac{\partial u}{\partial y} \right) \phi \tag{2b}$$

provides a Lax representation for (1).

Solution. Taking the derivative of Eq. (2b) with respect to x gives

$$\frac{\partial^2 \partial \phi}{\partial x \partial y} = \frac{\partial^2 u}{\partial x \partial y} - 2\alpha \frac{\partial \phi}{\partial x} - \gamma \frac{\partial^2 u}{\partial x \partial y} \phi - \gamma \frac{\partial u}{\partial y} \frac{\partial \phi}{\partial x}. \tag{3}$$

Taking the derivative of (2a) with respect to y gives

$$\frac{\partial^2 \partial \phi}{\partial y \partial x} = -\frac{\partial^2 u}{\partial y \partial x} - 2\beta \frac{\partial \phi}{\partial y} - \gamma \frac{\partial^2 u}{\partial y \partial x} \partial \phi - \gamma \frac{\partial u}{\partial x} \frac{\partial \phi}{\partial y}. \tag{4}$$

Subtracting (3) and (4) leads to

$$0 = 2\frac{\partial^2 u}{\partial y \partial x} - \left(2\alpha + \gamma \frac{\partial u}{\partial y} \right) \frac{\partial \phi}{\partial x} + \left(2\beta + \gamma \frac{\partial u}{\partial x} \right) \frac{\partial \phi}{\partial y}. \tag{5}$$

Inserting Eqs. (2a) and (2b) into Eq. (5) gives the Thomas equation (1).

Remark. *Equation (1) can be linearized by applying the transformation*

$$u(x, y) = \frac{1}{\gamma} \ln v(x, y).$$

Chapter 16

Hirota Technique

Problem 1. The *Hirota bilinear operators D_x and D_t* are defined as

$$D_t^n D_x^m f \circ g := \left(\frac{\partial}{\partial t} - \frac{\partial}{\partial t'} \right)^n \left(\frac{\partial}{\partial x} - \frac{\partial}{\partial x'} \right)^m f(t,x) g(t',x') \Bigg|_{x'=x, t'=t} \tag{1}$$

where $m, n = 0, 1, 2, \ldots$. The *Boussinesq equation* is given by

$$\frac{\partial^2 u}{\partial t^2} - \frac{\partial^2 u}{\partial x^2} - 3 \frac{\partial^2}{\partial x^2} u^2 - \frac{\partial^4 u}{\partial x^4} = 0 .$$

Express the Boussinesq equation using Hirota bilinear operators.

Remark. *The Hirota bilinear operator is linear. From (1) it follows that*

$$D_x^m f \circ f = 0 \quad \text{for odd } m ,$$

$$D_x^m f \circ g = (-1)^m D_x^m g \circ f ,$$

$$D_x f \circ g = \frac{\partial f}{\partial x} g - f \frac{\partial g}{\partial x} ,$$

$$D_t f \circ g = \frac{\partial f}{\partial t} g - f \frac{\partial g}{\partial t} .$$

Solution. We set

$$u = 2 \frac{\partial^2}{\partial x^2} \ln f \tag{2}$$

where $f > 0$. Obviously

$$\frac{\partial^2}{\partial x^2}(\ln f) = \frac{1}{2f^2}D_x^2 f \circ f$$

and

$$\frac{\partial^4}{\partial x^4}(\ln f) = \frac{1}{2f^2}D_x^4 f \circ f - 6\left(\frac{1}{2f^2}D_x^2 f \circ f\right)^2.$$

Then we obtain

$$u = \frac{1}{f^2}D_x^2 f \circ f$$

and

$$\frac{\partial^2 u}{\partial x^2} = \frac{1}{f^2}D_x^4 f \circ f - 3u^2.$$

Furthermore

$$2\frac{\partial^2 \ln f}{\partial t^2} = \frac{1}{f^2}D_t^2 f \circ f.$$

It follows that

$$2\frac{\partial^2 \ln f}{\partial t^2} - u - 3u^2 - \frac{\partial^2 u}{\partial x^2} = \frac{1}{f^2}((D_t^2 - D_x^2 - D_x^4)f \circ f). \tag{3}$$

Taking the second derivative of (3) with respect to x we find

$$2\frac{\partial^4 \ln f}{\partial t^2 \partial x^2} - \frac{\partial^2 u}{\partial x^2} - 3\frac{\partial^2}{\partial x^2}u^2 - \frac{\partial^4 u}{\partial x^4} = \frac{\partial^2}{\partial x^2}\left(\frac{1}{f^2}(D_t^2 - D_x^2 - D_x^4)f \circ f\right).$$

Hence

$$\frac{\partial^2 u}{\partial t^2} - \frac{\partial^2 u}{\partial x^2} - 3\frac{\partial^2}{\partial x^2}u^2 - \frac{\partial^4 u}{\partial x^4} = \frac{\partial^2}{\partial x^2}\left(\frac{1}{f^2}(D_t^2 - D_x^2 - D_x^4)f \circ f\right).$$

Thus, if f is a solution of

$$(D_t^2 - D_x^2 - D_x^4)f \circ f = 0,$$

then u as given by (2) is a solution of the Boussinesq equation.

Remark. *The Boussinesq equation admits a Lax representation and can be solved by the inverse scattering method. Furthermore, one finds auto-Bäcklund transformations, infinite hierarchies of conservation laws and infinite hierarchies of Lie–Bäcklund vector fields.*

Problem 2.   The system of partial differential equations

$$\left(\frac{\partial}{\partial t} + c_1 \frac{\partial}{\partial x}\right) u_1 = -u_1 u_2 \,, \tag{1a}$$

$$\left(\frac{\partial}{\partial t} + c_2 \frac{\partial}{\partial x}\right) u_2 = u_1 u_2 \tag{1b}$$

describes the *interaction of two waves* u_1 and u_2 propagating with constant velocities c_1 and c_2, respectively.

(i) Find a conservation law.

(ii) Let

$$u_1 := \frac{G_1}{F}\,, \qquad u_2 := \frac{G_2}{F}\,.$$

Show that system (1) can be written as

$$D_1 G_1 \circ F = -G_1 G_2 \,, \tag{2a}$$

$$D_2 G_2 \circ F = G_1 G_2 \tag{2b}$$

where

$$D_j := D_t + c_j D_x \,, \quad j = 1, 2 \,.$$

(iii) Expand F, G_1 and G_2 in power series, i.e.

$$F = 1 + \epsilon f_1 + \epsilon^2 f_2 + \cdots \,, \tag{3a}$$

$$G_1 = \epsilon g_1 + \epsilon^2 g_2 + \cdots \,, \tag{3b}$$

$$G_2 = \epsilon h_1 + \epsilon^2 h_2 + \cdots \tag{3c}$$

and construct a solution to system (1).

Solution.   (i) Adding (1a) and (1b) leads to the conservation law

$$\frac{\partial}{\partial t}(u_1 + u_2) + \frac{\partial}{\partial x}(c_1 u_1 + c_2 u_2) = 0 \,.$$

(ii) Since

$$\frac{\partial u_j}{\partial t} = \frac{1}{F^2}\left(\frac{\partial G_j}{\partial t} F - \frac{\partial F}{\partial t} G_j\right) \,,$$

$$\frac{\partial u_j}{\partial x} = \frac{1}{F^2}\left(\frac{\partial G_j}{\partial x} F - \frac{\partial F}{\partial x} G_j\right)$$

where $j = 1, 2$, we find (2a) and (2b).

(iii) Substituting (3) into system (2) and collecting terms with the same power in ϵ, we find

$$D_1 g_1 \circ 1 = 0, \tag{4a}$$

$$D_2 h_1 \circ 1 = 0, \tag{4b}$$

$$D_1(g_1 \circ f_1 + g_2 \circ 1) = -g_1 h_1, \tag{4c}$$

$$D_2(h_1 \circ f_1 + h_2 \circ 1) = g_1 h_1 \tag{4d}$$

and so on. From (4a) and (4b) we have

$$g_1(x, t) = g_1(x - c_1 t), \tag{5a}$$

$$h_1(x, t) = h_1(x - c_2 t) \tag{5b}$$

where g_1 and h_1 are arbitrary functions. Inserting (5a) and (5b) into (4c) and (4d) we find that all higher terms can be chosen to be zero if f_1 satisfies the equations

$$\left(\frac{\partial}{\partial t} + c_1 \frac{\partial}{\partial x} \right) f_1 = h_1, \tag{6a}$$

$$\left(\frac{\partial}{\partial t} + c_2 \frac{\partial}{\partial x} \right) f_1 = -g_1. \tag{6b}$$

The general solution to system (6) is given by

$$f_1(x, t) = F_1(x - c_1 t) + F_2(x - c_2 t)$$

where F_1 and F_2 are related to g_1 and h_1 by

$$\left(\frac{\partial}{\partial t} + c_1 \frac{\partial}{\partial x} \right) F_2(x - c_2 t) = h_1(x - c_2 t),$$

$$\left(\frac{\partial}{\partial t} + c_2 \frac{\partial}{\partial x} \right) F_1(x - c_1 t) = -g_1(x - c_1 t).$$

Therefore we have the exact solution of system (1)

$$u_1(x, t) = \frac{-(\partial/\partial t + c_2 \, \partial/\partial x) F_1(x - c_1 t)}{F_1(x - c_1 t) + F_2(x - c_2 t)},$$

$$u_2(x, t) = \frac{(\partial/\partial t + c_1 \, \partial/\partial x) F_2(x - c_2 t)}{F_1(x - c_1 t) + F_2(x - c_2 t)}.$$

Chapter 17

Painlevé Test

Problem 1. A partial differential equation has the *Painlevé property* when the solution of the partial differential equation is single-valued about the movable, singularity manifold. One requires that the solution be a single-valued functional of the data, i.e., arbitrary functions. To prove that a partial differential equation has the Painlevé property one expands a solution about a movable singular manifold

$$\phi(z_1, \ldots, z_n) = 0.$$

Let $u = u(z_1, \ldots, z_n)$ be a solution of the partial differential equation and assume that

$$u = \phi^n \sum_{j=0}^{\infty} u_j \phi^j \tag{1}$$

where ϕ and u_j are analytic functions of $z_1, \ldots, z_n$ in a neighbourhood of the manifold $\phi = 0$.

(i) *Burgers' equation* is given by

$$\frac{\partial u}{\partial t} + u \frac{\partial u}{\partial x} = \frac{\partial^2 u}{\partial x^2}. \tag{2}$$

Show that Burgers' equation has the Painlevé property.

(ii) Show that one can set $u_j = 0$ for $j \geq 2$. Find the partial differential equation for u_1 in this case.

Solution. (i) Inserting the expansion (1) into (2) yields

$$n = -1$$

and

$$(j-2)(j+1)\left(\frac{\partial \phi}{\partial x}\right)^2 = F_j\left(u_{j-1}, \ldots, u_0, \frac{\partial \phi}{\partial t}\frac{\partial \phi}{\partial x}\right).$$

In order that expansion (1) is valid F_2 has to vanish identically. We find

$$j = 0, \quad u_0 = -2\frac{\partial \phi}{\partial x}, \tag{3a}$$

$$j = 1, \quad \frac{\partial \phi}{\partial t} + u_1\frac{\partial \phi}{\partial x} = \frac{\partial^2 \phi}{\partial x^2}, \tag{3b}$$

$$j = 2, \quad \frac{\partial}{\partial x}\left(\frac{\partial \phi}{\partial t} + u_1\frac{\partial \phi}{\partial x} - \frac{\partial^2 \phi}{\partial x^2}\right) = 0. \tag{3c}$$

The relation at $j = 2$ (so-called *compatibility condition*) is satisfied identically owing to the relation (3b). Therefore the expansion (1) is valid for arbitrary functions ϕ and u_2.

(ii) If we set the arbitrary function u_2 equal to zero

$$u_2(x, t) = 0$$

and require that the function u_1 satisfies Burgers' Eq. (2), then $u_j = 0$ for $j \geq 2$. We obtain

$$u = -2\frac{1}{\phi}\frac{\partial \phi}{\partial x} + u_1 \tag{4}$$

where u and u_1 satisfy Burgers' equation and

$$\frac{\partial \phi}{\partial t} + u_1\frac{\partial \phi}{\partial x} = \frac{\partial^2 \phi}{\partial x^2}. \tag{5}$$

Since $u_1 = 0$ is a solution of Burgers' equation, we find that (5) takes the form (linear diffusion equation)

$$\frac{\partial \phi}{\partial t} = \frac{\partial^2 \phi}{\partial x^2}.$$

In this case Eq. (4) simplifies to

$$u = -2\frac{1}{\phi}\frac{\partial \phi}{\partial x}.$$

This is the *Cole–Hopf transformation*.

Remark.   *When we consider the* inviscid Burgers' equation

$$\frac{\partial u}{\partial t} = u \frac{\partial u}{\partial x},$$ (6)

we find an expansion of the form

$$u = u_0 + u_1 \phi^n + u_2 \phi^{2n} + \cdots$$

where $n = 1/2$. For the first two expansion coefficients, i.e., u_0 and u_1 we find

$$\phi^{-1/2} : u_0 \frac{\partial \phi}{\partial x} = \frac{\partial \phi}{\partial t},$$ (7a)

$$\phi^0 : \frac{\partial u_0}{\partial t} + u_0 \frac{\partial u_0}{\partial x} + \frac{1}{2} u_1^2 \frac{\partial \phi}{\partial x} = 0.$$ (7b)

If we require that u_0 satisfies the inviscid Burgers' Eq. (6), then we find from the infinite coupled system for the expansion coefficients that $u_1 = u_2 = \cdots = 0$. Then inserting (7a) into (7b) yields

$$\left(\frac{\partial \phi}{\partial x}\right)^2 \frac{\partial^2 \phi}{\partial t^2} + \left(\frac{\partial \phi}{\partial t}\right)^2 \frac{\partial^2 \phi}{\partial x^2} - 2 \frac{\partial \phi}{\partial x} \frac{\partial \phi}{\partial t} \frac{\partial^2 \phi}{\partial x \partial t} = 0.$$

This equation can be linearized by applying a Legendre transformation.

Problem 2.   The $SO(2,1)$ invariant *nonlinear σ-model* can be written as

$$(w + \bar{w}) \frac{\partial^2 w}{\partial x \partial t} - 2 \frac{\partial w}{\partial x} \frac{\partial w}{\partial t} = 0$$ (1)

where $\bar{w}$ denotes the complex conjugate of w. Show that the partial differential Eq. (1) has the Painlevé property.

Hint.   Introduce the real fields u and v, i.e., $w = u + iv$.

Solution.   Introducing the real fields u and v, we find that (1) can be written as

$$u \frac{\partial^2 u}{\partial x \partial t} - \frac{\partial u}{\partial x} \frac{\partial u}{\partial t} + \frac{\partial v}{\partial x} \frac{\partial v}{\partial t} = 0,$$ (2a)

$$u \frac{\partial^2 v}{\partial x \partial t} - \frac{\partial u}{\partial t} \frac{\partial v}{\partial x} - \frac{\partial u}{\partial x} \frac{\partial v}{\partial t} = 0,$$ (2b)

where we used $w + \bar{w} = 2u$. Inserting the ansatz

$$u = \phi^m \sum_{j=0}^{\infty} u_j \phi^j , \quad v = \phi^n \sum_{j=0}^{\infty} v_j \phi^j \tag{3}$$

into system (2) yields, for the first terms in the expansion

$$n = m = -1 \tag{4}$$

and

$$u_0^2 + v_0^2 = 0 , \quad 0 \cdot u_0 v_0 \frac{\partial \phi}{\partial x} \frac{\partial \phi}{\partial t} = 0 . \tag{5}$$

Thus u_0 or v_0 is arbitrary. Inserting the expansions (3) into (2) with the values given by (4) and (5) we find that at $j = 1$ u_1 or v_1 can be chosen arbitrarily. For $j = 2$, we find that u_2 or v_2 can be chosen arbitrarily. Thus (1) has the Painlevé property.

Chapter 18

Lie Algebras

Problem 1. A *Lie algebra* is defined as follows: A vector space L over a field F, with an operation $L \times L \to L$ denoted by

$$(x, y) \to [x, y]$$

and called the commutator of x and y, is called a Lie algebra over F if the following axioms are satisfied:

 (L1) The bracket operation is bilinear.

 (L2) $[x, x] = 0$ for all $x \in L$.

 (L3) $[x, [y, z]] + [y, [z, x]] + [z, [x, y]] = 0$ $(x, y, z \in L)$.

Remark. *Axiom (L3) is called the* Jacobi identity.

 Bilinearity means that

$$[\alpha x + \beta y, z] = \alpha[x, z] + \beta[y, z],$$
$$[x, \alpha y + \beta z] = \alpha[x, y] + \beta[x, z]$$

where $\alpha, \beta \in F$. A Lie algebra is called real when $F = \mathbf{R}$. A Lie algebra is called complex when $F = \mathbf{C}$. If $F = \mathbf{R}$ or $F = \mathbf{C}$, then $[x, x] = 0$ and $[x, y] = -[y, x]$ are equivalent.

 A Lie algebra is called abelian or commutative if $[x, y] = 0$ for all x, $y \in L$.

Let L_1 and L_2 be Lie algebras and

$$\Phi : L_1 \to L_2$$

a bijection such that for all $\alpha, \beta \in F$ and all $x, y \in L_1$

$$\Phi(\alpha x + \beta y) = \alpha\Phi(x) + \beta\Phi(y), \quad \Phi([x,y]) = [\Phi(x), \Phi(y)].$$

Then Φ is called an isomorphism and the Lie algebras L_1 and L_2 are isomorphic.

Classify all real two-dimensional Lie algebras, where $F = \mathbf{R}$. Denote the basis elements by X and Y.

Solution. There exists only two non-isomorphic two-dimensional Lie algebras. Let X and Y span a two-dimensional Lie algebra. Then their commutator is either 0 or $[X, Y] = Z \neq 0$. In the first case we have a two-dimensional abelian Lie algebra. In the second case we have

$$[X, Y] = Z = \alpha X + \beta Y, \quad \alpha, \beta \in F$$

where α and β are not both zero. Let us suppose $\alpha \neq 0$, then one has

$$[Z, \alpha^{-1}Y] = Z.$$

Hence, for all non-abelian two-dimensional Lie algebras one can choose a basis $\{Z, U\}$ such that

$$[Z, U] = Z.$$

Consequently, all non-abelian two-dimensional Lie algebras are isomorphic. An example is the set of differential operators

$$\left\{ \frac{d}{dx}, x\frac{d}{dx} \right\}$$

where $Z \to d/dx$ and $U \to xd/dx$.

Problem 2. Let L be the real vector space $\mathbf{R}^3$. Let $\mathbf{a}, \mathbf{b} \in L$. Define

$$\mathbf{a} \times \mathbf{b} := \begin{pmatrix} a_2 b_3 - a_3 b_2 \\ a_3 b_1 - a_1 b_3 \\ a_1 b_2 - a_2 b_1 \end{pmatrix}. \tag{1}$$

Remark. *Equation (1) is the* cross *or* vector product.

(i) Show that L is a Lie algebra.

(ii) Show that in general

$$\mathbf{a} \times (\mathbf{b} \times \mathbf{c}) \neq (\mathbf{a} \times \mathbf{b}) \times \mathbf{c}.$$

Solution. (i) Obviously, we have

$$(\mathbf{a} + \mathbf{b}) \times (\mathbf{c} + \mathbf{d}) = \mathbf{a} \times \mathbf{c} + \mathbf{a} \times \mathbf{d} + \mathbf{b} \times \mathbf{c} + \mathbf{b} \times \mathbf{d}.$$

From (1), we see that

$$\mathbf{a} \times \mathbf{b} = -\mathbf{b} \times \mathbf{a}.$$

Using definition (1) we obtain, after a lengthy calculation, that

$$\mathbf{a} \times (\mathbf{b} \times \mathbf{c}) + \mathbf{c} \times (\mathbf{a} \times \mathbf{b}) + \mathbf{b} \times (\mathbf{c} \times \mathbf{a}) = \mathbf{0}. \tag{2}$$

Equation (2) is called the *Jacobi identity*.

(ii) Let

$$\mathbf{a} := \begin{pmatrix} 1 \\ 0 \\ 0 \end{pmatrix}, \quad \mathbf{b} := \begin{pmatrix} 0 \\ 1 \\ 0 \end{pmatrix}, \quad \mathbf{c} := \begin{pmatrix} 0 \\ 1 \\ 0 \end{pmatrix}.$$

Then

$$\mathbf{a} \times (\mathbf{b} \times \mathbf{c}) \neq (\mathbf{a} \times \mathbf{b}) \times \mathbf{c}.$$

Problem 3. Let v, w be two elements of a Lie algebra L. Then

$$\mathrm{ad}v(w) := [v, w].$$

Let

$$x := \begin{pmatrix} 0 & 1 \\ 0 & 0 \end{pmatrix}, \quad h := \begin{pmatrix} 1 & 0 \\ 0 & -1 \end{pmatrix}, \quad y := \begin{pmatrix} 0 & 0 \\ 1 & 0 \end{pmatrix}$$

be an ordered basis for the Lie algebra $sl(2, \mathbf{R})$.

(i) Compute the matrices ad$x$, ad$h$ and ad$y$ relative to this basis. This is the so-called *adjoint representation*.

(ii) Calculate the eigenvalues and eigenvectors of ad$x$, ad$h$ and ad$y$.

(iii) Do the matrices

$$\mathrm{ad}x, \quad \mathrm{ad}h, \quad \mathrm{ad}y$$

form basis of a Lie algebra under the commutator? If so, is this Lie algebra isomorphic to $sl(2, \mathbf{R})$?

Solution. Since

$$[x, h] = -2x, \quad [x, y] = h, \quad [y, h] = 2y,$$

we find

$$\operatorname{ad}x(x) = 0, \quad \operatorname{ad}x(h) = -2x, \quad \operatorname{ad}x(y) = h,$$
$$\operatorname{ad}h(x) = 2x, \quad \operatorname{ad}h(h) = 0, \quad \operatorname{ad}h(y) = -2y,$$
$$\operatorname{ad}y(x) = -h \quad \operatorname{ad}y(h) = 2y, \quad \operatorname{ad}y(y) = 0.$$

From

$$(x, h, y)\operatorname{ad}x = (0, -2x, h)$$

we find

$$\operatorname{ad}x = \begin{pmatrix} 0 & -2 & 0 \\ 0 & 0 & 1 \\ 0 & 0 & 0 \end{pmatrix}.$$

Analogously,

$$\operatorname{ad}h = \begin{pmatrix} 2 & 0 & 0 \\ 0 & 0 & 0 \\ 0 & 0 & -2 \end{pmatrix}$$

and

$$\operatorname{ad}y = \begin{pmatrix} 0 & 0 & 0 \\ -1 & 0 & 0 \\ 0 & 2 & 0 \end{pmatrix}.$$

(iii) Since

$$[\operatorname{ad}x, \operatorname{ad}h] = -2\operatorname{ad}x,$$

$$[\operatorname{ad}x, \operatorname{ad}y] = \operatorname{ad}h,$$

$$[\operatorname{ad}y, \operatorname{ad}h] = 2\operatorname{ad}y,$$

we find that the set

$$\{\operatorname{ad}x, \operatorname{ad}h, \operatorname{ad}y\}$$

is closed under the commutator. Thus this set forms a basis of a Lie algebra L. Thus the vector space isomorphism

$$\phi : sl(2, \mathbf{R}) \to L$$

defined on the basis elements by

$$\phi(x) \to \mathrm{ad}x, \quad \phi(h) \to \mathrm{ad}h, \quad \phi(y) \to \mathrm{ad}y$$

is a Lie algebra isomorphism. Thus $sl(2, \mathbf{R})$ is isomorphic to L.

Problem 4.   (i) Show that the vector fields

$$\left\{ \frac{d}{dx}, x\frac{d}{dx}, x^2\frac{d}{dx} \right\} \tag{1}$$

form a Lie algebra L under the *commutator*. Recall that the commutator is given by

$$\left[f_1(x)\frac{d}{dx}, f_2(x)\frac{d}{dx} \right] g(x) := \left(f_1\frac{df_2}{dx} - f_2\frac{df_1}{dx} \right) \frac{dg}{dx} \tag{2}$$

where g is an arbitrary smooth function. Thus (2) can also be written as

$$\left[f_1(x)\frac{d}{dx}, f_2(x)\frac{d}{dx} \right] := \left(f_1\frac{df_2}{dx} - f_2\frac{df_1}{dx} \right) \frac{d}{dx} . \tag{3}$$

(ii) Determine the *centre* Z of the Lie algebra L. The centre of L is defined as

$$Z(L) := \{ z \in L | [z, x] = 0 \quad \text{for all } x \in L \} .$$

(iii) Determine

$$\mathrm{ad}\left(\frac{d}{dx} \right), \quad \mathrm{ad}\left(x\frac{d}{dx} \right), \quad \mathrm{ad}\left(x^2\frac{d}{dx} \right) .$$

(iv) Do the vector fields

$$\left\{ \frac{d}{dx}, x\frac{d}{dx}, x^2\frac{d}{dx}, x^3\frac{d}{dx} \right\} \tag{4}$$

form a Lie algebra? If not can the set be extended to form a Lie algebra.

Solution.   (i) Straightforward calculation yields

$$\left[\frac{d}{dx}, x\frac{d}{dx}\right] = \frac{d}{dx},$$

$$\left[\frac{d}{dx}, x^2\frac{d}{dx}\right] = 2x\frac{d}{dx},$$

$$\left[x\frac{d}{dx}, x^2\frac{d}{dx}\right] = x^2\frac{d}{dx}.$$

Since the Jacobi identity is satisfied for vector fields, we find that the vector fields given by (1) form a basis of a Lie algebra.

(ii) Now an arbitrary element of L can be written as

$$f(x)\frac{d}{dx}$$

where $f(x) = a + bx + cx^2$ with $a, b, c \in \mathbf{R}$. Then we find from the condition

$$\left[f(x)\frac{d}{dx}, \frac{d}{dx}\right] = 0$$

that

$$\frac{df}{dx} = 0.$$

From the condition

$$\left[f(x)\frac{d}{dx}, x\frac{d}{dx}\right] = 0$$

together with condition (5), we obtain

$$f(x) = 0.$$

Thus the centre Z of the Lie algebra L is given by $Z = \{0\}$.

(iii) Let

$$A := \frac{d}{dx}, \quad B := x\frac{d}{dx}, \quad C := x^2\frac{d}{dx}.$$

We introduce the order A, B, C. Then we have

$$\operatorname{ad}A(A) = 0,$$

$$\operatorname{ad}A(B) = A,$$

$$\operatorname{ad}A(C) = 2B.$$

Thus we find

$$\mathrm{ad}A = \begin{pmatrix} 0 & 1 & 0 \\ 0 & 0 & 2 \\ 0 & 0 & 0 \end{pmatrix}.$$

Analogously

$$\mathrm{ad}B = \begin{pmatrix} -1 & 0 & 0 \\ 0 & 0 & 0 \\ 0 & 0 & 1 \end{pmatrix},$$

$$\mathrm{ad}C = \begin{pmatrix} 0 & 0 & 0 \\ -2 & 0 & 0 \\ 0 & -1 & 0 \end{pmatrix}.$$

(iv) Let

$$A_k := x^k \frac{d}{dx}$$

where $k = 0, 1, 2, \ldots$. Since

$$[A_2, A_3] = x^4 \frac{d}{dx} \equiv A_4,$$

we find that the set (4) does not form a basis of a Lie algebra. Since

$$[A_j, A_k] = (k - j)A_{j+k-1},$$

we find that the vector fields given by the set

$$S := \{ A_k : k = 0, 1, 2, \ldots \}$$

form an infinite-dimensional Lie algebra under the commutator (3).

Problem 5.   Let L be any Lie algebra. If $x, y \in L$, define

$$\kappa(x, y) := \mathrm{tr}(\mathrm{ad}x\,\mathrm{ad}y)$$

where κ is called the *Killing form* and $\mathrm{tr}(\cdot)$ denotes the trace.

(i) Show that

$$\kappa([x, y], z) = \kappa(x, [y, z]).$$

(ii) If B_1 and B_2 are subspaces of a Lie algebra L, we define $[B_1, B_2]$ to be the subspace spanned by all products $[b_1, b_2]$ with $b_1 \in B_1$ and $b_2 \in B_2$. This is the set of sums

$$\sum_j [b_{1j}, b_{2j}]$$

where $b_{1j} \in B_1$ and $b_{2j} \in B_2$. Let

$$L^2 := [L, L], \quad L^3 := [L^2, L], \ldots, L^k := [L^{k-1}, L].$$

A Lie algebra L is called *nilpotent* if

$$L^n = \{0\}$$

for some positive integer n. Prove that if L is nilpotent, the Killing form of L is identically zero.

Solution. (i) For $m \times m$ matrices we have the properties

$$\operatorname{tr}(A + B) = \operatorname{tr} A + \operatorname{tr} B,$$

$$\operatorname{tr}(AB) = \operatorname{tr}(BA).$$

Moreover we apply

$$\operatorname{ad}[x, y] \equiv [\operatorname{ad}x, \operatorname{ad}y].$$

Using these properties we obtain

$$\kappa([x, y], z) = \operatorname{tr}(\operatorname{ad}[x, y]\operatorname{ad}z) = \operatorname{tr}([\operatorname{ad}x, \operatorname{ad}y]\operatorname{ad}z)$$

$$= \operatorname{tr}(\operatorname{ad}x \operatorname{ad}y \operatorname{ad}z - \operatorname{ad}y \operatorname{ad}x \operatorname{ad}z)$$

$$= \operatorname{tr}(\operatorname{ad}x \operatorname{ad}y \operatorname{ad}z) - \operatorname{tr}(\operatorname{ad}y \operatorname{ad}x \operatorname{ad}z)$$

$$= \operatorname{tr}(\operatorname{ad}x \operatorname{ad}y \operatorname{ad}z) - \operatorname{tr}(\operatorname{ad}x \operatorname{ad}z \operatorname{ad}y)$$

$$= \operatorname{tr}(\operatorname{ad}x[\operatorname{ad}y, \operatorname{ad}z]) = \operatorname{tr}(\operatorname{ad}x \operatorname{ad}[y, z])$$

$$= \kappa(x, [y, z]).$$

(ii) For any $u \in L$ we have

$$(\operatorname{ad}x \operatorname{ad}y)^1(u) = [x, [y, u]] \in L^3,$$

$$(\operatorname{ad}x \operatorname{ad}y)^2(u) = [x, [y, [x, [y, u]]]] \in L^5.$$

In general

$$(adx\,ady)^k(u) \in L^{2k+1}\,.$$

Since L is nilpotent, say

$$L^n = \{0\}\,,$$

we find that for $2k + 1 \geq n$

$$(adx\,ady)^k(u) = 0 \quad \text{for all } u \in L$$

or equivalently

$$(adx\,ady)^k = 0$$

where 0 is the zero matrix. Thus the matrix $(adx\,ady)$ is nilpotent. If λ is an eigenvalue of $(adx\,ady)$ with eigenvector $\mathbf{v}$, i.e.,

$$(adx\,ady)\mathbf{v} = \lambda\mathbf{v}\,,$$

then

$$\lambda^k\mathbf{v} = (adx\,ady)^k\mathbf{v} = 0\,.$$

Therefore

$$\lambda^k = 0$$

and thus $\lambda = 0$. Thus the eigenvalues $\lambda_1, \ldots, \lambda_m$ of $(adx\,ady)$ are all zero, and we find

$$\kappa(x, y) = \text{tr}(adx\,ady) = \sum_{j=1}^m \lambda_j = 0$$

since the trace of a square matrix is the sum of the eigenvalues. This is true for any $x, y \in L$ and the result follows.

Problem 6.   Let $gl(n, \mathbf{R})$ be the Lie algebra of all $n \times n$ matrices over $\mathbf{R}$. Let

$$x \in gl(n, \mathbf{R})$$

have n distinct eigenvalues $\lambda_1, \ldots, \lambda_n$ in $\mathbf{R}$. Prove that the eigenvalues of

$$adx$$

are the n^2 scalars

$$\lambda_i - \lambda_j$$

where $i, j = 1, 2, \ldots, n$.

Solution. Since the eigenvalues $\lambda_1, \ldots, \lambda_n$ of x are real and distinct, the corresponding eigenvectors $\mathbf{v}_1, \ldots, \mathbf{v}_n$ are linearly independent (not necessarily orthogonal), and thus form a basis for $\mathbf{R}^n$. The existence of a basis of eigenvectors of x means that x can be diagonalized, i.e., there exists a matrix $R \in GL(n, \mathbf{R})$ such that

$$R^{-1}xR = \mathrm{diag}(\lambda_1, \ldots, \lambda_n)$$

where $GL(n, \mathbf{R})$ denotes the general linear group $(GL(n, \mathbf{R}) \subset gl(n, \mathbf{R}))$. Now

$$\mathrm{ad}(R^{-1}xR)(E_{ij}) \equiv [R^{-1}xR, E_{ij}] = (\lambda_i - \lambda_j)E_{ij}$$

where E_{ij} are the matrices having 1 in the (i, j) position and 0 elsewhere. Thus

$$\lambda_i - \lambda_j, \quad i, j = 1, 2, \ldots, n$$

are eigenvalues of

$$\mathrm{ad}(R^{-1}xR).$$

Since the corresponding set of eigenvectors E_{ij} $(i, j = 1, 2, \ldots, n)$ is a basis of $gl(n, \mathbf{R})$ we conclude that these are the only eigenvalues of $\mathrm{ad}(R^{-1}xR)$. Now if λ is an eigenvalue of $\mathrm{ad}x$ with eigenvector $\mathbf{v}$, then

$$\mathrm{ad}(R^{-1}xR)(R^{-1}\mathbf{v}R) = R^{-1}xRR^{-1}\mathbf{v}R - R^{-1}\mathbf{v}RR^{-1}xR$$

$$= R^{-1}x\mathbf{v}R - R^{-1}\mathbf{v}xR$$

$$= R^{-1}(\mathrm{ad}x(\mathbf{v}))R.$$

Thus

$$\mathrm{ad}(R^{-1}xR)(R^{-1}\mathbf{v}R) = \lambda(R^{-1}\mathbf{v}R)$$

so that λ is an eigenvalue of $\mathrm{ad}(R^{-1}xR)$. Thus we find

$$\lambda = \lambda_i - \lambda_j \quad \text{for some } i, j \in \{1, \ldots, n\}.$$

Conversely we have

$$\mathrm{ad}(R^{-1}xR)E_{ij} = (\lambda_i - \lambda_j)E_{ij} \,.$$

It follows that

$$R^{-1}xRE_{ij} - E_{ij}R^{-1}xR = (\lambda_i - \lambda_j)E_{ij} \,.$$

Therefore

$$xRE_{ij}R^{-1} - RE_{ij}R^{-1}x = (\lambda_i - \lambda_j)RE_{ij}R^{-1}$$

or

$$\mathrm{ad}x(RE_{ij}R^{-1}) = (\lambda_i - \lambda_j)(RE_{ij}R^{-1}) \,.$$

Thus $\lambda_i - \lambda_j$ is indeed an eigenvalue of $\mathrm{ad}x$ for all $i, j \in \{1, \ldots, n\}$. This proves the theorem.

Problem 7.   Let

$$H(\mathbf{p}, \mathbf{q}) = \frac{1}{2}(p_1^2 + p_2^2 + p_3^2) + U(\|\mathbf{q}\|)$$

be a Hamilton function in $\mathbf{R}^6$, where $\| \cdot \|$ denotes the Euclidean norm, i.e.,

$$\|\mathbf{q}\| = \sqrt{q_1^2 + q_2^2 + q_3^2} \,.$$

(i) Show that

$$h_1(\mathbf{p}, \mathbf{q}) = q_2p_3 - q_3p_2 \,,$$

$$h_2(\mathbf{p}, \mathbf{q}) = q_3p_1 - q_1p_3 \,,$$

$$h_3(\mathbf{p}, \mathbf{q}) = q_1p_2 - q_2p_1$$

are first integrals.

(ii) Show that the functions h_1, h_2, h_3 form a Lie algebra under the Poisson bracket.

(iii) Is the Lie algebra simple? A Lie algebra is called *simple* when it contains no proper ideals.

Solution.   (i) The *Poisson bracket* of two smooth functions f and g is defined as

$$\{f(\mathbf{p}, \mathbf{q}), g(\mathbf{p}, \mathbf{q})\} := \sum_{j=1}^{N} \left(\frac{\partial f}{\partial q_j} \frac{\partial g}{\partial p_j} - \frac{\partial f}{\partial p_j} \frac{\partial g}{\partial q_j} \right)$$

where $N = 3$ in the present case. A function $I(\mathbf{p}, \mathbf{q})$ is called a *first integral* with respect to H if

$$\{I, H\} = 0.\tag{1}$$

Equation (1) is equivalent to $dH/dt = 0$. Since

$$\{h_1, H\} = 0,$$

$$\{h_2, H\} = 0,$$

$$\{h_3, H\} = 0,$$

we find that h_1, h_2 and h_3 are first integrals of H.

(ii) We obtain

$$\{h_1, h_2\} = h_3,\tag{2a}$$

$$\{h_2, h_3\} = h_1,\tag{2b}$$

$$\{h_3, h_1\} = h_2.\tag{2c}$$

We conclude that h_1, h_2, and h_3 form a basis of a Lie algebra.

(iii) A subspace I of a Lie algebra L is called an *ideal* of L if $x \in L$, $y \in I$ together imply $[x, y] \in I$. If L has no ideals except itself and $\{0\}$, and if moreover

$$[L, L] \neq 0$$

we call L simple. Obviously, if L is simple it implies that

$$Z(L) = \{0\}$$

where $Z(L)$ denotes the center of the Lie algebra L. Moreover

$$[L, L] = L.$$

From the commutation relations (2), we find that the Lie algebra is simple.

Problem 8.   Let L be a Lie algebra. If $H \in L$, then we call $A \in L$ a recursion element with respect to H with value $\mu \in \mathbf{C}$ if $A \neq 0$ and

$$[H, A] = \mu A.$$

If $\mu > 0$, then A is called a raising element with value μ and if $\mu < 0$, then A is called a lowering element with value μ. Assume that the elements of L are linear operators in a Hilbert space. Consider the eigenvalue equation

$$H\phi = \lambda\phi \tag{1}$$

and

$$[H, A] = \mu A. \tag{2}$$

Find $H(A\phi)$.

Solution.   From (2), we obtain

$$HA = AH + \mu A.$$

Thus

$$H(A\phi) = (\mu A + AH)\phi = \mu(A\phi) + A(H\phi).$$

Using (1), we find

$$H(A\phi) = (\mu + \lambda)(A\phi).$$

This is an eigenvalue equation. Consequently, if $A\phi \neq 0$, then $A\phi$ is an eigenfunction of H with eigenvalue $\lambda + \mu$. Thus, if $\mu \neq 0$ and we have one eigenfunction ϕ for H and a recursion element A for H, then we can generate infinitely many eigenfunctions.

Remark.   *If A and B are recursion elements with respect to H with values μ_1 and μ_2 and the product AB is well defined, then*

$$[H, AB] = [H, A]B + A[H, B] = (\mu_1 + \mu_2)AB.$$

Consequently AB is a recursion element with respect to H with value $\mu_1 + \mu_2$.

Problem 9.   The *Baker–Campbell–Hausdorff formula* is given by

$$\exp(A)B\exp(-A) = \sum_{n=0}^{\infty} \frac{[A, B]_n}{n!}$$

where the repeated commutator is defined by

$$[A, B]_n := [A, [A, B]_{n-1}]$$

with $[A, B]_0 = B$. Consider the nontrivial Lie algebra with two generators X and Y and the commutation relation

$$[X, Y] = Y.$$

Let

$$U = \exp(\alpha X + \beta Y), \quad V = \exp(aX) \exp(bY).$$

Find UXU^{-1}, UYU^{-1}, VXV^{-1}, and VYV^{-1}.

Solution. Since

$$[\alpha X + \beta Y, X]_n = -\frac{\beta}{\alpha} \alpha^n Y, \quad n > 0$$

and

$$[\alpha X + \beta Y, Y]_n = \alpha^n Y,$$

we obtain

$$UXU^{-1} = X - \frac{\beta}{\alpha}(\exp(\alpha) - 1)Y, \quad UYU^{-1} = \exp(\alpha)Y.$$

On the other hand, V transforms the generators to

$$VXV^{-1} = X - b\exp(a)Y, \quad VYV^{-1} = \exp(a)Y.$$

Comparing the two sets of similarity transformations, one obtains

$$a = \alpha, \quad b = \frac{\beta}{\alpha}(1 - \exp(-\alpha)).$$

Problem 10. Let G be a Lie group with Lie algebra g. For complex z near 1, we consider the function

$$f(z) := \frac{\ln(z)}{z - 1} = \sum_{n=0}^{\infty} \frac{(-1)^n}{n + 1}(z - 1)^n.$$

Then for X, Y near 0 in g we have

$$\exp X \exp Y = \exp C(X, Y)$$

where

$$C(X,Y) = X + Y + \sum_{n=1}^{\infty} \frac{(-1)^n}{n+1} \int_0^1 \left(\sum_{\substack{k,\ell \geq 0 \\ k+\ell \geq 1}} \frac{t^k}{k!\ell!} (\operatorname{ad}X)^k (\operatorname{ad}Y)^\ell \right)^n X \, dt$$

$$= X + Y + \sum_{n=1}^{\infty} \frac{(-1)^n}{n+1} \sum_{\substack{k_1,\ldots,k_n \geq 0 \\ \ell_1,\ldots,\ell_n \geq 0 \\ k_i + \ell_i \geq 1}}$$

$$\times \frac{(\operatorname{ad}X)^{k_1}(\operatorname{ad}Y)^{\ell_1} \cdots (\operatorname{ad}X)^{k_n}(\operatorname{ad}Y)^{\ell_n}}{(k_1 + \cdots + k_n + 1)k_1! \cdots k_n! \ell_1! \cdots \ell_n!} X$$

where $(\operatorname{ad}X)Y := [X, Y]$. Thus

$$C(X,Y) = X + Y + \frac{1}{2}[X,Y] + \frac{1}{12}([X,[X,Y]] - [Y,[Y,X]]) + \cdots.$$

Apply this equation to the Bose operators $X = b$ and $Y = b^\dagger$.

Solution.   The commutator of b and $b^\dagger$ is given by

$$[b, b^\dagger] = I$$

where I is the identity operator. Since

$$[b, I] = 0, \quad [b^\dagger, I] = 0,$$

we find

$$C(b, b^\dagger) = b + b^\dagger + I.$$

Problem 11.   The *Nahm equations* is a system of non-linear ordinary differential equations given by

$$\frac{dT_j}{dt} = \frac{1}{2} \sum_{k=1}^{3} \sum_{l=1}^{3} \epsilon_{jkl} [T_k, T_l](t) \tag{1}$$

for three $n \times n$ matrices T_j of complex-valued functions of the variable t. The indices j, k, l range over 1, 2, 3. The tensor ϵ_{jkl} is the totally anti-symmetric tensor with

$$\epsilon_{123} = 1.$$

Let $\{H_\alpha, \alpha = 1, \ldots, n-1\}$ be generators of the Cartan–Lie subalgebra $s\ell(n)$, and let $\{E_\alpha, E_{-\alpha}\}$ be step operators satisfying

$$[H_\alpha, E_{\pm\beta}] = \pm K_{\beta\alpha} E_{\pm\beta}, \tag{2a}$$

$$[E_\alpha, E_{-\beta}] = \delta_{\alpha\beta} H_\beta, \tag{2b}$$

where $\delta_{\alpha\beta}$ is the Kronecker delta. The $(n-1) \times (n-1)$ matrix $K_{\beta\alpha}$ is the *Cartan matrix* of the Lie algebra $s\ell(n)$. The Cartan matrix is given by

$$(K_{\beta\alpha}) = \begin{pmatrix} 2 & -1 & 0 & \cdots & 0 & 0 \\ -1 & 2 & -1 & \cdots & 0 & 0 \\ 0 & -1 & 2 & \cdots & 0 & 0 \\ \vdots & \vdots & \vdots & \ddots & \vdots & \vdots \\ 0 & 0 & 0 & \cdots & 2 & -1 \\ 0 & 0 & 0 & \cdots & -1 & 2 \end{pmatrix}.$$

Assume that

$$T_1(t) = \frac{1}{2} i \sum_{\alpha=1}^{n-1} q_\alpha(t)(E_\alpha + E_{-\alpha}), \quad T_2(t) = -\frac{1}{2} \sum_{\alpha=1}^{n-1} q_\alpha(t)(E_\alpha - E_{-\alpha}),$$

$$T_3(t) = \frac{1}{2} i \sum_{\alpha=1}^{n-1} p_\alpha(t) H_\alpha,$$

where q_α and p_α are smooth functions of t. Find the equations of motion for q_α and p_α.

Solution. Substituting T_1, T_2 and T_3 into the Nahm Eq. (1) and using the commutation relations (2a) and (2b) yields

$$\frac{dp_\alpha}{dt} = q_\alpha^2, \quad \frac{dq_\alpha}{dt} = \frac{1}{2} \sum_\beta p_\beta K_{\alpha\beta} q_\alpha.$$

If we set

$$\phi_\alpha = 2 \ln q_\alpha,$$

we obtain

$$\frac{d^2 \phi_\alpha}{dt^2} = \sum_{\beta=1}^{n-1} K_{\alpha\beta} \exp(\phi_\beta)$$

which are the *Toda-molecule equations*.

Problem 12.   Let A and B be $n \times n$ matrices over $\mathbf{C}$. It is known that if $[A, B] = 0$, then $\exp(A + B) = \exp(A) \exp(B)$. Let C and D be $n \times n$ matrices over $\mathbf{C}$. Assume that $\exp(C + D) = \exp(C) \exp(D)$. Can we conclude that $[C, D] = 0$?

Solution.   The answer is no. A counterexample is as follows. Let $\sigma = (\sigma_1, \sigma_2, \sigma_3)$ be the Pauli spin matrices. Consider the two unit vectors in $\mathbf{C}^3$

$$\mathbf{m} = (\cos(\alpha), \sin(\alpha), 0)^T, \quad \mathbf{n} = (\cos(\beta), \sin(\beta), 0)^T.$$

Then the sum

$$\mathbf{p} = \mathbf{m} + \mathbf{n}$$

is a unit vector if and only if

$$\cos(\alpha - \beta) = -\frac{1}{2}.$$

Thus $\alpha - \beta = 2\pi/3$. Moreover,

$$\mathbf{m} \times \mathbf{n} = (0, 0, \cos(\alpha)\sin(\beta) - \sin(\alpha)\cos(\beta))^T$$

where

$$\cos(\alpha)\sin(\beta) - \sin(\alpha)\cos(\beta) = -\sin(\alpha - \beta).$$

If $\alpha - \beta = 2\pi/3$, we have

$$-\sin(\alpha - \beta) = -\frac{1}{2}\sqrt{3}.$$

Now let

$$C = 2i\pi\sigma \cdot \mathbf{m}, \quad D = 2i\pi\sigma \cdot \mathbf{n}.$$

Thus $C + D = 2i\pi\sigma \cdot \mathbf{p}$. $\exp(C)$, $\exp(D)$ and $\exp(C + D)$ now represent rotations by 4π around $\mathbf{m}$, $\mathbf{n}$ and $\mathbf{p}$, respectively. Thus

$$\exp(C) = \exp(D) = \exp(C + D) = I$$

although

$$[C, D] = -i4\pi^2\sigma \cdot (\mathbf{m} \times \mathbf{n}) \neq 0.$$

Chapter 19

Differential Forms

Problem 1. We define differential p-forms of class C^∞ on an open set Ω of $\mathbf{R}^n$ to be the expressions

$$\omega := \sum_{j_1 < j_2 \cdots < j_p}^{n} c_{j_1 j_2 \cdots j_p}(\mathbf{x}) dx_{j_1} \wedge dx_{j_2} \wedge \cdots \wedge dx_{j_p}$$

where the functions $c_{j_1 j_2 \cdots j_p} \in C^\infty(\Omega)$ and the integers $j_1, \ldots, j_p$ lie between 1 and n. Two such differential forms may be added componentwise. One defines the *Graßmann product* (also called *exterior product* or *wedge product*) of a p-form and a q-form as follows: For any permutation σ of the indices $j_1, \ldots, j_p$,

$$dx_{\sigma(j_1)} \wedge dx_{\sigma(j_2)} \wedge \cdots \wedge dx_{\sigma(j_p)} = \operatorname{sgn}(\sigma) dx_{j_1} \wedge dx_{j_2} \wedge \cdots \wedge dx_{j_p}$$

where $\operatorname{sgn}(\sigma)$ denotes the sign of the permutation σ. Let

$$\omega' = \sum_{k_1 < k_2 \cdots < k_q}^{n} b_{k_1 k_2 \cdots k_q}(\mathbf{x}) dx_{k_1} \wedge dx_{k_2} \wedge \cdots \wedge dx_{k_q} .$$

Then

$$\omega \wedge \omega' := \sum_{\substack{j_1 < j_2 < \cdots < j_p \\ k_1 < k_2 < \cdots < k_q}}^{n} c_{j_1 j_2 \cdots j_p}(\mathbf{x}) b_{k_1 k_2 \cdots k_q}(\mathbf{x})$$

$$\times dx_{j_1} \wedge \cdots \wedge dx_{j_p} \wedge dx_{k_1} \wedge \cdots \wedge dx_{k_q} .$$

The exterior derivative of a p-form is defined by

$$d\omega = \sum_{j_1 < j_2 \cdots < j_p}^{n} dc_{j_1 j_2 \cdots j_p}(\mathbf{x}) \wedge dx_{j_1} \wedge dx_{j_2} \wedge \cdots \wedge dx_{j_p}$$

where, for a smooth function f, we have

$$df = \sum_{j=1}^{n} \frac{\partial f}{\partial x_j} dx_j.$$

Let $\Omega = \mathbf{R}^3$ and

$$\alpha := f_1(\mathbf{x}) dx_1 + f_2(\mathbf{x}) dx_2 + f_3(\mathbf{x}) dx_3.$$

Calculate $d\alpha$. Let $d\alpha = 0$. Find the condition on the functions f_1, f_2 and f_3.

Solution.　Applying

$$dx_j \wedge dx_k = -dx_k \wedge dx_j$$

yields

$$d\alpha = \left(\frac{\partial f_2}{\partial x_1} - \frac{\partial f_1}{\partial x_2}\right) dx_1 \wedge dx_2 + \left(\frac{\partial f_3}{\partial x_2} - \frac{\partial f_2}{\partial x_3}\right) dx_2 \wedge dx_3$$

$$+ \left(\frac{\partial f_1}{\partial x_3} - \frac{\partial f_3}{\partial x_1}\right) dx_3 \wedge dx_1.$$

Thus the condition $d\alpha = 0$ leads to

$$\frac{\partial f_2}{\partial x_1} - \frac{\partial f_1}{\partial x_2} = 0 \qquad \frac{\partial f_3}{\partial x_2} - \frac{\partial f_2}{\partial x_3} = 0, \qquad \frac{\partial f_1}{\partial x_3} - \frac{\partial f_3}{\partial x_1} = 0$$

since $dx_1 \wedge dx_2$, $dx_2 \wedge dx_3$ and $dx_3 \wedge dx_1$ are linearly independent.

Problem 2.　Let $f : \mathbf{R}^n \to \mathbf{R}$ be a C^1 function with $f(\mathbf{x}) \neq 0$ for all $\mathbf{x} \in \mathbf{R}^n$. Let ω be a differential one-form defined on $\mathbf{R}^n$. Assume that

$$d(f\omega) = 0. \tag{1}$$

Show that

$$\omega \wedge d\omega = 0. \tag{2}$$

Solution.　From (1) it follows that

$$d(f\omega) = (df) \wedge \omega + f d\omega = 0.$$

Taking the exterior product with ω yields

$$\omega \wedge (df) \wedge \omega + \omega \wedge (f d\omega) = 0.$$

Applying the associative law for differential forms and that

$$\omega \wedge \omega = 0$$

for differential one-forms gives

$$\omega \wedge (f d\omega) = 0.$$

Thus

$$f(\omega \wedge d\omega) = 0.$$

Since $f(\mathbf{x}) \neq 0$ for all $\mathbf{x} \in \mathbf{R}^n$, Eq. (2) follows.

Problem 3. Consider a differential p-form ω defined on $\mathbf{R}^n$. If there is a $(p-1)$-form ψ defined on $\mathbf{R}^n$ such that

$$\omega = d\psi$$

then ω is called an *exact differential form.*

(i) Show that if ω_1 and ω_2 are exact p-differential forms then $\omega_1 \wedge \omega_2$ is also exact.

(ii) Give an example.

Solution. (i) Since ω_1 and ω_2 are exact we have

$$\omega_1 = d\psi_1, \quad \omega_2 = d\psi_2.$$

Since ω_1 and ω_2 are differential p-forms we find that $\omega_1 \wedge \omega_2$ is a $2p$ differential form. We have to find a $2p - 1$ differential form α defined on $\mathbf{R}^n$ such that

$$d\alpha = \omega_1 \wedge \omega_2.$$

Thus

$$d\alpha = d\psi_1 \wedge d\psi_2.$$

Since

$$d(\psi_1 \wedge d\psi_2) = d\psi_1 \wedge d\psi_2$$

where we used that

$$dd\psi_2 = 0,$$

we find

$$\alpha = \psi_1 \wedge d\psi_2.$$

(ii) Consider

$$\omega_1 = dx_1 \wedge dx_2, \quad \omega_2 = dx_3 \wedge dx_4$$

defined on $\mathbf{R}^4$. Then we can choose ψ_1 and ψ_2 as

$$\psi_1 = x_1 dx_2, \quad \psi_2 = x_3 dx_4.$$

Therefore

$$\psi_1 \wedge d\psi_2 = \psi_1 \wedge \omega_2 = x_1 dx_2 \wedge dx_3 \wedge dx_4.$$

Remark. ψ_1 *and* ψ_2 *are not unique, for example we could also choose*

$$\psi_1 = \frac{1}{2}(x_1 dx_2 - x_2 dx_1), \quad \psi_2 = \frac{1}{2}(x_3 dx_4 - x_4 dx_3).$$

Problem 4.   The differential form

$$\alpha = \frac{dz_1 \wedge dz_2}{z_1 z_2}$$

where $z_j = x_j + iy_j$ $(x_j, y_j \in \mathbf{R})$ is defined on

$$\mathbf{C}^2 \setminus \{(z_1, z_2) : z_1 = 0 \vee z_2 = 0\}.$$

Find the real and imaginary part of α.

Solution.   Since

$$dz_j = dx_j + idy_j, \quad j = 1, 2$$

and

$$z_1 z_2 = x_1 x_2 - y_1 y_2 + i(x_1 y_2 + x_2 y_1),$$

we obtain

$$\frac{dz_1 \wedge dz_2}{z_1 z_2} = \frac{(x_1 x_2 - y_1 y_2)(dx_1 \wedge dx_2 - dy_1 \wedge dy_2) + (x_1 y_2 + x_2 y_1)(dx_1 \wedge dy_2 + dy_1 \wedge dx_2)}{x_1^2 x_2^2 + y_1^2 y_2^2 + x_1^2 y_2^2 + x_2^2 y_1^2}$$

$$+ i\frac{(x_1 x_2 - y_1 y_2)(dx_1 \wedge dy_2 + dy_1 \wedge dx_2) - (x_1 y_2 + x_2 y_1)(dx_1 \wedge dx_2 - dy_1 \wedge dy_2)}{x_1^2 x_2^2 + y_1^2 y_2^2 + x_1^2 y_2^2 + x_2^2 y_1^2}.$$

Problem 5. Let

$$g := \sum_{j,k=1}^{m} g_{jk}(\mathbf{x}) dx_j \otimes dx_k$$

be the *metric tensor field* of a Riemannian or pseudo-Riemannian real C^∞ manifold M ($\dim M = m < \infty$). Let $(x_1, x_2, \ldots, x_m)$ be the local coordinate system in a local coordinate neighbourhood (U, ψ). Since $(dx_j)_p$ ($j = 1, \ldots, m$) is a basis of T_p^* (dual space of the tangent vector space T_p) at each point p of U, we can express an r-form ω_p ($p \in U$) uniquely in the form

$$\omega_p = \sum_{j_1 < \cdots < j_r}^{m} a_{j_1 \cdots j_r}(p)(dx_{j_1})_p \wedge \cdots \wedge (dx_{j_r})_p .$$

The *Hodge star operator* is an f-linear mapping which transforms an r-form into its dual $(m - r)$-form. The *-operator which is applied to an r-form defined on an arbitrary Riemannian or pseudo-Riemannian manifold M with metric tensor field g, is defined by

$$*(dx_{j_1} \wedge dx_{j_2} \wedge \cdots \wedge dx_{j_r})$$

$$:= \sum_{k_1,\ldots,k_m=1}^{m} g^{j_1 k_1} \cdots g^{j_r k_r} \frac{1}{(m-r)!} \cdot \frac{g}{\sqrt{|g|}} \varepsilon_{k_1 \cdots k_m} dx_{k_{r+1}} \wedge \cdots \wedge dx_{k_m}$$

where $\varepsilon_{k_1 \cdots k_m}$ is the totally antisymmetric tensor with

$$\varepsilon_{12 \cdots m} = +1 ,$$

$$g \equiv \det(g_{ij}) ,$$

$$\sum_{j=1}^{m} g^{ij} g_{jk} = \delta_k^i$$

where δ_k^i denotes the Kronecker delta.

(i) Let $M = \mathbf{R}^2$ and

$$g = dx_1 \otimes dx_1 + dx_2 \otimes dx_2 .$$

Calculate $(*dx_1)$ and $(*dx_2)$.

(ii) Let $M = \mathbf{R}^3$ and

$$g = dx_1 \otimes dx_1 + dx_2 \otimes dx_2 + dx_3 \otimes dx_3 .$$

Calculate

$$*dx_1, \quad *dx_2, \quad *dx_3,$$

$$*(dx_1 \wedge dx_2), \quad *(dx_2 \wedge dx_3), \quad *(dx_3 \wedge dx_1).$$

(iii) Let $M = \mathbf{R}^4$ and

$$g = dx_1 \otimes dx_1 + dx_2 \otimes dx_2 + dx_3 \otimes dx_3 - dx_4 \otimes dx_4.$$

This is the *Minkowski metric.* Calculate

$$*(dx_1 \wedge dx_2), \quad *(dx_2 \wedge dx_3), \quad *(dx_3 \wedge dx_1),$$

$$*(dx_1 \wedge dx_4), \quad *(dx_2 \wedge dx_4), \quad *(dx_3 \wedge dx_4),$$

$$*dx_1, \quad\quad *dx_2, \quad\quad *dx_3, \quad *dx_4.$$

Solution.   (i) Since

$$g_{11} = g_{22} = 1, \quad g_{12} = g_{21} = 0,$$

we find

$$g^{11} = g^{22} = 1, \quad g^{12} = g^{21} = 0.$$

Therefore

$$*dx_1 = \sum_{k_1,k_2=1}^{2} g^{1k_1} \frac{1}{(2-1)!} \frac{g}{\sqrt{|g|}} \varepsilon_{k_1 k_2} dx_{k_2}.$$

We have

$$g = \det(g_{ij}) = 1$$

and

$$\varepsilon_{12} = 1, \quad \varepsilon_{21} = -1, \quad \varepsilon_{11} = \varepsilon_{22} = 0.$$

It follows that

$$*dx_1 = \sum_{j_2=1}^{2} g^{11} \varepsilon_{1j_2} dx_{j_2} = dx_2.$$

Similarly,

$$*dx_2 = -dx_1.$$

(ii) Since

$$g_{11} = g_{22} = g_{33} = 1 \quad \text{and} \quad g_{jk} = 0 \quad \text{for } j \neq k,$$

we obtain

$$g^{11} = g^{22} = g^{33} = 1 \quad \text{and} \quad g^{jk} = 0 \quad \text{for } j \neq k.$$

Therefore we find for $*dx_1$

$$*dx_1 = \sum_{k_1,k_2,k_3=1}^{3} g^{1k_1} \frac{1}{(3-1)!} \varepsilon_{k_1 k_2 k_3} dx_{k_2} \wedge dx_{k_3}$$

$$= \sum_{k_2,k_3=1}^{3} \frac{1}{2} \varepsilon_{1k_2 k_3} dx_{k_2} \wedge dx_{k_3}.$$

Thus

$$*dx_1 = dx_2 \wedge dx_3.$$

Analogously

$$*dx_2 = dx_3 \wedge dx_1,$$

$$*dx_3 = dx_1 \wedge dx_2.$$

Furthermore

$$*(dx_1 \wedge dx_2) = \sum_{k_1,k_2,k_3=1}^{3} g^{1k_1} g^{2k_2} \varepsilon_{k_1 k_2 k_3} dx_{k_3} = \sum_{k_3=1}^{3} \varepsilon_{12k_3} dx_{k_3} = dx_3.$$

Analogously

$$*(dx_2 \wedge dx_3) = dx_1,$$

$$*(dx_3 \wedge dx_1) = dx_2.$$

(iii) In the present case we have

$$g_{11} = g_{22} = g_{33} = 1, \quad g_{44} = -1.$$

Therefore

$$g^{11} = g^{22} = g^{33} = 1, \quad g^{44} = -1.$$

Moreover

$$g = \det(g_{jk}) = -1.$$

Consequently, we find

$$*(dx_1 \wedge dx_2) = -dx_3 \wedge dx_4,$$

$$*(dx_2 \wedge dx_3) = -dx_1 \wedge dx_4,$$

$$*(dx_3 \wedge dx_1) = -dx_2 \wedge dx_4,$$

$$*(dx_1 \wedge dx_4) = dx_2 \wedge dx_3,$$

$$*(dx_2 \wedge dx_4) = dx_3 \wedge dx_1,$$

$$*(dx_3 \wedge dx_4) = dx_1 \wedge dx_2,$$

and

$$*dx_1 = -dx_2 \wedge dx_3 \wedge dx_4,$$

$$*dx_2 = -dx_3 \wedge dx_1 \wedge dx_4,$$

$$*dx_3 = -dx_1 \wedge dx_2 \wedge dx_4,$$

$$*dx_4 = -dx_1 \wedge dx_2 \wedge dx_3.$$

Problem 6.   In electrodynamics we have the differential two-forms

$$\beta = E_1 dx_1 \wedge dt + E_2 dx_2 \wedge dt + E_3 dx_3 \wedge dt$$

$$+ B_3 dx_1 \wedge dx_2 + B_1 dx_2 \wedge dx_3 + B_2 dx_3 \wedge dx_1$$

and

$$*\beta = \frac{E_1}{c} dx_2 \wedge dx_3 + \frac{E_2}{c} dx_3 \wedge dx_1 + \frac{E_3}{c} dx_1 \wedge dx_2$$

$$- B_3 c dx_3 \wedge dt - B_1 c dx_1 \wedge dt - B_2 c dx_2 \wedge dt$$

where c is a positive constant (speed of light).

(i) Calculate $d\beta$, $d(*\beta)$ and give an interpretation of $d\beta = 0$ and $d(*\beta) = 0$.

(ii) Calculate $\beta \wedge \beta$ and $\beta \wedge (*\beta)$ and give an interpretation.

Solution. (i) We set $x_4 \equiv ct$. Since

$$dE_k = \frac{\partial E_k}{\partial x_1} dx_1 + \frac{\partial E_k}{\partial x_2} dx_2 + \frac{\partial E_k}{\partial x_3} dx_3 + \frac{\partial E_k}{\partial x_4} dx_4,$$

$$dB_k = \frac{\partial B_k}{\partial x_1} dx_1 + \frac{\partial B_k}{\partial x_2} dx_2 + \frac{\partial B_k}{\partial x_3} dx_3 + \frac{\partial B_k}{\partial x_4} dx_4$$

for $k = 1, 2, 3$ and $dx_j \wedge dx_k = -dx_k \wedge dx_j$, we find

$$d\beta = \left(\frac{\partial B_1}{\partial x_1} + \frac{\partial B_2}{\partial x_2} + \frac{\partial B_3}{\partial x_3} \right) dx_1 \wedge dx_2 \wedge dx_3$$

$$+ \left(\frac{\partial B_3}{\partial t} - \frac{\partial E_1}{\partial x_2} + \frac{\partial E_2}{\partial x_1} \right) dx_1 \wedge dx_2 \wedge dt$$

$$+ \left(\frac{\partial B_2}{\partial t} - \frac{\partial E_3}{\partial x_1} + \frac{\partial E_1}{\partial x_3} \right) dx_3 \wedge dx_1 \wedge dt$$

$$+ \left(\frac{\partial B_1}{\partial t} - \frac{\partial E_2}{\partial x_3} + \frac{\partial E_3}{\partial x_2} \right) dx_2 \wedge dx_3 \wedge dt.$$

From the condition $d\beta = 0$, we obtain

$$0 = \left(\frac{\partial B_1}{\partial x_1} + \frac{\partial B_2}{\partial x_2} + \frac{\partial B_3}{\partial x_3} \right), \tag{1a}$$

$$0 = \left(\frac{\partial B_3}{\partial t} - \frac{\partial E_1}{\partial x_2} + \frac{\partial E_2}{\partial x_1} \right), \tag{1b}$$

$$0 = \left(\frac{\partial B_2}{\partial t} - \frac{\partial E_3}{\partial x_1} + \frac{\partial E_1}{\partial x_3} \right), \tag{1c}$$

$$0 = \left(\frac{\partial B_1}{\partial t} - \frac{\partial E_2}{\partial x_3} + \frac{\partial E_3}{\partial x_2} \right). \tag{1d}$$

Calculating $d(*\beta)$ yields

$$d(*\beta) = \frac{1}{c} \left(\frac{\partial E_1}{\partial x_1} + \frac{\partial E_2}{\partial x_2} + \frac{\partial E_3}{\partial x_3} \right) dx_1 \wedge dx_2 \wedge dx_3$$

$$+ \left(\frac{1}{c} \frac{\partial E_3}{\partial t} - c \frac{\partial B_2}{\partial x_1} + c \frac{\partial B_1}{\partial x_2} \right) dx_1 \wedge dx_2 \wedge dt$$

$$+ \left(\frac{1}{c} \frac{\partial E_2}{\partial t} - c \frac{\partial B_1}{\partial x_3} + c \frac{\partial B_3}{\partial x_1} \right) dx_3 \wedge dx_1 \wedge dt$$

$$+ \left(\frac{1}{c} \frac{\partial E_1}{\partial t} - c \frac{\partial B_3}{\partial x_2} + c \frac{\partial B_2}{\partial x_3} \right) dx_2 \wedge dx_3 \wedge dt \,.$$

From the condition $d(*\beta) = 0$, we obtain

$$0 = \left(\frac{\partial E_1}{\partial x_1} + \frac{\partial E_2}{\partial x_2} + \frac{\partial E_3}{\partial x_3} \right) , \qquad (2a)$$

$$0 = \left(\frac{1}{c} \frac{\partial E_3}{\partial t} - c \frac{\partial B_2}{\partial x_1} + c \frac{\partial B_1}{\partial x_2} \right) , \qquad (2b)$$

$$0 = \left(\frac{1}{c} \frac{\partial E_2}{\partial t} - c \frac{\partial B_1}{\partial x_3} + c \frac{\partial B_3}{\partial x_1} \right) , \qquad (2c)$$

$$0 = \left(\frac{1}{c} \frac{\partial E_1}{\partial t} - c \frac{\partial B_3}{\partial x_2} + c \frac{\partial B_2}{\partial x_3} \right) . \qquad (2d)$$

System (1) and (2) are *Maxwell's equations* in free space, i.e.,

$$\text{div } \mathbf{B} = 0, \qquad -\frac{\partial \mathbf{B}}{\partial t} = \nabla \times \mathbf{E},$$

$$\text{div } \mathbf{E} = 0, \qquad \frac{1}{c^2} \frac{\partial \mathbf{E}}{\partial t} = \nabla \times \mathbf{B}.$$

(ii) Straightforward calculation yields

$$\beta \wedge \beta = 2(B_1 E_1 + B_2 E_2 + B_3 E_3) dx_1 \wedge dx_2 \wedge dx_3 \wedge dt , \qquad (3)$$

$$\beta \wedge (*\beta) = \left(\frac{E_1^2}{c} + \frac{E_2^2}{c} + \frac{E_3^2}{c} - B_1^2 c - B_2^2 c - B_3^2 c \right)$$

$$\times dx_1 \wedge dx_2 \wedge dx_3 \wedge dt . \qquad (4)$$

Equation (4) describes the energy density of the electromagnetic field. Both (3) and (4) are invariant under the Lorentz transformation.

Remark.   *Observe that*

$$* * \beta = -\beta .$$

Problem 7.   Within the techniques of differential forms the *vector potential* and *scalar potential* is given by the one-form

$$\alpha = A_1 dx_1 + A_2 dx_2 + A_3 dx_3 - U dt$$

and the electromagnetic field by the two-form

$$\beta = E_1 dx_1 \wedge dt + E_2 dx_2 \wedge dt + E_3 dx_3 \wedge dt$$

$$+ B_3 dx_1 \wedge dx_2 + B_1 dx_2 \wedge dx_3 + B_2 dx_3 \wedge dx_1 .$$

Find the relations which follow from

$$d\alpha = \beta .$$

Solution. We set $ct \equiv x_4$. Since

$$dx_j \wedge dx_k = -dx_k \wedge dx_j$$

and

$$dA_k = \frac{\partial A_k}{\partial x_1} dx_1 + \frac{\partial A_k}{\partial x_2} dx_2 + \frac{\partial A_k}{\partial x_3} dx_3 + \frac{\partial A_k}{\partial t} dt ,$$

$$dU = \frac{\partial U}{\partial x_1} dx_1 + \frac{\partial U}{\partial x_2} dx_2 + \frac{\partial U}{\partial x_3} dx_3 + \frac{\partial U}{\partial t} dt ,$$

we find

$$d\alpha = \left(\frac{\partial A_2}{\partial x_1} - \frac{\partial A_1}{\partial x_2} \right) dx_1 \wedge dx_2 + \left(\frac{\partial A_3}{\partial x_2} - \frac{\partial A_2}{\partial x_3} \right) dx_2 \wedge dx_3$$

$$+ \left(\frac{\partial A_1}{\partial x_3} - \frac{\partial A_3}{\partial x_1} \right) dx_3 \wedge dx_1 + \left(-\frac{\partial A_1}{\partial t} - \frac{\partial U}{\partial x_1} \right) dx_1 \wedge dt$$

$$+ \left(-\frac{\partial A_2}{\partial t} - \frac{\partial U}{\partial x_2} \right) dx_2 \wedge dt + \left(-\frac{\partial A_3}{\partial t} - \frac{\partial U}{\partial x_3} \right) dx_3 \wedge dt .$$

Comparing the six basis elements

$$dx_1 \wedge dx_2 , \quad dx_2 \wedge dx_3 , \quad dx_3 \wedge dx_1 , \quad dx_1 \wedge dt , \quad dx_2 \wedge dt , \quad dx_3 \wedge dt ,$$

we find

$$B_3 = \frac{\partial A_2}{\partial x_1} - \frac{\partial A_1}{\partial x_2} ,$$

$$B_1 = \frac{\partial A_3}{\partial x_2} - \frac{\partial A_2}{\partial x_3} ,$$

$$B_2 = \frac{\partial A_1}{\partial x_3} - \frac{\partial A_3}{\partial x_1} ,$$

$$E_1 = -\frac{\partial A_1}{\partial t} - \frac{\partial U}{\partial x_1} ,$$

$$E_2 = -\frac{\partial A_2}{\partial t} - \frac{\partial U}{\partial x_2},$$

$$E_3 = -\frac{\partial A_3}{\partial t} - \frac{\partial U}{\partial x_3}.$$

It follows that

$$\mathbf{B} = \nabla \times \mathbf{A},$$

$$\mathbf{E} = -\nabla U - \frac{\partial \mathbf{A}}{\partial t}.$$

Problem 8.   The basic quantity in electromagnetism is the differential two-form

$$\beta = E_1(\mathbf{x}, t)dx_1 \wedge dt + E_2(\mathbf{x}, t)dx_2 \wedge dt + E_3(\mathbf{x}, t)dx_3 \wedge dt$$

$$+ B_3(\mathbf{x}, t)dx_1 \wedge dx_2 + B_1(\mathbf{x}, t)dx_2 \wedge dx_3 + B_2(\mathbf{x}, t)dx_3 \wedge dx_1.$$

The system ' (prime) and without prime are connected by the *Lorentz transformation*

$$x_1' = \gamma(x_1 - vt), \quad t' = \gamma\left(t - v\frac{x_1}{c^2}\right), \quad x_2' = x_2, \quad x_3' = x_3 \qquad (1)$$

where

$$\gamma := \frac{1}{\sqrt{1 - v^2/c^2}}$$

and v with $0 \leq v < c$ is a constant. If β is a "physical quantity", then

$$\beta' = \beta \qquad (2)$$

where

$$\beta = E_1(\mathbf{x}, t)dx_1 \wedge dt + \cdots + B_3(\mathbf{x}, t)dx_1 \wedge dx_2$$

and

$$\beta' = E_1'(\mathbf{x}'(\mathbf{x}, t), t'(\mathbf{x}, t))dx_1'(\mathbf{x}, t) \wedge dt'(\mathbf{x}, t)$$

$$+ \cdots + B_3'(\mathbf{x}'(\mathbf{x}, t), t'(\mathbf{x}, t))dx_1'(\mathbf{x}, t) \wedge dx_2'(\mathbf{x}, t).$$

Find the transformation law for $\mathbf{B}$ and $\mathbf{E}$ from the condition (2).

Solution.   From the Lorentz transformation (2), it follows that

$$dx_1'(\mathbf{x}, t) = \gamma(dx_1 - v dt)\,, \qquad dt'(\mathbf{x}, t) = \gamma\left(dt - \left(\frac{v}{c^2}\right) dx_1\right)\,,$$

$$dx_2'(\mathbf{x}, t) = dx_2\,, \qquad\qquad dx_3'(\mathbf{x}, t) = dx_3\,.$$

Consequently,

$$dx_1'(\mathbf{x}, t) \wedge dt'(\mathbf{x}, t) = \gamma^2 dx_1 \wedge dt - \gamma^2\left(\frac{v^2}{c^2}\right) dx_1 \wedge dt = dx_1 \wedge dt\,,$$

$$dx_2'(\mathbf{x}, t) \wedge dt'(\mathbf{x}, t) = dx_2 \wedge \gamma\left(dt - \left(\frac{v}{c^2}\right) dx_1\right)$$

$$= \gamma dx_2 \wedge dt + \gamma\left(\frac{v}{c^2}\right) dx_1 \wedge dx_2\,,$$

$$dx_3'(\mathbf{x}, t) \wedge dt'(\mathbf{x}, t) = dx_3 \wedge \gamma\left(dt - \left(\frac{v}{c^2}\right) dx_1\right)$$

$$= \gamma dx_3 \wedge dt - \gamma\left(\frac{v}{c^2}\right) dx_3 \wedge dx_1\,,$$

$$dx_1'(\mathbf{x}, t) \wedge dx_2'(\mathbf{x}, t) = \gamma(dx_1 - v dt) \wedge dx_2\,,$$

$$= \gamma dx_1 \wedge dx_2 + \gamma v dx_2 \wedge dt\,,$$

$$dx_2'(\mathbf{x}, t) \wedge dx_3'(\mathbf{x}, t) = dx_2 \wedge dx_3\,,$$

$$dx_3'(\mathbf{x}, t) \wedge dx_1'(\mathbf{x}, t) = dx_3 \wedge \gamma(dx_1 - v dt) = \gamma dx_3 \wedge dx_1 - \gamma v dx_3 \wedge dt\,.$$

Inserting these equations into (4) yields

$$E_1(\mathbf{x}, t) dx_1 \wedge dt + E_2(\mathbf{x}, t) dx_2 \wedge dt + \cdots + B_3(\mathbf{x}, t) dx_1 \wedge dx_2$$

$$= E_1'(\mathbf{x}'(\mathbf{x}, t), t'(\mathbf{x}, t)) dx_1 \wedge dt$$

$$+ E_2'(\mathbf{x}'(\mathbf{x}, t), t'(\mathbf{x}, t))\left(\gamma dx_2 \wedge dt + \gamma\left(\frac{v}{c^2}\right) dx_1 \wedge dx_2\right)$$

$$+ E_3'(\mathbf{x}'(\mathbf{x}, t), t'(\mathbf{x}, t))\left(\gamma dx_3 \wedge dt - \gamma\left(\frac{v}{c^2}\right) dx_3 \wedge dx_1\right)$$

$$+ B_3'(\mathbf{x}'(\mathbf{x}, t), t'(\mathbf{x}, t))(\gamma dx_1 \wedge dx_2 + \gamma v dx_2 \wedge dt)$$

$$+ B_1'(\mathbf{x}'(\mathbf{x}, t), t'(c, t)) dx_2 \wedge dx_3$$

$$+ B_2'(\mathbf{x}'(\mathbf{x}, t), t'(\mathbf{x}, t))(\gamma dx_3 \wedge dx_1 - \gamma v dx_3 \wedge dt)\,.$$

Comparing the basis elements of the two-forms

$$dx_1 \wedge dt, \quad dx_2 \wedge dt, \quad dx_3 \wedge dt, \quad dx_1 \wedge dx_2, \quad dx_2 \wedge dx_3, \quad dx_3 \wedge dx_1,$$

we find the transformation law for **E** and **B**

$$E_1(\mathbf{x}, t) = E_1'(\mathbf{x}'(\mathbf{x}, t), t'(\mathbf{x}, t)),$$

$$B_1(\mathbf{x}, t) = B_1'(\mathbf{x}'(\mathbf{x}, t), t'(\mathbf{x}, t)),$$

$$E_2(\mathbf{x}, t) = \gamma E_2'(\mathbf{x}'(\mathbf{x}, t), t'(\mathbf{x}, t)) + \gamma v B_3'(\mathbf{x}'(\mathbf{x}, t), t'(\mathbf{x}, t)),$$

$$E_3(\mathbf{x}, t) = \gamma E_3'(\mathbf{x}'(\mathbf{x}, t), t'(\mathbf{x}, t)) - \gamma v B_2'(\mathbf{x}'(\mathbf{x}, t), t'(\mathbf{x}, t)),$$

$$B_3(\mathbf{x}, t) = \gamma B_3'(\mathbf{x}'(\mathbf{x}, t), t'(\mathbf{x}, t)) + \gamma \left(\frac{v}{c^2}\right) E_2'(\mathbf{x}'(\mathbf{x}, t), t'(\mathbf{x}, t)),$$

$$B_2(\mathbf{x}, t) = \gamma B_2'(\mathbf{x}'(\mathbf{x}, t), t'(\mathbf{x}, t)) - \gamma \left(\frac{v}{c^2}\right) E_3'(\mathbf{x}'(\mathbf{x}, t), t'(\mathbf{x}, t)).$$

Problem 9.   Let the volume V of the system and the temperature T of the system be the independent variables of the given thermodynamic system. Furthermore, let the external pressure P and the internal energy U of the system be the dependent variables. All the objects under consideration are smooth. Prove that

Theorem.   *Let $\omega := dU + P dV$ be a one-form in a two-dimensional space (V, T). Assume that $\partial P/\partial T \neq 0$ and that P and U are related by the equation*

$$\frac{\partial U}{\partial V} = T \frac{\partial P}{\partial T} - P \tag{1}$$

which is the so-called thermodynamical equation of state. Then

(i) *The one-form ω is not closed.*

(ii) *There exists a one-form $\delta \equiv f(V, T)\omega$ such that $d\delta = 0$.*

(iii) *The function $f(V, T)$ is given by*

$$f(V, T) = \frac{g(K(V, T))}{T} \tag{2}$$

where g is a smooth function of $K(V, T)$ and $K(V, T)$ must satisfy the partial differential equations

$$\frac{\partial K}{\partial V} - \frac{1}{T} \frac{\partial U}{\partial V} - \frac{P}{T} = 0, \tag{3}$$

$$\frac{\partial K}{\partial T} - \frac{1}{T}\frac{\partial U}{\partial T} = 0 \,. \tag{4}$$

In particular we can choose $g(K) = 1$.

Solution. (i) The exterior derivative of the one-form

$$\omega = dU(V,T) + P(V,T)dV = \frac{\partial U}{\partial T}dT + \left(\frac{\partial U}{\partial V} + P\right)dV$$

yields

$$d\omega = dP \wedge dV = \frac{\partial P}{\partial T}dT \wedge dV \neq 0\,,$$

since $\partial P/\partial T \neq 0$. $\partial P/\partial T \neq 0$ is a reasonable assumption to make.

(ii) Owing to $\omega \wedge d\omega = 0$, the *theorem of Frobenius* tells us that there is a function $f(V,T)$ such that $d(f\omega) = 0$. The theorem of Frobenius can be given in this special form because there are only two independent variables. This is locally trivially true (though not necessarily globally, the Frobenius theorem itself being a local result) whether or not (1) holds, since ω is a one-form in only two variables.

(iii) The condition $d(f\omega) = 0$ yields

$$0 = d(f\omega) = (df)\wedge\omega + fd\omega = \left(\left(\frac{\partial U}{\partial V} + P\right)\frac{\partial f}{\partial T} - \frac{\partial U}{\partial T}\frac{\partial f}{\partial V} + \frac{\partial P}{\partial T}f\right)dT\wedge dV\,.$$

Thus we obtain a linear partial differential equation of first order

$$\left(\frac{\partial U}{\partial V} + P\right)\frac{\partial f}{\partial T} - \frac{\partial U}{\partial T}\frac{\partial f}{\partial V} = -\frac{\partial P}{\partial T}f\,. \tag{5}$$

Inserting (2) through (4) into (5), we find that (5) is satisfied identically. When we interpret that ω represents an "infinitesimal quantity of heat" the equation $\omega = dU + PdV$ is the *first law of thermodynamics*. If we assume that the internal energy U and pressure P (external pressure) are related by (1) (the so-called thermodynamical equation of state, where the right-hand side of the equation can be computed from the equation of state), then there is a function $f(V,T) = g(K(V,T))/T$ such that $g\omega/T$ is a closed form. $K(V,T)$ must satisfy (3) and (4). Let $g = 1$. The quantity ω/T is usually called the "infinitesimal entropy". We write $\omega = TdS$. This means that

$$dS = \frac{dU + PdV}{T}\,. \tag{6}$$

S is the first integral of the exterior differential equation $dU + PdV = 0$.
Since

$$dS = \frac{\partial S}{\partial V}dV + \frac{\partial S}{\partial T}dT, \quad dU = \frac{\partial U}{\partial V}dV + \frac{\partial U}{\partial T}dT,$$

we can write (6) as

$$\frac{\partial S}{\partial V} - \frac{1}{T}\frac{\partial U}{\partial V} - \frac{P}{T} = 0 \tag{7}$$

and

$$\frac{\partial S}{\partial T} - \frac{1}{T}\frac{\partial U}{\partial T} = 0. \tag{8}$$

Comparing (3) and (4) with (7) and (8), we must put $K = S$. Consequently,
the most general local integrating factor for ω is

$$f(V, T) = \frac{1}{T} \times \text{ arbitrary smooth function of } S.$$

To go further and prove that physically fT is actually a constant, one must
invoke the zeroth law and assume that T is the temperature. The given
derivation can also be considered from a converse point of view. From (6),
we are able to derive the thermodynamical equation of state. Because of
the fact that $ddS = 0$, it follows that

$$0 = d\left(\frac{dU + PdV}{T}\right).$$

Consequently,

$$0 = \frac{1}{T^2}\left(\frac{\partial U}{\partial V} + P - T\frac{\partial P}{\partial T}\right)dV \wedge dT.$$

Therefore

$$\frac{\partial U}{\partial V} + P - T\frac{\partial P}{\partial T} = 0.$$

Thus we obtain the thermodynamical equation of state.

Problem 10.   The differential form

$$\alpha = \frac{dz}{z}$$

with $z = x + iy$ $(x, y \in \mathbf{R})$ is defined on $\mathbf{C} \setminus \{0\}$.

(i) Find $d\alpha$.

(ii) Calculate

$$\oint_C \alpha$$

where C is the unit circle around the origin in the complex plane $\mathbf{C}$.

Solution.   (i) We have

$$d\alpha = d\left(\frac{1}{z}\right) \wedge dz = -\frac{1}{z^2} dz \wedge dz = 0$$

since $dz \wedge dz = 0$.

(ii) From $z = x + iy$ it follows that

$$dz = dx + idy.$$

Thus the differential form α takes the form

$$\alpha = \frac{dx + idy}{x + iy} = \frac{xdx + ydy + i(xdy - ydx)}{x^2 + y^2}.$$

Introducing

$$z = r \exp(i\phi)$$

with $r = 1$, we have

$$dz = ri \exp(i\phi)d\phi.$$

Thus

$$\oint_C \frac{dz}{z} = i \int_0^{2\pi} d\phi = 2\pi i.$$

Problem 11.   Let $\Omega = \mathbf{R}^3 \setminus \{(0,0,0)\}$. Consider

$$\alpha = \frac{\lambda}{r}(x_1 dx_2 \wedge dx_3 + x_2 dx_3 \wedge dx_1 + x_3 dx_1 \wedge dx_2)$$

where $\lambda > 0$ and $r^2 := x_1^2 + x_2^2 + x_3^2$. Consider the map f given by

$$x_1(u,v) = \sin u \cos v, \quad x_2(u,v) = \sin u \sin v, \quad x_3(u,v) = \cos u$$

where $0 \le u < \pi$ and $0 \le v < 2\pi$. Calculate

$$\int_{S^2} f^* \alpha.$$

Solution.   Since

$$f^*(dx_1 \wedge dx_2) = d(\sin u \cos v) \wedge d(\sin u \sin v) = \sin u \cos u du \wedge dv\,,$$

$$f^*(dx_2 \wedge dx_3) = d(\sin u \sin v) \wedge d(\cos u) = \sin^2 u \cos v du \wedge dv\,,$$

$$f^*(dx_3 \wedge dx_1) = d(\cos u) \wedge d(\sin u \cos v) = \sin^2 u \sin v du \wedge dv\,,$$

we find

$$f^*\alpha = \lambda \sin u du \wedge dv$$

since $f(r) = 1$. Consequently

$$\int_{S^2} f^*\alpha = \lambda \int_{u=0}^{\pi} \int_{v=0}^{2\pi} \sin u du dv = 4\pi\lambda$$

where

$$S^2 := \{(x_1, x_2, x_3) : x_1^2 + x_2^2 + x_3^2 = 1\}\,.$$

Problem 12.   Let M be an oriented n-manifold and let D be a regular domain in M. Let ω be a differential form of degree $(n-1)$ of compact support. Then

$$\int_D d\omega = \int_{\partial D} i^*\omega$$

where the boundary ∂D of D is considered as an oriented submanifold with the orientation induced by that of M. $i : \partial D \to M$ is the canonical injection map.

Remark.   *Equation* (1) *is called* Stokes theorem.

Let $M = \mathbf{R}^3$ and

$$K := \left\{ (x, y, z) : \frac{x^2}{a^2} + \frac{y^2}{b^2} + \frac{z^2}{c^2} \leq 1, \, a, b, c > 0 \right\}$$

and

$$\omega := \frac{x^3}{a^2} dy \wedge dz + \frac{y^3}{b^2} dz \wedge dx + \frac{z^3}{c^2} dx \wedge dy\,.$$

Calculate ∂K and $d\omega$. The orientation is x, y, z. Show that

$$\int_{\partial K} \omega = \int_K d\omega\,. \tag{1}$$

Solution.   We find that the boundary of K is given by

$$\partial K = \left\{ (x, y, z) : \frac{x^2}{a^2} + \frac{y^2}{b^2} + \frac{z^2}{c^2} = 1, \, a, b, c > 0 \right\}.$$

The exterior derivative of ω leads to the differential three-form

$$d\omega = \frac{3x^2}{a^2} dx \wedge dy \wedge dz + \frac{3y^2}{b^2} dy \wedge dz \wedge dx + \frac{3z^2}{c^2} dz \wedge dx \wedge dy$$

$$= 3 \left(\frac{x^2}{a^2} + \frac{y^2}{b^2} + \frac{z^2}{c^2} \right) dx \wedge dy \wedge dx \,.$$

To calculate the right-hand side of (1), we introduce *spherical coordinates*

$$x(r, \phi, \theta) = ar \cos \phi \cos \theta \,,$$

$$y(r, \phi, \theta) = br \sin \phi \cos \theta \,,$$

$$z(r, \phi, \theta) = cr \sin \theta \,,$$

where $0 \leq \phi < 2\pi, -\pi/2 \leq \theta < \pi/2$ and $0 \leq r \leq 1$. Then

$$dx \wedge dy \wedge dz = abcr^2 \cos \theta dr \wedge d\phi \wedge d\theta$$

and

$$\frac{x^2}{a^2} + \frac{y^2}{b^2} + \frac{z^2}{c^2} = r^2 (\cos^2 \phi \cos^2 \theta + \sin^2 \phi \cos^2 \theta + \sin^2 \theta) = r^2 \,.$$

Consequently

$$\int_K d\omega = \int_0^1 \int_{-\pi/2}^{\pi/2} \int_0^{2\pi} 3abcr^4 \cos \theta dr d\theta d\phi = \frac{4\pi}{5} 3abc \,.$$

Now we calculate the left-hand side of (1). We set

$$x(\phi, \theta) = a \cos \phi \cos \theta \,,$$

$$y(\phi, \theta) = b \sin \phi \cos \theta \,,$$

$$z(\phi, \theta) = c \sin \theta \,,$$

where $0 \leq \phi < 2\pi$ and $-\pi/2 \leq \theta < \pi/2$. It follows that

$$dx \wedge dy = ab \cos \theta \sin \theta d\phi \wedge d\theta \,,$$

$$dy \wedge dz = cb \cos \phi \cos^2 \theta d\phi \wedge d\theta \,,$$

$$dz \wedge dx = ac \sin \phi \cos^2 \theta d\phi \wedge d\theta \,.$$

Therefore

$$\int_{\partial K} \omega = I_1 + I_2 + I_3$$

where

$$I_1 = \int_{-\pi/2}^{\pi/2} \int_0^{2\pi} abc \cos^4 \phi \cos^5 \theta \, d\phi d\theta = \frac{4}{5} abc\pi,$$

$$I_2 = \int_{-\pi/2}^{\pi/2} \int_0^{2\pi} abc \sin^4 \phi \cos^5 \theta \, d\phi d\theta = \frac{4}{5} abc\pi,$$

$$I_3 = \int_{-\pi/2}^{\pi/2} \int_0^{2\pi} abc \sin^4 \theta \cos \theta \, d\phi d\theta = \frac{4}{5} abc\pi.$$

It follows that

$$\int_{\partial K} \omega = \frac{4\pi}{5} 3abc.$$

Problem 13.   *Poincaré's lemma* tells us that if ω is a p-differential form on M ($\dim M = n$) for which there exists a $(p-1)$-differential form α such that $d\alpha = \omega$, then $d\omega = 0$.

The converse of Poincaré's lemma tells us that if ω is a p-differential form on an open set $U \subset M$ (which is contractible to a point) such that $d\omega = 0$, then there exists a $(p-1)$ differential form α such that $\omega = d\alpha$. The exception is $p = 0$. Then $\omega = f$ and the vanishing of df simply means f is constant.

Let

$$\omega = a(\mathbf{x}) dx_{i_1} \wedge dx_{i_2} \wedge \cdots \wedge dx_{i_p}$$

and let λ be a real parameter with $\lambda \in [0, 1]$. We introduce the linear operator T_λ which is defined as

$$T_\lambda \omega := \int_0^1 \lambda^{p-1} \left(x_1 \frac{\partial}{\partial x_1} + x_2 \frac{\partial}{\partial x_2} + \cdots + x_n \frac{\partial}{\partial x_n} \right) \rfloor a(\lambda \mathbf{x}) dx_{i_1}$$
$$\wedge dx_{i_2} \wedge \cdots \wedge dx_{i_p} d\lambda$$

where $\rfloor$ denotes the *contraction* (also called the *interior product* of a differential form and a vector field). We have

$$\frac{\partial}{\partial x_j} \rfloor dx_k = \delta_{jk}. \tag{1}$$

Since the operator T_λ is linear it must only be defined for a monom. We have

$$d(T_\lambda \omega) = \omega .$$

(i) Apply the converse of Poincaré's lemma to

$$\omega = dx_1 \wedge dx_2$$

defined on $\mathbf{R}^2$.

(ii) Apply the converse of Poincaré's lemma to

$$\omega = dx_1 \wedge dx_2 \wedge dx_3$$

defined on $\mathbf{R}^3$.

Solution. (i) We have $d\omega = 0$ and $\mathbf{R}^2$ is contractible to $0 \in \mathbf{R}^2$. From the definition (1), we obtain

$$T_\lambda \omega = \left(x_1 \frac{\partial}{\partial x_1} + x_2 \frac{\partial}{\partial x_2} + \cdots + x_n \frac{\partial}{\partial x_n} \right) \rfloor (dx_{i_1} \wedge dx_{i_2} \wedge \cdots \wedge dx_{i_p})$$

$$\times \int_0^1 \lambda^{p-1} a(\lambda \mathbf{x}) d\lambda .$$

We have $n = 2$ and $p = 2$. Since

$$\left(x_1 \frac{\partial}{\partial x_1} + x_2 \frac{\partial}{\partial x_2} \right) \rfloor (dx_1 \wedge dx_2) = -x_2 dx_1 + x_1 dx_2 ,$$

we find

$$T_\lambda \omega = (x_1 dx_2 - x_2 dx_1) \int_0^1 \lambda d\lambda = \frac{1}{2}(x_1 dx_2 - x_2 dx_1) .$$

(ii) We have $d\omega = 0$ and $\mathbf{R}^3$ is contractible to $0 \in \mathbf{R}^3$. Since

$$\left(x_1 \frac{\partial}{\partial x_1} + x_2 \frac{\partial}{\partial x_2} + x_3 \frac{\partial}{\partial x_3} \right) \rfloor (dx_1 \wedge dx_2 \wedge dx_3)$$

$$= x_1 dx_2 \wedge dx_3 - x_2 dx_1 \wedge dx_3 + x_3 dx_1 \wedge dx_2$$

and

$$\int_0^1 \lambda^2 d\lambda = \frac{1}{3} ,$$

we find

$$T_\lambda \omega = \frac{1}{3}(x_1 dx_2 \wedge dx_3 + x_2 dx_3 \wedge dx_1 + x_3 dx_1 \wedge dx_2)$$

where we used $dx_1 \wedge dx_3 = -dx_3 \wedge dx_1$.

Problem 14.　Let

$$\tilde{\alpha} := \sum_{j=1}^{n} (a_j(x,t)dx + A_j(x,t)dt) \otimes X_j$$

be a *Lie algebra-valued differential form* where $\{X_1, X_2, \ldots, X_n\}$ forms a basis of a Lie algebra. The exterior derivative is defined as

$$d\tilde{\alpha} := \sum_{j=1}^{n} \left(-\frac{\partial a_j}{\partial t} + \frac{\partial A_j}{\partial x} \right) dx \wedge dt \otimes X_j \,.$$

The commutator is defined as

$$[\tilde{\alpha}, \tilde{\alpha}] = \sum_{k=1}^{n} \sum_{j=1}^{n} (a_k A_j - a_j A_k) dx \wedge dt \otimes [X_k, X_j] \,.$$

The covariant exterior derivative is defined as

$$D_{\tilde{\alpha}} \tilde{\alpha} := d\tilde{\alpha} + \frac{1}{2}[\tilde{\alpha}, \tilde{\alpha}] \,.$$

(i) Calculate the covariant derivative of $\tilde{\alpha}$ and find the equation which follows from the condition

$$D_{\tilde{\alpha}} \tilde{\alpha} = 0 \,.$$

(ii) Study the case where the basis of the Lie algebra satisfies the commutation relations

$$[X_1, X_2] = 2X_2, \quad [X_3, X_1] = 2X_3, \quad [X_2, X_3] = X_1. \tag{1}$$

(iii) Let

$$a_1 = -\eta, \quad a_2 = \frac{1}{2}\frac{\partial u}{\partial x}, \quad a_3 = -\frac{1}{2}\frac{\partial u}{\partial x}, \tag{2a}$$

$$A_1 = -\frac{1}{4\eta}\cos u, \quad A_2 = A_3 = -\frac{1}{4\eta}\sin u, \tag{2b}$$

where η is an arbitrary constant with $\eta \neq 0$. Find the equation of motion.

Solution. (i) Since

$$[X_k, X_j] = \sum_{i=1}^{n} C_{kj}^i X_i ,$$

it follows that

$$D_{\tilde{a}}\tilde{\alpha} = \sum_{i=1}^{n} \left(\left(-\frac{\partial a_i}{\partial t} + \frac{\partial A_i}{\partial x} \right) + \frac{1}{2} \sum_{k=1}^{n} \sum_{j=1}^{n} (a_k A_j - a_j A_k) C_{kj}^i \right) dx \wedge dt \otimes X_i .$$

Since the X_j, $j = 1, \ldots, n$ form a basis of the Lie algebra, the condition $D_{\tilde{a}}\tilde{\alpha} = 0$ yields

$$\left(-\frac{\partial a_i}{\partial t} + \frac{\partial A_i}{\partial x} \right) + \frac{1}{2} \sum_{k=1}^{n} \sum_{j=1}^{n} (a_k A_j - a_j A_k) C_{kj}^i = 0$$

for $i = 1, \ldots, n$. Since $C_{kj}^i = -C_{jk}^i$, it follows that

$$\left(-\frac{\partial a_i}{\partial t} + \frac{\partial A_i}{\partial x} \right) + \sum_{k<j}^{n} (a_k A_j - a_j A_k) C_{kj}^i = 0 .$$

(ii) Consider now a special case where $n = 3$ and X_1, X_2 and X_3 satisfy the commutation relations (1). We find

$$-\frac{\partial a_1}{\partial t} + \frac{\partial A_1}{\partial x} + a_2 A_3 - a_3 A_2 = 0 , \tag{3a}$$

$$-\frac{\partial a_2}{\partial t} + \frac{\partial A_2}{\partial x} + 2(a_1 A_2 - a_2 A_1) = 0 , \tag{3b}$$

$$-\frac{\partial a_3}{\partial t} + \frac{\partial A_3}{\partial x} - 2(a_1 A_3 - a_3 A_1) = 0 . \tag{3c}$$

A convenient choice of a basis $\{X_1, X_2, X_3\}$ is given by

$$X_1 = \begin{pmatrix} 1 & 0 \\ 0 & -1 \end{pmatrix} , \quad X_2 = \begin{pmatrix} 0 & 1 \\ 0 & 0 \end{pmatrix} , \quad X_3 = \begin{pmatrix} 0 & 0 \\ 1 & 0 \end{pmatrix} .$$

Consequently, the Lie algebra under consideration is $sl(2, \mathbf{R})$.

(iii) Inserting (2) into (3a) through (3c) yields the sine-Gordon equation

$$\frac{\partial^2 u}{\partial x \partial t} = \sin u .$$

Remark.   *Other one-dimensional soliton equations (such as the Korteweg–de Vries equation, the nonlinear Schrödinger equation, the Liouville equation) can be derived with this method.*

Problem 15.   Let

$$\tilde{\alpha} := \sum_{a=1}^{3} \alpha_a \otimes T_a \tag{1}$$

be a Lie algebra-valued differential one-form. Here $\otimes$ denotes the tensor product and T_a $(a = 1, 2, 3)$ are the generators given by

$$T_1 := \frac{1}{2} \begin{pmatrix} 0 & -i \\ -i & 0 \end{pmatrix}, \quad T_2 := \frac{1}{2} \begin{pmatrix} 0 & -1 \\ 1 & 0 \end{pmatrix}, \quad T_3 := \frac{1}{2} \begin{pmatrix} -i & 0 \\ 0 & i \end{pmatrix}.$$

Remark. T_1, T_2 and T_3 form a basis of the Lie algebra $su(2)$.

The quantities α_a are differential one-forms

$$\alpha_a := \sum_{j=1}^{4} A_{aj}(\mathbf{x}) dx_j$$

where $\mathbf{x} = (x_1, x_2, x_3, x_4)$. The quantity $\tilde{\alpha}$ is usually called the *connection* or *vector potential*. The actions of the Hodge operator $*$, and the exterior derivative d, may be consistently defined by

$$*\tilde{\alpha} := \sum_{a=1}^{3} (*\alpha_a) \otimes T_a$$

and

$$d\tilde{\alpha} := \sum_{a=1}^{3} (d\alpha_a) \otimes T_a .$$

The bracket $[\,,\,]$ of Lie algebra-valued differential forms, say $\tilde{\beta}$ and $\tilde{\gamma}$, is defined as

$$[\tilde{\beta}, \tilde{\gamma}] := \sum_{a=1}^{n} \sum_{b=1}^{n} (\beta_a \wedge \gamma_b) \otimes [T_a, T_b] .$$

The covariant exterior derivative of a Lie algebra-valued differential p-form $\tilde{\gamma}$ with respect to a Lie algebra-valued one-form $\tilde{\beta}$ is defined as

$$D_{\tilde{\beta}} \tilde{\gamma} := d\tilde{\gamma} - g[\tilde{\beta}, \tilde{\gamma}]$$

where

$$g = \begin{cases} -1 & p \text{ even}, \\ -\dfrac{1}{2} & p \text{ odd}. \end{cases}$$

Therefore

$$D_{\tilde{\beta}}\tilde{\beta} = d\tilde{\beta} + \frac{1}{2}[\tilde{\beta}, \tilde{\beta}], \quad D_{\tilde{\beta}}(D_{\tilde{\beta}}\tilde{\beta}) = 0.$$

The last equation is called the *Bianchi identity.* Let $\tilde{\alpha}$ be the Lie algebra valued-differential form given by Eq. (1). The *Yang-Mills equations* are given by

$$D_{\tilde{\alpha}}(*D_{\tilde{\alpha}}\tilde{\alpha}) = 0. \tag{2}$$

The quantity $D_{\tilde{\alpha}}\tilde{\alpha}$ is called the *curvature form* or *field strength tensor.* Equation (2) is a coupled system of nonlinear partial differential equations of second order. In addition we have to impose gauge conditions. Let

$$g = dx_1 \otimes dx_1 + dx_2 \otimes dx_2 + dx_3 \otimes dx_3 - dx_4 \otimes dx_4.$$

(i) Write down Eq. (2) explicitly.

(ii) Impose the gauge conditions

$$A_{a4} = 0, \tag{3a}$$

$$\sum_{i=1}^{3} \frac{\partial A_{ai}}{\partial x_i} = 0, \tag{3b}$$

$$\frac{\partial A_{ai}}{\partial x_j} = 0, \quad i, j = 1, 2, 3, \tag{3c}$$

where $a = 1, 2, 3$.

Solution.   (i) The commutation relations of T_1, T_2 and T_3 are given by

$$[T_1, T_2] = T_3, \quad [T_2, T_3] = T_1, \quad [T_3, T_1] = T_2.$$

Since

$$D_{\tilde{\alpha}}\tilde{\alpha} = d\tilde{\alpha} + \frac{1}{2}[\tilde{\alpha}, \tilde{\alpha}],$$

$$\frac{1}{2}[\tilde{\alpha}, \tilde{\alpha}] = (\alpha_1 \wedge \alpha_2) \otimes T_3 + (\alpha_2 \wedge \alpha_3) \otimes T_1 + (\alpha_3 \wedge \alpha_1) \otimes T_2,$$

$$d\tilde{\alpha} = \sum_{a=1}^{3} d\alpha_a \otimes T_a,$$

and

$$*(D_{\tilde{a}}\tilde{\alpha}) = (*d\alpha_1 + *(\alpha_2 \wedge \alpha_3)) \otimes T_1 + (*d\alpha_2 + *(\alpha_3 \wedge \alpha_1)) \otimes T_2$$
$$+ (*d\alpha_3 + *(\alpha_1 \wedge \alpha_2)) \otimes T_3,$$

we obtain

$$D_{\tilde{a}} * (D_{\tilde{a}}\tilde{\alpha})$$
$$= d(*d\alpha_1 + *(\alpha_2 \wedge \alpha_3)) \otimes T_1 + d(*d\alpha_2 + *(\alpha_3 \wedge \alpha_1)) \otimes T_2$$
$$+ d(*d\alpha_3 + *(\alpha_1 \wedge \alpha_2)) \otimes T_3 + (\alpha_2 \wedge (*d\alpha_3 + *(\alpha_1 \wedge \alpha_2))$$
$$- \alpha_3 \wedge (*d\alpha_2 + *(\alpha_3 \wedge \alpha_1))) \otimes T_1 + (\alpha_3 \wedge (*d\alpha_1 + *(\alpha_2 \wedge \alpha_3))$$
$$- \alpha_1 \wedge (*d\alpha_3 + *(\alpha_1 \wedge \alpha_2))) \otimes T_2$$
$$+ (\alpha_1 \wedge (*d\alpha_2 + *(\alpha_3 \wedge \alpha_1)) - \alpha_2 \wedge (*d\alpha_1 + *(\alpha_2 \wedge \alpha_3))) \otimes T_3.$$

Then from the condition (2) it follows that

$$d(*d\alpha_1 + *(\alpha_2 \wedge \alpha_3)) + (\alpha_2 \wedge (*d\alpha_3 + *(\alpha_1 \wedge \alpha_2))$$
$$- \alpha_3 \wedge (*d\alpha_2 + *(\alpha_3 \wedge \alpha_1))) = 0,$$
$$d(*d\alpha_2 + *(\alpha_3 \wedge \alpha_1)) + (\alpha_3 \wedge (*d\alpha_1 + *(\alpha_2 \wedge \alpha_3))$$
$$- \alpha_1 \wedge (*d\alpha_3 + *(\alpha_1 \wedge \alpha_2))) = 0,$$
$$d(*d\alpha_3 + *(\alpha_1 \wedge \alpha_2)) + (\alpha_1 \wedge (*d\alpha_2 + *(\alpha_3 \wedge \alpha_1))$$
$$- \alpha_2 \wedge (*d\alpha_1 + *(\alpha_2 \wedge \alpha_3))) = 0.$$

Now $*(\alpha_a \wedge \alpha_b)$ is a two-form and therefore $d * (\alpha_a \wedge \alpha_b)$ is a three-form. We find 12 coupled partial differential equations.

(ii) Let us now impose the gauge conditions (3). Thus we have eliminated the space dependence of the fields. Taking into account Eqs. (3), we arrive at the autonomous system of ordinary differential equations of second order

$$\frac{d^2 A_{ai}}{dt^2} + \sum_{b=1}^{3} \sum_{j=1}^{3} (A_{bj} A_{bj} A_{ai} - A_{aj} A_{bj} A_{bi}) = 0 \qquad (4)$$

together with

$$\sum_{j=1}^{3}\sum_{b=1}^{3}\sum_{c=1}^{3}\varepsilon_{abc}A_{bj}\frac{dA_{cj}}{dt} = 0.$$

System (4) can be viewed as a Hamilton system with

$$H\left(\frac{A_{ai}, dA_{ai}}{dt}\right) = \frac{1}{2}\sum_{a=1}^{3}\sum_{i=1}^{3}\left(\frac{dA_{ai}}{dt}\right)^2 + \frac{1}{4}\sum_{a=1}^{3}\sum_{i=1}^{3}(A_{ai}^2)^2$$

$$- \frac{1}{4}\sum_{i=1}^{3}\sum_{j=1}^{3}\left(\sum_{a=1}^{3}A_{ai}A_{aj}\right)^2.$$

To simplify further we put $A_{ai}(t) = O_{ai}f_a(t)$ (no summation), where $O = (O_{ai})$ is a time-independent 3×3 orthogonal matrix. We put

$$\frac{df_a}{dt} \equiv p_a, \quad f_a \equiv q_a$$

where $a = 1, 2, 3$. Then we obtain

$$\frac{d^2q_1}{dt^2} + q_1(q_2^2 + q_3^2) = 0, \quad \frac{d^2q_2}{dt^2} + q_2(q_1^2 + q_3^2) = 0, \quad \frac{d^2q_3}{dt^2} + q_3(q_1^2 + q_2^2) = 0$$

with the Hamilton function

$$H(\mathbf{p}, \mathbf{q}) = \frac{1}{2}(p_1^2 + p_2^2 + p_3^2) + \frac{1}{2}(q_1^2 q_2^2 + q_1^2 q_3^2 + q_2^2 q_3^2).$$

Problem 16.   Find the closed plane curve of a given length L which encloses a maximum area.

Solution.   Given the curve $(x(t), y(t))$, where $t \in [t_0, t_1]$. We assume that $x(t)$ and $y(t)$ are continuously differentiable and $x(t_0) = x(t_1)$, $y(t_0) = y(t_1)$. Let A be the area enclosed and L be the given length. Then we have

$$L = \int_{t_0}^{t_1}\left(\left(\frac{dx}{dt}\right)^2 + \left(\frac{dy}{dt}\right)^2\right)^{1/2} dt.$$

Since the exterior derivative d of $(xdy - ydx)/2$ is given by

$$d\frac{1}{2}(xdy - ydx) = dx \wedge dy,$$

we can apply *Stokes theorem* and find the area

$$A = \int_{t_0}^{t_1} \left(x\frac{dy}{dt} - y\frac{dx}{dt} \right) dt .$$

To find the maximum area we apply the *Lagrange multiplier method* and consider

$$H = \int_{t_0}^{t_1} \left(\left(x\frac{dy}{dt} - y\frac{dx}{dt} \right) + \lambda \left(\left(\frac{dx}{dt}\right)^2 + \left(\frac{dy}{dt}\right)^2 \right)^{1/2} \right) dt$$

where λ is the Lagrange multiplier. Thus we consider

$$\phi(x, y, \dot{x}, \dot{y}) = \frac{1}{2}(x\dot{y} - y\dot{x}) + \lambda(\dot{x}^2 + \dot{y}^2)^{1/2}$$

where ϕ satisfies the Euler–Lagrange equation. Thus, by partial differentiation

$$\frac{\partial \phi}{\partial x} = \frac{\dot{y}}{2}, \quad \frac{\partial \phi}{\partial y} = -\frac{\dot{x}}{2},$$

$$\frac{\partial \phi}{\partial \dot{x}} = -\frac{y}{2} + \frac{\lambda\dot{x}}{(\dot{x}^2 + \dot{y}^2)^{1/2}}, \quad \frac{\partial \phi}{\partial \dot{y}} = \frac{x}{2} + \frac{\lambda\dot{y}}{(\dot{x}^2 + \dot{y}^2)^{1/2}}.$$

The *Euler–Lagrange equations* are

$$\frac{\partial \phi}{\partial x} - \frac{d}{dt}\frac{\partial \phi}{\partial \dot{x}} = 0, \quad \frac{\partial \phi}{\partial y} - \frac{d}{dt}\frac{\partial \phi}{\partial \dot{y}} = 0.$$

Thus

$$\frac{\dot{y}}{2} - \frac{d}{dt}\left(-\frac{y}{2} + \frac{\lambda\dot{x}}{(\dot{x}^2 + \dot{y}^2)^{1/2}} \right) = 0, \quad -\frac{\dot{x}}{2} - \frac{d}{dt}\left(\frac{x}{2} + \frac{\lambda\dot{y}}{(\dot{x}^2 + \dot{y}^2)^{1/2}} \right) = 0.$$

If we choose a special parameter s, the arc length along the competing curves, then

$$(\dot{x}^2 + \dot{y}^2)^{1/2} = 1$$

and

$$ds = (\dot{x}^2 + \dot{y}^2)^{1/2}dt$$

so that Euler–Lagrange equations reduce to

$$\frac{dy}{ds} - \lambda\frac{d^2x}{ds^2} = 0, \quad \frac{dx}{ds} + \lambda\frac{d^2y}{ds^2} = 0.$$

Integration of these linear differential equations yields

$$y - \lambda \frac{dx}{ds} = C_1, \quad x + \lambda \frac{dy}{ds} = C_2$$

where C_1 and C_2 are constants of integration. Elimination of y yields

$$\lambda^2 \frac{d^2 x}{ds^2} + x = C_2$$

with the solution

$$x(s) = a \sin\left(\frac{s}{\lambda}\right) + b \cos\left(\frac{s}{\lambda}\right) + C_2$$

where a and b are constants of integration. For y, we find

$$y(s) = a \cos\left(\frac{s}{\lambda}\right) - b \sin\left(\frac{s}{\lambda}\right) + C_1 .$$

Thus $(x(s), y(s))$ describe a circle.

Chapter 20

Lie Derivative

Problem 1. Let M be an n-dimensional C^∞ differentiable manifold with local coordinates x_j, $j = 1, \ldots, n$. Real valued C^∞ vector fields and real valued C^∞ differential forms on M can be considered. The components of the vector field X are denoted by $X_j \partial/\partial x_j$. This means

$$X := X_1(\mathbf{x})\frac{\partial}{\partial x_1} + X_2(\mathbf{x})\frac{\partial}{\partial x_2} + \cdots + X_n(\mathbf{x})\frac{\partial}{\partial x_n} \tag{1}$$

in local coordinates. The *Lie derivative* of a differential form α with respect to X is defined by the derivative of α along the integral curve $t \mapsto \Phi_t$ of X, i.e.,

$$L_X \alpha := \lim_{t \to 0} \frac{\Phi_t^* \alpha - \alpha}{t}.$$

It can be shown that this can be written as

$$L_X \alpha := d(X \rfloor \alpha) + X \rfloor (d\alpha)$$

where $d\alpha$ is the exterior derivative of the differential form α and $X \rfloor \alpha$ is the contraction of α by X. In local coordinates we have

$$\frac{\partial}{\partial x_j} \rfloor dx_k = \delta_{jk}$$

where δ_{jk} denotes the Kronecker delta. Furthermore, we have the *product rule*

$$X \rfloor (\alpha \wedge \beta) = (X \rfloor \alpha) \wedge \beta + (-1)^r \alpha \wedge (X \rfloor \beta) \tag{2}$$

where α is an r-form. The linear operators $d(.)$, $X\rfloor$ and $L_X(.)$ are coordinate-free operators.

(i) Let $M = \mathbf{R}^n$ and

$$\omega := dx_1 \wedge dx_2 \wedge \cdots \wedge dx_n .$$

Find

$$X \rfloor \omega$$

where the vector field X is given by (1). Find

$$d(X \rfloor \omega) .$$

Give an interpretation of the result.

(ii) Consider the autonomous nonlinear system of ordinary differential equations

$$\frac{dx_1}{dt} = x_1 - x_1 x_2 , \qquad \frac{dx_2}{dt} = -x_2 + x_1 x_2$$

where $x_1 > 0$ and $x_2 > 0$. This equation is a *Lotka–Volterra model*. The corresponding vector field is

$$X = (x_1 - x_1 x_2)\frac{\partial}{\partial x_1} + (-x_2 + x_1 x_2)\frac{\partial}{\partial x_2} .$$

Let

$$\Omega = \frac{dx_1 \wedge dx_2}{x_1 x_2} .$$

Calculate $L_X \Omega$.

Solution. (i) Applying the product rule (2), we obtain

$$X \rfloor \omega = \sum_{j=1}^{n}(-1)^{j+1}X_j dx_1 \wedge \cdots \wedge \widehat{dx_j} \wedge \cdots \wedge dx_n \tag{3}$$

where the circumflex indicates omission. From (3), we obtain

$$d(X \rfloor \omega) = \left(\sum_{j=1}^{n}\frac{\partial X_j}{\partial x_j}\right) dx_1 \wedge dx_2 \wedge \cdots \wedge dx_n .$$

This is the divergence of the vector field X. Since $d\omega = 0$, we have

$$L_X \omega = d(X \rfloor \omega) .$$

(ii) Since $d\Omega = 0$, we have

$$L_X\Omega = d(X\rfloor\Omega).$$

Thus

$$L_X\Omega = d\left[\left(\frac{x_1 - x_1 x_2}{x_1 x_2}\right)dx_2 - \left(\frac{-x_2 + x_1 x_2}{x_1 x_2}\right)dx_1\right]$$

$$= d\left(\frac{1}{x_2}dx_2 + \frac{1}{x_1}dx_1\right) = 0$$

where we have used

$$ddx_1 = 0, \quad ddx_2 = 0.$$

Since

$$L_X\Omega = 0$$

we say that Ω is *invariant* under X.

Problem 2.   (i) Let $M = \mathbf{R}^n$. Let X_1, X_2 and Y be smooth vector fields defined on M. Assume that

$$[X_1, Y] = fY, \quad [X_2, Y] = gY$$

where f and g are smooth functions defined on M. Let $\Omega := dx_1 \wedge dx_2 \wedge \cdots \wedge dx_n$ be the volume form on M. Show that

$$L_Y(X_1\rfloor X_2\rfloor Y\rfloor\Omega) = (\text{div}Y)(X_1\rfloor X_2\rfloor Y\rfloor\Omega)$$

where $\text{div}Y$ denotes the divergence of the vector field Y, i.e.,

$$\text{div}Y := \sum_{j=1}^{n} \frac{\partial Y_j}{\partial x_j}.$$

(ii) Let X, Y be smooth vector fields and let f and g be smooth functions. Assume that

$$L_Y g = 0, \quad [X, Y] = fY. \tag{1}$$

Calculate

$$L_Y(L_X g).$$

Hint. Apply the identity

$$L_{[X,Y]} \equiv [L_X, L_Y].\tag{2}$$

Solution. (i) Straightforward calculation shows that

$$
\begin{aligned}
L_Y(X_1 \rfloor X_2 \rfloor Y \rfloor \Omega) &= [Y, X_1] \rfloor X_2 \rfloor Y \rfloor \Omega + X_1 \rfloor L_Y(X_2 \rfloor Y \rfloor \Omega) \\
&= -fY \rfloor X_2 \rfloor Y \rfloor \Omega + X_1 \rfloor L_Y(X_2 \rfloor Y \rfloor \Omega) \\
&= X_1 \rfloor ([Y, X_2] \rfloor Y \rfloor \Omega + X_2 \rfloor L_Y(Y \rfloor \Omega)) \\
&= -X_1 \rfloor (gY) \rfloor Y \rfloor \Omega + X_1 \rfloor X_2 \rfloor L_Y(Y \rfloor \Omega) \\
&= X_1 \rfloor X_2 \rfloor (\mathrm{div} Y) Y \rfloor \Omega \\
&= (\mathrm{div} Y)(X_1 \rfloor X_2 \rfloor Y \rfloor \Omega)
\end{aligned}
$$

where we have used that $Y \rfloor Y \rfloor \Omega = 0$.

(ii) Applying identity (2) and equation (1), we find

$$L_Y(L_X g) = L_X(L_Y g) - L_{[X,Y]}g = -L_{[X,Y]}g = -L_{fY}g = -fL_Y g = 0.$$

Therefore the function $L_X g$ is invariant under the vector field Y.

Problem 3. (i) Let

$$T := \sum_{i,j=1}^{n} a_{ij}(\mathbf{x}) \frac{\partial}{\partial x_i} \otimes dx_j$$

be a smooth $(1,1)$ tensor field. Let

$$V = \sum_{k=1}^{n} V_k(\mathbf{x}) \frac{\partial}{\partial x_k}$$

be a smooth vector field. Find

$$L_V T = 0 \tag{1}$$

where $L_V(\cdot)$ denotes the Lie derivative.

(ii) Assume that $L_V T = 0$. Show that

$$L_V \sum_{j=1}^{n} a_{jj} = 0. \tag{2}$$

Hint. Recall that the Lie derivative is linear and obeys the product rule, i.e.,

$$L_V\left(a_{ij}\frac{\partial}{\partial x_i}\otimes dx_j\right)\equiv (L_V a_{ij})\frac{\partial}{\partial x_i}\otimes dx_j$$

$$+a_{ij}\left(L_V\frac{\partial}{\partial x_i}\right)\otimes dx_j+a_{ij}\frac{\partial}{\partial x_i}\otimes(L_V dx_j).$$

Solution.   (i) First we calculate $L_V T$. Applying the linearity of the Lie derivative, we find

$$L_V T = L_V\sum_{i,j=1}^{n}\left(a_{ij}\frac{\partial}{\partial x_i}\otimes dx_j\right)=\sum_{i,j=1}^{n}L_V\left(a_{ij}\frac{\partial}{\partial x_i}\otimes dx_j\right).$$

Applying the product rule for the Lie derivative we find

$$L_V T = \sum_{i,j=1}^{n}(L_V a_{ij})\frac{\partial}{\partial x_i}\otimes dx_j+\sum_{i,j=1}^{n}a_{ij}\left(L_V\frac{\partial}{\partial x_i}\right)\otimes dx_j$$

$$+\sum_{i,j=1}^{n}a_{ij}\frac{\partial}{\partial x_i}\otimes L_V dx_j$$

$$=\sum_{i,j,k=1}^{n}\left(V_k\frac{\partial a_{ij}}{\partial x_k}\frac{\partial}{\partial x_i}\otimes dx_j+a_{ij}\left[V_k\frac{\partial}{\partial x_k},\frac{\partial}{\partial x_i}\right]\otimes dx_j\right.$$

$$\left.+a_{ij}\frac{\partial}{\partial x_i}\otimes\delta_{kj}dV_k\right)$$

$$=\sum_{i,j,k=1}^{n}V_k\frac{\partial a_{ij}}{\partial x_k}\frac{\partial}{\partial x_i}\otimes dx_j-\sum_{i,j,k=1}^{n}a_{ij}\frac{\partial V_k}{\partial x_i}\frac{\partial}{\partial x_k}\otimes dx_j$$

$$+\sum_{i,j,k=1}^{n}a_{ij}\frac{\partial}{\partial x_i}\otimes\delta_{kj}dV_k$$

$$=\sum_{i,j,k=1}^{n}V_k\frac{\partial a_{ij}}{\partial x_k}\frac{\partial}{\partial x_i}\otimes dx_j-\sum_{i,j,k=1}^{n}a_{kj}\frac{\partial V_i}{\partial x_k}\frac{\partial}{\partial x_i}\otimes dx_j$$

$$+\sum_{i,k=1}^{n}a_{ik}\frac{\partial}{\partial x_i}\otimes dV_k$$

$$= \sum_{i,j,k=1}^{n} V_k \frac{\partial a_{ij}}{\partial x_k} \frac{\partial}{\partial x_i} \otimes dx_j - \sum_{i,j,k=1}^{n} a_{kj} \frac{\partial V_i}{\partial x_k} \frac{\partial}{\partial x_i} \otimes dx_j$$

$$+ \sum_{i,j,k=1}^{n} a_{ik} \frac{\partial}{\partial x_i} \otimes \frac{\partial V_k}{\partial x_j} dx_j \, .$$

Consequently,

$$L_V T = \sum_{i,j=1}^{n} \left(\sum_{k=1}^{n} \left(V_k \frac{\partial a_{ij}}{\partial x_k} - a_{kj} \frac{\partial V_i}{\partial x_k} + a_{ik} \frac{\partial V_k}{\partial x_j} \right) \right) \frac{\partial}{\partial x_i} \otimes dx_j \, . \qquad (3)$$

Condition (1) yields

$$\sum_{k=1}^{n} \left(V_k \frac{\partial a_{ij}}{\partial x_k} - a_{kj} \frac{\partial V_i}{\partial x_k} + a_{ik} \frac{\partial V_k}{\partial x_j} \right) = 0 \quad \text{for all } i, j = 1, \ldots, n \, .$$

(ii) From (3), it follows that for $i = j$

$$\sum_{k=1}^{n} \left(V_k \frac{\partial a_{jj}}{\partial x_k} - a_{kj} \frac{\partial V_j}{\partial x_k} + a_{jk} \frac{\partial V_k}{\partial x_j} \right) = 0$$

for $j = 1, \ldots, n$. Since

$$L_V \sum_{j=1}^{n} a_{jj} = \sum_{k=1}^{n} \sum_{j=1}^{n} V_k \frac{\partial a_{jj}}{\partial x_k}$$

we find (2).

Problem 4. Let $M = \mathbf{R}^n$ (or any open subset of $\mathbf{R}^n$). X is a vector field on M and α an r-form on M $(r < n)$. We call α a *conformal invariant r-form* of X if

$$L_X \alpha = g \alpha \, .$$

Here $L_X \alpha$ is the Lie derivative of α with respect to X and g an arbitrary smooth function. Prove that

Theorem. *Let $M = \mathbf{R}^n$. The volume form is given by*

$$\omega := dx_1 \wedge \cdots \wedge dx_n \, .$$

Let X and Y be two vector fields such that

$$[X, Y] = fY$$

and

$$\alpha := Y \,\rfloor\, \omega\,.$$

Then

$$L_X \alpha = (f + \mathrm{div} X)\alpha\,.$$

Solution.   Applying the rules for the Lie derivative we find

$$L_X(Y \,\rfloor\, \omega) = [X, Y] \,\rfloor\, \omega + Y \,\rfloor\, (L_X \omega)$$

$$= (fY) \,\rfloor\, \omega + Y \,\rfloor\, (X \,\rfloor\, d\omega + d(X \,\rfloor\, \omega))$$

$$= f(Y \,\rfloor\, \omega) + Y \,\rfloor\, (d(X \,\rfloor\, \omega))$$

$$= f(Y \,\rfloor\, \omega) + Y \,\rfloor\, ((\mathrm{div} X)\omega)$$

$$= (f + \mathrm{div} X)(Y \,\rfloor\, \omega)\,.$$

Hence we have

$$g = f + \mathrm{div} X\,.$$

Problem 5.   The nonlinear partial differential equation

$$\frac{\partial^2 u}{\partial x \partial t} = \sin u \tag{1}$$

is the so-called one-dimensional *sine-Gordon equation*. Show that

$$V = (4u_{xxx} + 2(u_x)^3)\frac{\partial}{\partial u}$$

is a Lie–Bäcklund vector field of (1).

Solution.   Taking the derivative of (1) with respect to x we find that (1) defines the manifolds

$$u_{xt} = \sin u\,,$$

$$u_{xxt} = u_x \cos u\,,$$

$$u_{xxxt} = u_{xx} \cos u - (u_x)^2 \sin u\,, \tag{2}$$

$$u_{xxxxt} = u_{xxx} \cos u - 3u_x u_{xx} \sin u - (u_x)^3 \cos u\,.$$

The *prolongated vector field* $\bar{V}$ of V is given by

$$\bar{V} = V + (4u_{xxxt} + 6(u_x)^2 u_{xt})\frac{\partial}{\partial u_t} + (4u_{xxxxt} + 12u_x u_{xx} u_{xt} + 6(u_x)^2 u_{xxt})\frac{\partial}{\partial u_{xt}}\,.$$

It follows that the Lie derivative of $u_{xt} - \sin u$ with respect to $\bar{V}$ is given by

$$L_{\bar{V}}(u_{xt} - \sin u) = 4u_{xxxxt} + 12u_x u_{xx} u_{xt}$$

$$+ 6(u_x)^2 u_{xxt} - (4u_{xxx} + 2(u_x)^3) \cos u. \qquad (3)$$

Inserting (2) into (3) yields

$$L_{\bar{V}}(u_{xt} - \sin u) = 0.$$

Consequently, the vector field V is a Lie–Bäcklund symmetry vector field of the sine-Gordon equation.

Chapter 21

Metric Tensor Fields

Problem 1. Let

$$g = dx \otimes dx + \cos(u(x,t))dx \otimes dt + \cos(u(x,t))dt \otimes dx + dt \otimes dt$$

be a metric tensor field, where u is a smooth function of x and t.

(i) Calculate the *Riemann curvature scalar R*.

(ii) Find the equation which follows from the condition

$$R = -2\,.$$

Solution.   We set $1 \equiv x$ and $2 \equiv t$. Then

$$g_{11} = g_{22} = 1$$

and

$$g_{12} = g_{21} = \cos(u(x,t))\,.$$

We write g as a 2×2 matrix

$$g = \begin{pmatrix} g_{11} & g_{12} \\ g_{21} & g_{22} \end{pmatrix}.$$

The inverse of g is given by

$$g^{-1} = \begin{pmatrix} g^{11} & g^{12} \\ g^{21} & g^{22} \end{pmatrix}$$

where

$$g^{11} = g^{22} = \frac{1}{\sin^2 u}$$

and

$$g^{12} = g^{21} = -\frac{\cos u}{\sin^2 u} \, .$$

Next we have to calculate the *Christoffel symbols*. They are defined as

$$\Gamma^a_{mn} := \frac{1}{2} g^{ab} (g_{bm,n} + g_{bn,m} - g_{mn,b})$$

where the summation convention is used and

$$g_{bm,1} := \frac{\partial g_{bm}}{\partial x} \, , \quad g_{bm,2} := \frac{\partial g_{bm}}{\partial t} \, .$$

Obviously

$$g_{11,1} = g_{11,2} = g_{22,1} = g_{22,2} = 0 \, ,$$

$$g_{12,1} = g_{21,1} = -\frac{\partial u}{\partial x} \sin u \, ,$$

$$g_{12,2} = g_{21,2} = -\frac{\partial u}{\partial t} \sin u \, .$$

Therefore

$$\Gamma^1_{11} = \frac{\partial u}{\partial x} \frac{\cos u}{\sin u} \, ,$$

$$\Gamma^1_{12} = \Gamma^1_{21} = 0 \, ,$$

$$\Gamma^1_{22} = -\frac{\partial u}{\partial t} \frac{1}{\sin u} \, ,$$

$$\Gamma^2_{11} = -\frac{\partial u}{\partial x} \frac{1}{\sin u} \, ,$$

$$\Gamma^2_{12} = \Gamma^2_{21} = 0 \, ,$$

$$\Gamma^2_{22} = \frac{\partial u}{\partial t} \frac{\cos u}{\sin u} \, .$$

Now we have to calculate the *Riemann curvature tensor* which is defined by

$$R^r_{msq} := \Gamma^r_{mq,s} - \Gamma^r_{ms,q} + \Gamma^r_{ns} \Gamma^n_{mq} - \Gamma^r_{nq} \Gamma^n_{ms} \, .$$

Obviously $R^b_{mss} = 0$, i.e.,

$$R^1_{111} = R^1_{122} = R^1_{211} = R^1_{222} = R^2_{111} = R^2_{122} = R^2_{211} - R^2_{222} = 0.$$

Moreover

$$R^1_{112} = -R^1_{121} = -\frac{\partial^2 u}{\partial x \partial t} \frac{\cos u}{\sin u},$$

$$R^1_{212} = -R^1_{221} = -\frac{\partial^2 u}{\partial x \partial t} \frac{1}{\sin u},$$

$$R^2_{112} = -R^2_{121} = \frac{\partial^2 u}{\partial x \partial t} \frac{1}{\sin u},$$

$$R^2_{212} = -R^2_{221} = \frac{\partial^2 u}{\partial x \partial t} \frac{\cos u}{\sin u}.$$

The *Ricci tensor* R_{mq} is defined by

$$R_{mq} := R^a_{maq} = -R^a_{mqa},$$

i.e., the Ricci tensor is constructed by contraction. We find

$$R_{11} = R^1_{111} + R^2_{121} = -\frac{\partial^2 u}{\partial x \partial t} \frac{1}{\sin u},$$

$$R_{12} = R^1_{112} + R^2_{122} = -\frac{\partial^2 u}{\partial x \partial t} \frac{\cos u}{\sin u},$$

$$R_{21} = R^1_{211} + R^2_{221} = -\frac{\partial^2 u}{\partial x \partial t} \frac{\cos u}{\sin u},$$

$$R_{22} = R^1_{212} + R^2_{222} = -\frac{\partial^2 u}{\partial x \partial t} \cdot \frac{1}{\sin u}.$$

From R_{nq}, we obtain R^m_q via

$$R^m_q = g^{mn} R_{nq}.$$

We find

$$R^1_1 = R^2_2 = -\frac{1}{\sin u} \frac{\partial^2 u}{\partial x \partial t},$$

$$R^2_1 = R^1_2 = 0.$$

Finally, the *curvature scalar* R is given by

$$R = R^m_m.$$

Consequently

$$R = -\frac{2}{\sin u}\frac{\partial^2 u}{\partial x \partial t}.$$

If $R = -2$, then

$$\frac{\partial^2 u}{\partial x \partial t} = \sin u.$$

This is the so-called *sine-Gordon equation*.

Problem 2. Given the metric tensor field

$$g = g_{11}(q_1, q_2)dq_1 \otimes dq_1 + g_{22}(q_1, q_2)dq_2 \otimes dq_2.$$

(i) Calculate the *Riemann curvature scalar R*.

(ii) Simplify g to the special case

$$g_{11}(q_1, q_2) = g_{22}(q_1, q_2) = E - V(q_1, q_2).$$

Solution. (i) First we have to calculate the *Christoffel symbols*. Since

$$\Gamma^1_{11} := \frac{1}{2}\sum_{k=1}^{2} g^{1k}\left(\frac{\partial g_{k1}}{\partial q_1} + \frac{\partial g_{1k}}{\partial q_1} - \frac{\partial g_{11}}{\partial q_k}\right)$$

and $g_{12} = g_{21} = 0$, we have

$$\Gamma^1_{11} = \frac{1}{2}g^{11}\left(\frac{\partial g_{11}}{\partial q_1} + \frac{\partial g_{11}}{\partial q_1} - \frac{\partial g_{11}}{\partial q_1}\right) = \frac{1}{2}g^{11}\frac{\partial g_{11}}{\partial q_1}.$$

Analogously

$$\Gamma^1_{12} = \frac{1}{2}\sum_{k=1}^{2} g^{1k}\left(\frac{\partial g_{k2}}{\partial q_1} + \frac{\partial g_{1k}}{\partial q_2} - \frac{\partial g_{12}}{\partial q_k}\right) = \frac{1}{2}g^{11}\frac{\partial g_{11}}{\partial q_2},$$

$$\Gamma^1_{21} = \frac{1}{2}\sum_{k=1}^{2} g^{1k}\left(\frac{\partial g_{k1}}{\partial q_2} + \frac{\partial g_{2k}}{\partial q_1} - \frac{\partial g_{22}}{\partial q_k}\right) = \frac{1}{2}g^{11}\frac{\partial g_{11}}{\partial q_2},$$

$$\Gamma^1_{22} = \frac{1}{2}\sum_{k=1}^{2} g^{1k}\left(\frac{\partial g_{k2}}{\partial q_2} + \frac{\partial g_{2k}}{\partial q_2} - \frac{\partial g_{22}}{\partial q_k}\right) = -\frac{1}{2}g^{11}\frac{\partial g_{22}}{\partial q_1},$$

$$\Gamma^2_{11} = \frac{1}{2}\sum_{k=1}^{2} g^{2k}\left(\frac{\partial g_{k1}}{\partial q_1} + \frac{\partial g_{1k}}{\partial q_1} - \frac{\partial g_{11}}{\partial q_k}\right) = -\frac{1}{2}g^{22}\frac{\partial g_{11}}{\partial q_2},$$

$$\Gamma_{12}^2 = \frac{1}{2}\sum_{k=1}^2 g^{2k}\left(\frac{\partial g_{k2}}{\partial q_1} + \frac{\partial g_{1k}}{\partial q_2} - \frac{\partial q_{12}}{\partial q_k}\right) = \frac{1}{2}g^{22}\frac{\partial g_{22}}{\partial q_1},$$

$$\Gamma_{21}^2 = \frac{1}{2}\sum_{k=1}^2 g^{2k}\left(\frac{\partial g_{k1}}{\partial q_2} + \frac{\partial g_{2k}}{\partial q_1} - \frac{\partial g_{21}}{\partial q_k}\right) = \frac{1}{2}g^{22}\frac{\partial g_{22}}{\partial q_1},$$

$$\Gamma_{22}^2 = \frac{1}{2}\sum_{k=1}^2 g^{2k}\left(\frac{\partial g_{k2}}{\partial q_2} + \frac{\partial g_{2k}}{\partial q_2} - \frac{\partial g_{22}}{\partial q_k}\right) = \frac{1}{2}g^{22}\frac{\partial g_{22}}{\partial q_2},$$

where

$$g^{11} = \frac{1}{g_{11}}, \quad g^{22} = \frac{1}{g_{22}}.$$

Next we have to calculate the curvature. We find

$$R_{12} = \sum_{h=1}^2\left(\frac{\partial\Gamma_{12}^h}{\partial q_h} - \frac{\partial\Gamma_{h1}^h}{\partial q_2} + \sum_{\ell=1}^2(\Gamma_{h\ell}^h\Gamma_{12}^\ell - \Gamma_{1h}^\ell\Gamma_{2\ell}^h)\right) = 0,$$

$$R_{21} = 0 \quad \text{for symmetry reasons},$$

$$R_{11} = \sum_{h=1}^2\left(\frac{\partial\Gamma_{11}^h}{\partial q_h} - \frac{\partial\Gamma_{h1}^h}{\partial q_1} + \sum_{\ell=1}^2(\Gamma_{h\ell}^h\Gamma_{11}^\ell - \Gamma_{1h}^\ell\Gamma_{1\ell}^h)\right)$$

$$= \frac{\partial\Gamma_{11}^2}{\partial q_2} - \frac{\partial\Gamma_{21}^2}{\partial q_1} + (\Gamma_{21}^2\Gamma_{11}^1 + \Gamma_{22}^2\Gamma_{11}^2 - \Gamma_{11}^2\Gamma_{12}^1 - \Gamma_{12}^2\Gamma_{12}^2),$$

$$R_{22} = \sum_{h=1}^2\left(\frac{\partial\Gamma_{22}^h}{\partial q_h} - \frac{\partial\Gamma_{h2}^h}{\partial q_2} + \sum_{\ell=1}^2(\Gamma_{h\ell}^h\Gamma_{22}^\ell - \Gamma_{2h}^\ell\Gamma_{2\ell}^h)\right)$$

$$= \frac{\partial\Gamma_{22}^1}{\partial q_1} - \frac{\Gamma_{12}^1}{\partial q_2} + \Gamma_{11}^1\Gamma_{22}^1 + \Gamma_{12}^1\Gamma_{22}^2 - \Gamma_{21}^1\Gamma_{21}^1 - \Gamma_{21}^2\Gamma_{22}^1.$$

Now the Riemann curvature scalar R is given by

$$R = g^{11}R_{11} + g^{22}R_{22}.$$

Consequently,

$$R = g^{11}\left(\frac{\partial\Gamma_{11}^2}{\partial q_2} - \frac{\partial\Gamma_{21}^2}{\partial q_1}\right) + g^{11}(\Gamma_{11}^1\Gamma_{21}^2 + \Gamma_{22}^2\Gamma_{11}^2 - \Gamma_{11}^2\Gamma_{12}^1 - \Gamma_{12}^2\Gamma_{12}^2)$$

$$+ g^{22}\left(\frac{\partial\Gamma_{22}^1}{\partial q_1} - \frac{\partial\Gamma_{12}^1}{\partial q_2}\right) + g^{22}(\Gamma_{22}^2\Gamma_{12}^1 + \Gamma_{11}^1\Gamma_{22}^1 - \Gamma_{22}^1\Gamma_{21}^2 - \Gamma_{21}^1\Gamma_{21}^1)$$

where

$$\frac{\partial \Gamma_{11}^2}{\partial q_2} = -\frac{1}{2}\frac{\partial}{\partial q_2}\left(g^{22}\frac{\partial g_{11}}{\partial q_2}\right) = -\frac{1}{2}\left(\frac{\partial g^{22}}{\partial q_2}\right)\left(\frac{\partial g_{11}}{\partial q_2}\right) - \frac{1}{2}g^{22}\frac{\partial^2 g_{11}}{\partial q_2^2},$$

$$\frac{\partial \Gamma_{21}^2}{\partial q_1} = \frac{1}{2}\frac{\partial}{\partial q_1}\left(g^{22}\frac{\partial g_{22}}{\partial q_1}\right) = \frac{1}{2}\left(\frac{\partial g^{22}}{\partial q_1}\right)\left(\frac{\partial g_{22}}{\partial q_1}\right) + \frac{1}{2}g^{22}\frac{\partial^2 g_{22}}{\partial q_1^2},$$

$$\frac{\partial \Gamma_{22}^1}{\partial q_1} = -\frac{1}{2}\frac{\partial}{\partial q_1}\left(g^{11}\frac{\partial g_{22}}{\partial q_1}\right) = -\frac{1}{2}\left(\frac{\partial g^{11}}{\partial q_1}\right)\left(\frac{\partial g_{22}}{\partial q_1}\right) - \frac{1}{2}g^{11}\frac{\partial^2 g_{22}}{\partial q_1^2},$$

$$\frac{\partial \Gamma_{12}^1}{\partial q_2} = \frac{1}{2}\frac{\partial}{\partial q_2}\left(g^{11}\frac{\partial g_{11}}{\partial q_2}\right) = \frac{1}{2}\left(\frac{\partial g^{11}}{\partial q_2}\right)\left(\frac{\partial g_{11}}{\partial q_2}\right) + \frac{1}{2}g^{11}\frac{\partial^2 g_{11}}{\partial q_2^2}.$$

Therefore

$$R = \frac{g^{11}g^{22}}{2}\left(-\frac{\partial^2 g_{11}}{\partial q_2^2} - \frac{\partial^2 g_{22}}{\partial q_1^2} - \frac{\partial^2 g_{22}}{\partial q_1^2} - \frac{\partial^2 g_{11}}{\partial q_2^2}\right)$$

$$- \frac{1}{2}\left(g^{11}\frac{\partial g^{22}}{\partial q_2}\frac{\partial g_{11}}{\partial q_2} + g^{11}\frac{\partial g^{22}}{\partial q_1}\frac{\partial g_{22}}{\partial q_1} + g^{22}\frac{\partial g^{11}}{\partial q_1}\frac{\partial g_{22}}{\partial q_1} + g^{22}\frac{\partial g^{11}}{\partial q_2}\frac{\partial g_{11}}{\partial q_2}\right).$$

(ii) If

$$g_{11} = g_{22} = E - V(q_1, q_2),$$

we obtain

$$g^{11} = \frac{1}{E - V}, \quad g^{22} = \frac{1}{E - V}.$$

Therefore

$$R = \frac{1}{(E-V)^3}\left(\left(\frac{\partial^2 V}{\partial q_1^2} + \frac{\partial^2 V}{\partial q_2^2}\right)(E - V) + \frac{\partial V}{\partial q_1}\frac{\partial V}{\partial q_1} + \frac{\partial V}{\partial q_2}\frac{\partial V}{\partial q_2}\right).$$

Remark. *The motion of a classical mechanical system with Hamilton function*

$$H(\mathbf{p}, \mathbf{q}) = \frac{1}{2}\alpha^{jk}p_j p_k + V(q_1, q_2, \dots, q_N)$$

(summation convention; $j, k = 1, 2, \dots, N$) can be represented as a geodesic flow *on a Riemannian manifold with metric*

$$g_{jk} = (E - V(\mathbf{q}))\alpha_{jk}.$$

The matrix (α^{jk}) is symmetric and the matrix (α_{kl}) is its inverse

$$\alpha^{jk}\alpha_{kl} = \delta_l^j$$

where δ_i^j is the Kronecker symbol. In our case we have $N = 2$ and $\alpha_{jk} = \delta_{jk}$. The geodesic flow on a closed Riemannina manifold of negative curvature is a so-called C-flow and is therefore ergodic.

Problem 3.   Let $a > b > 0$ and define

$$f : \mathbf{R}^2 \to \mathbf{R}^3$$

by

$$f(\theta, \phi) = ((a + b\cos\phi)\cos\theta, \ (a + b\cos\phi)\sin\theta, \ b\sin\phi)$$

where $0 \le \phi < 2\pi$ and $0 \le \theta < 2\pi$. The function f is a *parametrized torus* T^2 on $\mathbf{R}^3$. Let

$$g = dx_1 \otimes dx_1 + dx_2 \otimes dx_2 + dx_3 \otimes dx_3 .$$

The parametrized torus is doubly-periodic, i.e.,

$$f(\theta + 2k\pi, \phi) = f(\theta, \phi), \quad f(\theta, \phi + 2k\pi) = f(\theta, \phi), \quad k \in \mathbf{Z}.$$

(i) Calculate $g|_{T^2}$.

(ii) Calculate the Christoffel symbols Γ_{ab}^m from $g|_{T^2}$.

(iii) Give the differential equations of the geodesics.

Solution.   (i) Since

$$x_1(\phi, \theta) = (a + b\cos\phi)\cos\theta,$$
$$x_2(\phi, \theta) = (a + b\cos\phi)\sin\theta,$$
$$x_3(\phi, \theta) = b\sin\phi,$$

and

$$dx_j = \frac{\partial x_j}{\partial\theta}d\theta + \frac{\partial x_j}{\partial\phi}d\phi, \quad j = 1, 2, 3,$$

we find

$$dx_1 \otimes dx_1 + dx_2 \otimes dx_2 + dx_3 \otimes dx_3 = (a + b\cos\phi)^2 d\theta \otimes d\theta + b^2 d\phi \otimes d\phi$$

where we used that $\sin^2\theta + \cos^2\theta = 1$ and $\sin^2\phi + \cos^2\phi = 1$. We identify θ with the first coordinate and ϕ with the second coordinate. Thus

$$g_{11} = (a + b\cos(\phi))^2, \quad g_{12} = 0, \quad g_{21} = 0, \quad g_{22} = b^2$$

and therefore

$$g^{11} = \frac{1}{(a + b\cos(\phi))^2}, \quad g^{12} = 0, \quad g^{21} = 0, \quad g^{22} = \frac{1}{b^2}.$$

(ii) We set $y^1 = \theta$ and $y^2 = \phi$. The *Christoffel symbols* are given by

$$\Gamma^\alpha_{\beta\gamma} = \frac{1}{2}g^{\alpha\delta}\left(\frac{\partial g_{\beta\delta}}{\partial y^\gamma} + \frac{\partial g_{\gamma\delta}}{\partial y^\beta} - \frac{\partial g_{\beta\gamma}}{\partial y^\delta}\right)$$

where we make use the summation convention (summation over δ, $\delta = 1$, 2, 3). Thus we find

$$\Gamma^1_{11} = 0, \quad \Gamma^1_{12} = -\frac{b\sin\phi}{a + b\cos\phi}, \quad \Gamma^1_{21} = -\frac{b\sin\phi}{a + b\cos\phi}, \quad \Gamma^1_{22} = 0,$$

$$\Gamma^2_{11} = \frac{(a + b\cos\phi)\sin\phi}{b}, \quad \Gamma^2_{12} = 0, \quad \Gamma^2_{21} = 0, \quad \Gamma^2_{22} = 0.$$

(iii) The differential equations for the *geodesics* are determined by

$$\frac{d^2 y^\alpha}{ds^2} + \Gamma^\alpha_{\beta\gamma}\frac{dy^\beta}{ds}\frac{dy^\gamma}{ds} = 0$$

where we used the summation convention (summation over β and γ, $\beta = 1$, 2, 3, $\gamma = 1, 2, 3$). Thus we find

$$\frac{d^2\theta}{ds^2} - \frac{2b\sin\phi}{a + b\cos\phi}\frac{d\theta}{ds}\frac{d\phi}{ds} = 0,$$

$$\frac{d^2\phi}{ds^2} + \frac{(a + b\cos\phi)\sin\phi}{b}\left(\frac{d\theta}{ds}\right)^2 = 0.$$

Chapter 22

Killing Vector Fields

Problem 1. Let M be a smooth manifold and g be a metric tensor field. Then V is called a *Killing vector field* with respect to g if

$$L_V g = 0$$

where $L_V(\cdot)$ denotes the Lie derivative. Let

$$g = \frac{1}{2} \sum_{j,k=1}^{3} g_{jk}(\mathbf{x}) dx_j \otimes dx_k$$

be a metric tensor field in $\mathbf{R}^3$ and let

$$V(\mathbf{x}) = V_1(\mathbf{x}) \frac{\partial}{\partial x_1} + V_2(\mathbf{x}) \frac{\partial}{\partial x_2} + V_3(\mathbf{x}) \frac{\partial}{\partial x_3}.$$

Assume that $g_{jk} = \delta_{jk}$ where δ_{jk} is the Kronecker symbol, i.e.,

$$\delta_{jk} := \begin{cases} 1 & j = k, \\ 0 & j \neq k. \end{cases}$$

(i) Calculate

$$L_V g.$$

(ii) Give an interpretation of

$$L_V g = 0.$$

Solution. (i) The Lie derivative is linear and satisfies the product rule

$$L_V g = L_V(dx_1 \otimes dx_1) + L_V(dx_2 \otimes dx_2) + L_V(dx_3 \otimes dx_3),$$

$$L_V(dx_j \otimes dx_k) = (L_V dx_j) \otimes dx_k + dx_j \otimes (L_V dx_k).$$

Furthermore the Lie derivative satisfies

$$L_V dx_j = d(V \rfloor dx_j) = dV_j = \sum_{k=1}^{3} \frac{\partial V_j}{\partial x_k} dx_k .$$

Here $\rfloor$ denotes the contraction, i.e.,

$$\frac{\partial}{\partial x_j} \rfloor dx_k = \delta_{jk} .$$

It follows that

$$L_V g = \frac{1}{2} \sum_{j,k=1}^{3} \left(\frac{\partial V_j}{\partial x_k} + \frac{\partial V_k}{\partial x_j} \right) dx_j \otimes dx_k . \tag{1}$$

Since $dx_j \otimes dx_k$ $(j, k = 1, 2, 3)$ are basic elements, the right-hand side of (1) can also be written as a 3×3 matrix, namely

$$\frac{1}{2} \begin{pmatrix} 2\dfrac{\partial V_1}{\partial x_1} & \dfrac{\partial V_1}{\partial x_2} + \dfrac{\partial V_2}{\partial x_1} & \dfrac{\partial V_1}{\partial x_3} + \dfrac{\partial V_3}{\partial x_1} \\[2mm] \dfrac{\partial V_1}{\partial x_2} + \dfrac{\partial V_2}{\partial x_1} & 2\dfrac{\partial V_2}{\partial x_2} & \dfrac{\partial V_2}{\partial x_3} + \dfrac{\partial V_3}{\partial x_2} \\[2mm] \dfrac{\partial V_1}{\partial x_3} + \dfrac{\partial V_3}{\partial x_1} & \dfrac{\partial V_2}{\partial x_3} + \dfrac{\partial V_3}{\partial x_2} & 2\dfrac{\partial V_3}{\partial x_3} \end{pmatrix} .$$

(ii) The physical meaning is as follows: Consider a deformable body B in $\mathbf{R}^3$, $B \subset \mathbf{R}^3$. A displacement with or without deformation of the body B is a diffeomorphism of $\mathbf{R}^3$ defined in a neighbourhood of B. All such diffeomorphisms form a local group generated by the so-called displacement vector field $V(\mathbf{x})$. Consequently, the strain tensor field can be considered as the Lie derivative of the metric tensor field g in $\mathbf{R}^3$ with respect to the vector field $V(\mathbf{x})$. The metric tensor field gives rise to the distance between two points, namely

$$(ds)^2 = \sum_{j,k=1}^{3} g_{jk}(\mathbf{x}) dx_j dx_k .$$

Then the strain tensor field measures the variation of the distance between two points under a displacement generated by $V(\mathbf{x})$. The equation (invariance condition)

$$L_V g = 0$$

tells us that two points of the body B do not change during the displacement.

Summarize. The strain tensor is the Lie derivative of the metric tensor field with respect to the deformation (or exactly the displacement vector field).

Problem 2. Let

$$g = dx_1 \otimes dx_1 + dx_2 \otimes dx_2$$

be the metric tensor field of the two-dimensional Euclidean space. Assume that

$$L_V g = 0$$

where

$$V = V_1(\mathbf{x}) \frac{\partial}{\partial x_1} + V_2(\mathbf{x}) \frac{\partial}{\partial x_2}$$

is a smooth vector field defined on the two-dimensional Euclidean space. Show that

$$L_V (dx_1 \wedge dx_2) = 0 \,. \tag{2}$$

Solution. From $L_V g = 0$, it follows that

$$\frac{\partial V_1}{\partial x_1} = 0, \quad \frac{\partial V_2}{\partial x_2} = 0, \tag{2a}$$

$$\frac{\partial V_1}{\partial x_2} + \frac{\partial V_2}{\partial x_1} = 0 \,. \tag{2b}$$

Now

$$L_V (dx_1 \wedge dx_2) = (L_V dx_1) \wedge dx_2 + dx_1 \wedge (L_V dx_2)$$
$$= dV_1 \wedge dx_2 + dx_1 \wedge dV_2$$

$$= \frac{\partial V_1}{\partial x_1} dx_1 \wedge dx_2 + \frac{\partial V_2}{\partial x_2} dx_1 \wedge dx_2$$

$$= \left(\frac{\partial V_1}{\partial x_1} + \frac{\partial V_2}{\partial x_2} \right) dx_1 \wedge dx_2 \,. \tag{3}$$

Inserting (2a) into (3) results in (1).

Problem 3. Let E^3 be the three-dimensional Euclidean space with metric tensor field

$$g = dx_1 \otimes dx_1 + dx_2 \otimes dx_2 + dx_3 \otimes dx_3 \,. \tag{1}$$

Let

$$M \equiv S^2 = \{ (x_1, x_2, x_3) : x_1^2 + x_2^2 + x_3^2 = 1 \} \,.$$

This means the manifold M is the unit sphere.

(i) Calculate the metric tensor field $\tilde{g}$ for M. The parametrization is given by

$$x_1(u_1, u_2) = \cos u_1 \sin u_2 \,, \quad x_2(u_1, u_2) = \sin u_1 \sin u_2 \,, \quad x_3(u_1, u_2) = \cos u_2$$

where $u_1 \in [0, 2\pi)$ and $u_2 \in [0, \pi)$.

(ii) Find the vector fields V such that

$$L_V \tilde{g} = 0 \,.$$

(iii) Show that these vector fields form a Lie algebra under the commutator.

Solution. (i) Since

$$dx_1 = -\sin u_1 \sin u_2 du_1 + \cos u_1 \cos u_2 du_2 \,,$$
$$dx_2 = \cos u_1 \sin u_2 du_1 + \sin u_1 \cos u_2 du_2 \,,$$
$$dx_3 = -\sin u_2 du_2 \,,$$

we obtain from (1) that

$$\tilde{g} = \sin^2 u_2 du_1 \otimes du_1 + du_2 \otimes du_2 \,.$$

(ii) From the condition

$$L_V \tilde{g} = 0$$

with

$$V = V_1(u_1, u_2)\frac{\partial}{\partial u_1} + V_2(u_1, u_2)\frac{\partial}{\partial u_2},$$

we find

$$\left(2V_2 \sin u_2 \cos u_2 + 2\frac{\partial V_1}{\partial u_1}\sin^2 u_2\right) du_1 \otimes du_1$$

$$+ \left(\sin^2 u_2\frac{\partial V_1}{\partial u_2} + \frac{\partial V_2}{\partial u_1}\right) du_1 \otimes du_2$$

$$+ \left(\sin^2 u_2\frac{\partial V_1}{\partial u_2} + \frac{\partial V_2}{\partial u_1}\right) du_2 \otimes du_1 + 2\frac{\partial V_2}{\partial u_2}du_2 \otimes du_2 = 0.$$

This leads to the system of linear partial differential equations

$$2V_2 \sin u_2 \cos u_2 + 2\frac{\partial V_1}{\partial u_1}\sin^2 u_2 = 0, \tag{2a}$$

$$\sin^2 u_2\frac{\partial V_1}{\partial u_2} + \frac{\partial V_2}{\partial u_1} = 0, \tag{2b}$$

$$\frac{\partial V_2}{\partial u_2} = 0. \tag{2c}$$

From (2c) we conclude that V_2 depends only on u_1. Thus the solution of system (2) leads to the three independent vector fields

$$V^I = \frac{\partial}{\partial u_1},$$

$$V^{II} = \cos u_1 \cot u_2\frac{\partial}{\partial u_1} + \sin u_1\frac{\partial}{\partial u_2},$$

$$V^{III} = -\sin u_1 \cot u_2\frac{\partial}{\partial u_1} + \cos u_1\frac{\partial}{\partial u_2}.$$

(iii) We find

$$[V^I, V^{II}] = -\sin u_1 \cot u_2\frac{\partial}{\partial u_1} + \cos u_1\frac{\partial}{\partial u_2} = V^{III},$$

$$[V^I, V^{III}] = -\cos u_1 \cot u_2\frac{\partial}{\partial u_1} - \sin u_1\frac{\partial}{\partial u_2} = -V^{II},$$

$$[V^{II}, V^{III}] = \frac{\partial}{\partial u_1} = V^I.$$

Thus the vector fields V^I, V^{II} and V^{III} form a basis of a Lie algebra under the commutator.

Problem 4. The *Gödel metric tensor field* is given by

$$g = a^2 \left[dx \otimes dx - \frac{1}{2} e^{2x} dy \otimes dy + dz \otimes dz - c^2 dt \otimes dt \right.$$

$$\left. - c e^x dy \otimes dt - c e^x dt \otimes dy \right]$$

where a and c are constants.

(i) Show that the vector fields

$$Y = \frac{\partial}{\partial y}, \quad Z = \frac{\partial}{\partial z}, \quad T = \frac{\partial}{\partial t},$$

$$A = \frac{\partial}{\partial x} - y \frac{\partial}{\partial y},$$

$$B = y \frac{\partial}{\partial x} + \left(c e^{-2x} - \frac{1}{2} y^2 \right) \frac{\partial}{\partial y} - 2 e^{-x} \frac{\partial}{\partial t},$$

are Killing vector fields of the Gödel metric tensor.

(ii) Calculate the commutators.

(iii) Let g be a metric tensor field. Assume that V and W are Killing vector fields. Show that $[V, W]$ is also a Killing vector field.

Solution. (i) First we recall that the Lie derivative is a linear operation. Since the functions g_{jk} do not depend on y, z and t it is obvious that Y, Z and T are Killing vector fields. Applying the product rule and

$$L_{\partial/\partial x} e^{2x} dt = 2 e^{2x} dt, \quad L_{\partial/\partial x} e^{2x} dy = 2 e^{2x} dy,$$

$$L_{y\partial/\partial y} dy = dy, \quad L_{y\partial/\partial y} e^x dy = e^x dy,$$

we find that

$$L_A g = 0.$$

Now we evaluate the Lie derivative of g with respect to B. Applying the product rule and the rules

$$L_{fV}\alpha = f(L_V\alpha) + df \wedge (V \rfloor \alpha),$$

$$L_V(f\alpha) = (Vf)\alpha + f(L_V\alpha),$$

we obtain

$$L_B g = 0.$$

Remark.   *The Gödel metric tensor field contains* closed timelike lines; *that is, an observer can influence his own past.*

(ii) For the commutators we obtain

$$[Y, Z] = 0,$$

$$[Y, T] = 0,$$

$$[Z, T] = 0,$$

$$[Y, A] = -\frac{\partial}{\partial y} = -Y,$$

$$[Y, B] = \frac{\partial}{\partial x} - y\frac{\partial}{\partial y} = A,$$

$$[Z, A] = 0,$$

$$[Z, B] = 0,$$

$$[T, A] = 0,$$

$$[T, B] = 0,$$

$$[A, B] = -y\frac{\partial}{\partial x} + \left(\frac{1}{2}y^2 - ce^{-2x}\right)\frac{\partial}{\partial y} + 2e^{-x}\frac{\partial}{\partial t} = -B.$$

(iii) By assumption we have

$$L_V g = 0, \quad L_W g = 0.$$

Using the identity

$$L_{[V,W]}g \equiv [L_V, L_W]g \equiv L_V(L_W g) - L_W(L_V g),$$

it follows that

$$L_{[V,W]}g = 0.$$

Chapter 23

Inequalities

Problem 1. Let f and g be two integrable functions. Assume that

$$\int_I f(x)dx = \int_I g(x)dx \tag{1}$$

and $f(x) > 0$, $g(x) > 0$ for $x \in I$, where $I \subset \mathbf{R}$. Show that

$$\int_I f(x)\ln f(x)dx \geq \int_I f(x)\ln g(x)dx\,.$$

Solution. We have

$$\int_I f\ln f dx - \int_I f\ln g dx \equiv \int_I f(\ln f - \ln g)dx \equiv \int_I f\ln \frac{f}{g}dx$$

$$\equiv \int_I g\left(\frac{f}{g}\ln\left(\frac{f}{g}\right)\right)dx\,.$$

Thus

$$\int_I f\ln f dx - \int_I f\ln g dx \equiv \int_I g\left(\frac{f}{g}\ln\left(\frac{f}{g}\right) - \frac{f}{g} + 1\right)dx \geq 0$$

where we have used (1) and the fact that

$$\frac{f}{g}\ln\left(\frac{f}{g}\right) - \frac{f}{g} + 1 \geq 0\,.$$

Problem 2. (i) Let $0 < \mu < 1$ and $a \geq 0$, $b \geq 0$. Show that

$$a^\mu b^{1-\mu} \leq \mu a + (1-\mu)b\,. \tag{1}$$

(ii) Let p and q be two positive numbers which satisfy the condition

$$\frac{1}{p} + \frac{1}{q} = 1.$$

Let $c, d \in \mathbf{C}$. Show that

$$|cd| \leq \frac{|c|^p}{p} + \frac{|d|^q}{q}. \tag{2}$$

Solution. (i) For $a = b$, the inequality (1) is obviously satisfied. Without loss of generality we can assume that $b > a > 0$. Applying the theorem of the mean, we obtain

$$b^{1-\mu} - a^{1-\mu} = (1 - \mu)(b - a)\xi^{-\mu}$$

with $a < \xi < b$. Since

$$\xi^{-\mu} < a^{-\mu},$$

we obtain

$$b^{1-\mu} - a^{1-\mu} \leq (1 - \mu)(b - a)a^{-\mu}. \tag{3}$$

Multiplying the left and right-hand sides of (3) by a^{μ} leads to (1).

(ii) If we set

$$\mu = \frac{1}{p}, \quad 1 - \mu = \frac{1}{q}, \quad a = |c|^p, \quad b = |d|^q$$

in inequality (1) we obtain (2).

Problem 3. Let $a_1, \ldots, a_n$ and $b_1, \ldots, b_n$ be arbitrary complex numbers. Let p and q be two real positive numbers which satisfy

$$\frac{1}{p} + \frac{1}{q} = 1.$$

Show that

$$\sum_{k=1}^{n} |a_k b_k| \leq \left[\sum_{k=1}^{n} |a_k|^p \right]^{1/p} \left[\sum_{k=1}^{n} |b_k|^q \right]^{1/q}.$$

Solution. Let

$$A^p := \sum_{k=1}^{n} |a_k|^p, \quad B^q := \sum_{k=1}^{n} |b_k|^q.$$

We can assume that both A and B are positive. Let

$$\bar{a}_k := \frac{a_k}{A}, \quad \bar{b}_k := \frac{b_k}{B}.$$

Using the result from Problem 2, we obtain

$$|\bar{a}_k \bar{b}_k| \leq \frac{|\bar{a}_k|^p}{p} + \frac{|\bar{b}_k|^q}{q}.$$

Taking the sum over k on both sides yields

$$\sum_{k=1}^{n} |\bar{a}_k \bar{b}_k| \leq \sum_{k=1}^{n} \left(\frac{|\bar{a}_k|^p}{p} + \frac{|\bar{b}_k|^q}{q} \right) = \frac{1}{p} + \frac{1}{q} = 1.$$

Consequently,

$$\sum_{k=1}^{n} |a_k b_k| \leq AB.$$

Problem 4. Let B be an $n \times n$ matrix over $\mathbf{R}$. Let B^T be the transpose. Show that

$$\operatorname{tr}(B^{2m}) \leq \operatorname{tr}(B^m B^{mT}) \tag{1}$$

where tr denotes the trace.

Solution. Let A be an $n \times n$ matrix over $\mathbf{R}$. Then

$$(A^T A)_{jl} = \sum_{k=1}^{n} a_{kj} a_{kl}$$

and therefore

$$(A^T A)_{jj} = \sum_{k=1}^{n} a_{kj} a_{kj} \geq 0. \tag{2}$$

It follows that

$$\operatorname{tr}(A^T A) \geq 0.$$

We set

$$A^T := B^m - B^{mT}.$$

Hence $A = B^{mT} - B^m$. Using (2), we obtain

$$\operatorname{tr}((B^m - B^{mT})(B^{mT} - B^m)) \geq 0.$$

It follows that

$$\mathrm{tr}(B^m B^{mT}) + \mathrm{tr}(B^{mT} B^m) - \mathrm{tr}(B^{mT} B^{mT}) - \mathrm{tr}(B^m B^m) \geq 0.$$

Since

$$\mathrm{tr}(B^m B^{mT}) \equiv \mathrm{tr}(B^{mT} B^m),$$

$$\mathrm{tr}(B^m B^m) \equiv \mathrm{tr}(B^{mT} B^{mT}),$$

inequality (1) follows.

Problem 5.   Show that for two $n \times n$ positive-semidefinite real matrices A and B and $0 \leq \alpha \leq 1$, we have

$$\mathrm{tr}(A^\alpha B^{1-\alpha}) \leq (\mathrm{tr}(A))^\alpha (\mathrm{tr}(B))^{1-\alpha}.$$

Solution.   Since A and B are positive-semidefinite their eigenvalues are real and nonnegative. We order the eigenvalues of A and B in decreasing order

$$\lambda_1 \geq \lambda_2 \geq \cdots \geq \lambda_n \geq 0, \quad \mu_1 \geq \mu_2 \geq \cdots \geq \mu_n \geq 0.$$

For an arbitrary orthonormal set of vectors $\mathbf{x}_i$, we have

$$\sum_{i=1}^{k} (\mathbf{x}_i, B^{1-\alpha} \mathbf{x}_i) \leq \sum_{i=1}^{k} \mu_i^{1-\alpha}, \quad k = 1, 2, \ldots, n$$

where $(\ ,\ )$ denotes the scalar product. Choosing $\mathbf{x}_i$ to be the eigenvectors of A and summing by parts gives

$$\mathrm{tr}(A^\alpha B^{1-\alpha}) = \sum_{i=1}^{n} \lambda_i^\alpha (\mathbf{x}_i, B^{1-\alpha} \mathbf{x}_i).$$

Thus

$$\mathrm{tr}(A^\alpha B^{1-\alpha}) \leq \sum_{i=1}^{n} \lambda_i^\alpha \mu_i^{1-\alpha} \leq \left(\sum_{i=1}^{n} \lambda_i \right)^\alpha \left(\sum_{i=1}^{n} \mu_i \right)^{1-\alpha}$$

$$= (\mathrm{tr}(A))^\alpha (\mathrm{tr}(B))^{1-\alpha}$$

where the last inequality is just *Hölder's inequality* for positive real numbers. We also used that

$$\mathrm{tr}(A) = \sum_{i=1}^{n} \lambda_i, \quad \mathrm{tr}(B) = \sum_{i=1}^{n} \mu_i.$$

Problem 6.   Let A be a bounded linear operator in a Banach space. Let

$$f_{n,m}(A) := \left(\sum_{k=0}^{m} \frac{1}{k!} \left(\frac{A}{n} \right)^k \right)^n. \tag{1}$$

Show that

$$\| e^A - f_{n,m}(A) \| \leq \frac{1}{n^m (m+1)!} \| A \|^{m+1} e^{\| A \|}.$$

Solution.   Using the properties of the norm we have

$$\| e^A - f_{n,m}(A) \| \leq \| e^{A/n} - h \| \cdot \| (e^{A/n})^{n-1} + (e^{A/n})^{n-2} h + \cdots + h^{n-1} \|$$

$$\leq n \| e^{A/n} - h \| \cdot e^{(n-1) \| A \|/n} \tag{2}$$

where

$$h = \sum_{k=0}^{m} \frac{1}{k!} \left(\frac{A}{n} \right)^k.$$

Using Taylor's theorem we find

$$\| e^{A/n} - h \| = \left\| \sum_{k=m+1}^{\infty} \frac{1}{k!} \left(\frac{A}{n} \right)^k \right\| \leq \sum_{k=m+1}^{\infty} \frac{1}{k!} \left(\frac{\| A \|}{n} \right)^k$$

$$= e^{\| A \|/n} - \sum_{k=0}^{m} \frac{1}{k!} \left(\frac{\| A \|}{n} \right)^k$$

$$= \frac{1}{(m+1)!} \left(\frac{\| A \|}{n} \right)^{m+1} e^{\theta \| A \|/n}, \quad 0 < \theta < 1. \tag{3}$$

Inserting (3) into (2), we obtain (1). $f_{n,m}$ converges to e^A if $n \to \infty$ or if $m \to \infty$.

Chapter 24

Ising Model and Heisenberg Model

Problem 1.   (i) Find the spectrum for the four-point *Ising model*

$$\hat{H} = J\sum_{j=1}^{4} \sigma_j\sigma_{j+1}$$

with cyclic boundary conditions

$$\sigma_5 = \sigma_1$$

and $J < 0$. The quantity J is called the exchange constant.

(ii) Calculate the Helmholtz free energy F for the Ising model.

Solution.   (i) We have

$$\sigma_1 = \sigma_z \otimes I \otimes I \otimes I\,,$$

$$\sigma_2 = I \otimes \sigma_z \otimes I \otimes I\,,$$

$$\sigma_3 = I \otimes I \otimes \sigma_z \otimes I\,,$$

$$\sigma_4 = I \otimes I \otimes I \otimes \sigma_z\,,$$

where

$$\sigma_z := \begin{pmatrix} 1 & 0 \\ 0 & -1 \end{pmatrix}$$

and I is the 2×2 unit matrix.

Thus we find the diagonal matrices

$$\sigma_1\sigma_2 = \text{diag}(1,1,1,1,-1,-1,-1,-1,-1,-1,-1,-1,1,1,1,1),$$

$$\sigma_2\sigma_3 = \text{diag}(1,1,-1,-1,-1,-1,1,1,1,1,-1,-1,-1,-1,1,1),$$

$$\sigma_3\sigma_4 = \text{diag}(1,-1,-1,1,1,-1,-1,1,1,-1,-1,1,1,-1,-1,1),$$

$$\sigma_4\sigma_1 = \text{diag}(1,-1,1,-1,1,-1,1,-1,-1,1,-1,1,-1,1,-1,1).$$

It follows that

$$\sum_{j=1}^{4}\sigma_j\sigma_{j+1} = \text{diag}(4,0,0,0,0,-4,0,0,0,0,-4,0,0,0,0,4).$$

Consequently the spectrum of $\hat{H}$ is given by

$$4J \quad \text{twofold degenerate},$$

$$0 \quad \text{12 times degenerate},$$

$$-4J \quad \text{twofold degenerate}.$$

(ii) The *Helmholtz free energy* is given by

$$F(\beta) := -\frac{1}{\beta}\ln Z(\beta) = -\frac{1}{\beta}\ln \text{tr}\, e^{-\beta\hat{H}}$$

where $\beta := \frac{1}{kT}$ and Z denotes the *partition function*. Thus from the result of (i), we find that the partition function is given by

$$Z(\beta) := \text{tr}\, e^{-\beta\hat{H}} = 2e^{4\beta J} + 12 + 2e^{-4\beta J}.$$

Thus the Helmholtz free energy is given by

$$F(\beta) = -\frac{1}{\beta}\ln(2e^{4\beta J} + 12 + 2e^{-4\beta J}).$$

Problem 2. Calculate the eigenvalues and eigenvectors for the two-point *Heisenberg model*

$$\hat{H} = J\sum_{j=1}^{2}\mathbf{S}_j \cdot \mathbf{S}_{j+1}$$

with cyclic boundary conditions, i.e., $\mathbf{S}_3 \equiv \mathbf{S}_1$ where J is the so-called exchange constant ($J > 0$ or $J < 0$).

Solution.   It follows that

$$\hat{H} = J(\mathbf{S}_1 \cdot \mathbf{S}_2 + \mathbf{S}_2 \cdot \mathbf{S}_3) \equiv J(\mathbf{S}_1 \cdot \mathbf{S}_2 + \mathbf{S}_2 \cdot \mathbf{S}_1) \,.$$

Therefore

$$\hat{H} = J(S_{1x}S_{2x} + S_{1y}S_{2y} + S_{1z}S_{2z} + S_{2x}S_{1x} + S_{2y}S_{1y} + S_{2z}S_{1z}) \,.$$

Since $S_{1x} := S_x \otimes I$, $S_{2x} = I \otimes S_x$ etc., it follows that

$$\hat{H} = J[(S_x \otimes I)(I \otimes S_x) + (S_y \otimes I)(I \otimes S_y) + (S_z \otimes I)(I \otimes S_z)$$
$$+ (I \otimes S_x)(S_x \otimes I) + (I \otimes S_y)(S_y \otimes I) + (I \otimes S_z)(S_z \otimes I)] \,.$$

We find

$$\hat{H} = J[(S_x \otimes S_x) + (S_y \otimes S_y) + (S_z \otimes S_z) + (S_x \otimes S_x) + (S_y \otimes S_y) + (S_z \otimes S_z)] \,.$$

Therefore

$$\hat{H} = 2J((S_x \otimes S_x) + (S_y \otimes S_y) + (S_z \otimes S_z)) \,.$$

Since

$$S_x := \frac{1}{2}\sigma_x = \frac{1}{2}\begin{pmatrix} 0 & 1 \\ 1 & 0 \end{pmatrix}, \quad S_y := \frac{1}{2}\sigma_y = \frac{1}{2}\begin{pmatrix} 0 & -i \\ i & 0 \end{pmatrix},$$

$$S_z := \frac{1}{2}\sigma_z = \frac{1}{2}\begin{pmatrix} 1 & 0 \\ 0 & -1 \end{pmatrix},$$

we obtain

$$S_x \otimes S_x = \frac{1}{4}\begin{pmatrix} 0 & 1 \\ 1 & 0 \end{pmatrix} \otimes \begin{pmatrix} 0 & \cdot 1 \\ 1 & 0 \end{pmatrix} = \frac{1}{4}\begin{pmatrix} 0 & 0 & 0 & 1 \\ 0 & 0 & 1 & 0 \\ 0 & 1 & 0 & 0 \\ 1 & 0 & 0 & 0 \end{pmatrix}$$

etc. Then the Hamilton operator $\hat{H}$ is given by

$$\hat{H} = \frac{J}{2}\begin{pmatrix} 1 & 0 & 0 & 0 \\ 0 & -1 & 2 & 0 \\ 0 & 2 & -1 & 0 \\ 0 & 0 & 0 & 1 \end{pmatrix} \,.$$

The eigenvalues and eigenvectors can now be calculated. We set

$$|\uparrow\rangle := \begin{pmatrix} 1 \\ 0 \end{pmatrix} \text{ spin up}, \quad |\downarrow\rangle := \begin{pmatrix} 0 \\ 1 \end{pmatrix} \text{ spin down} \,.$$

Then

$$| \uparrow\uparrow\rangle := | \uparrow\rangle \otimes | \uparrow\rangle, \quad | \uparrow\downarrow\rangle := | \uparrow\rangle \otimes | \downarrow\rangle,$$

$$| \downarrow\uparrow\rangle := | \downarrow\rangle \otimes | \uparrow\rangle, \quad | \downarrow\downarrow\rangle := | \downarrow\rangle \otimes | \downarrow\rangle.$$

Consequently,

$$| \uparrow\uparrow\rangle = \begin{pmatrix} 1 \\ 0 \\ 0 \\ 0 \end{pmatrix}, \quad | \uparrow\downarrow\rangle = \begin{pmatrix} 0 \\ 1 \\ 0 \\ 0 \end{pmatrix}, \quad | \downarrow\uparrow\rangle = \begin{pmatrix} 0 \\ 0 \\ 1 \\ 0 \end{pmatrix}, \quad | \downarrow\downarrow\rangle = \begin{pmatrix} 0 \\ 0 \\ 0 \\ 1 \end{pmatrix}.$$

One sees at once that $| \uparrow\uparrow\rangle$ and $| \downarrow\downarrow\rangle$ are eigenvectors of the Hamilton operator with eigenvalues $J/2$ and $J/2$, respectively. This means the eigenvalue $J/2$ is degenerate. The eigenvalues of the matrix

$$\frac{J}{2} \begin{pmatrix} -1 & 2 \\ 2 & -1 \end{pmatrix}$$

are given by $J/2$, $-3J/2$. The eigenvectors are linear combinations of $| \uparrow\downarrow\rangle$ and $| \downarrow\uparrow\rangle$, i.e.,

$$\frac{1}{\sqrt{2}}(| \uparrow\downarrow\rangle + | \downarrow\uparrow\rangle), \quad \frac{1}{\sqrt{2}}(| \uparrow\downarrow\rangle - | \downarrow\uparrow\rangle).$$

Problem 3.   Consider the spin Hamilton operator

$$\hat{H} = a \sum_{j=1}^{4} \sigma_3(j)\sigma_3(j+1) + b \sum_{j=1}^{4} \sigma_1(j) \tag{1}$$

with cyclic boundary conditions, i.e.,

$$\sigma_3(5) \equiv \sigma_3(1).$$

Here a, b are real constants and σ_1, σ_2 and σ_3 are the Pauli matrices.

(i) Calculate the matrix representation of $\hat{H}$. Recall that

$$\sigma_a(1) = \sigma_a \otimes I \otimes I \otimes I, \quad \sigma_a(2) = I \otimes \sigma_a \otimes I \otimes I,$$

$$\sigma_a(3) = I \otimes I \otimes \sigma_a \otimes I, \quad \sigma_a(4) = I \otimes I \otimes I \otimes \sigma_a. \tag{2}$$

(ii) Show that the Hamilton operator $\hat{H}$ admits the C_{4v} symmetry group.

(iii) Calculate the matrix representation of $\hat{H}$ for the subspace which belongs to the representation A_1.

(iv) Show that the time-evolution of any expectation value of an observable, Ω, can be expressed as the sum of a time-independent and a time-dependent term

$$\langle\Omega\rangle(t) = \sum_n |c_n(0)|^2 \langle n|\Omega|n\rangle + \sum_{\substack{m,n \\ n\neq m}} c_m^*(0)c_n(0)e^{i(E_m - E_n)t/\hbar}\langle m|\Omega|n\rangle \quad (3)$$

where $c_n(0)$ are the coefficients of expansion of the initial state in terms of the energy eigenstates $|n\rangle$ of $\hat{H}$, E_n are the eigenvalues of $\hat{H}$.

Solution.   (i) The Pauli matrices are given by

$$\sigma_1 := \begin{pmatrix} 0 & 1 \\ 1 & 0 \end{pmatrix}, \quad \sigma_2 := \begin{pmatrix} 0 & -i \\ i & 0 \end{pmatrix}, \quad \sigma_3 := \begin{pmatrix} 1 & 0 \\ 0 & -1 \end{pmatrix}.$$

The matrix representation of the first term of the right-hand side of (1) has been calculated in Problem 1. We find the diagonal matrix

$$\sum_{j=1}^{4} \sigma_3(j)\sigma_3(j+1) = \mathrm{diag}(4,0,0,0,0,-4,0,0,0,0,-4,0,0,0,0,4).$$

The second term leads to non-diagonal terms. Using (2), we find the symmetric 16×16 matrix for $\hat{H}$

$$\begin{pmatrix}
4a & b & b & 0 & b & 0 & 0 & 0 & b & 0 & 0 & 0 & 0 & 0 & 0 & 0 \\
b & 0 & 0 & b & 0 & b & 0 & 0 & 0 & b & 0 & 0 & 0 & 0 & 0 & 0 \\
b & 0 & 0 & b & 0 & 0 & b & 0 & 0 & 0 & b & 0 & 0 & 0 & 0 & 0 \\
0 & b & b & 0 & 0 & 0 & 0 & b & 0 & 0 & 0 & b & 0 & 0 & 0 & 0 \\
b & 0 & 0 & 0 & 0 & b & b & 0 & 0 & 0 & 0 & 0 & b & 0 & 0 & 0 \\
0 & b & 0 & 0 & b & -4a & 0 & b & 0 & 0 & 0 & 0 & 0 & b & 0 & 0 \\
0 & 0 & b & 0 & b & 0 & 0 & b & 0 & 0 & 0 & 0 & 0 & 0 & b & 0 \\
0 & 0 & 0 & b & 0 & b & b & 0 & 0 & 0 & 0 & 0 & 0 & 0 & 0 & b \\
b & 0 & 0 & 0 & 0 & 0 & 0 & 0 & 0 & b & b & 0 & b & 0 & 0 & 0 \\
0 & b & 0 & 0 & 0 & 0 & 0 & 0 & b & 0 & 0 & b & 0 & b & 0 & 0 \\
0 & 0 & b & 0 & 0 & 0 & 0 & 0 & b & 0 & -4a & b & 0 & 0 & b & 0 \\
0 & 0 & 0 & b & 0 & 0 & 0 & 0 & 0 & b & b & 0 & 0 & 0 & 0 & b \\
0 & 0 & 0 & 0 & b & 0 & 0 & 0 & b & 0 & 0 & 0 & 0 & b & b & 0 \\
0 & 0 & 0 & 0 & 0 & b & 0 & 0 & 0 & b & 0 & 0 & b & 0 & 0 & b \\
0 & 0 & 0 & 0 & 0 & 0 & b & 0 & 0 & 0 & b & 0 & b & 0 & 0 & b \\
0 & 0 & 0 & 0 & 0 & 0 & 0 & b & 0 & 0 & 0 & b & 0 & b & b & 4a
\end{pmatrix}.$$

(ii) To find the discrete symmetries of the Hamilton operator (1), we study the discrete coordinate transformations $\tau : \{1,2,3,4\} \rightarrow \{1,2,3,4\}$ i.e., permutations of the set $\{1,2,3,4\}$ which preserve $\hat{H}$. Any permutation preserves the second term of $\hat{H}$. The first term is preserved by cyclic permutations of 1234 or 4321. Therefore we find the following set of symmetries

$$E : (1,2,3,4) \rightarrow (1,2,3,4) , \quad C_2 : (1,2,3,4) \rightarrow (3,4,1,2) ,$$

$$C_4 : (1,2,3,4) \rightarrow (2,3,4,1) , \quad C_4^3 : (1,2,3,4) \rightarrow (4,1,2,3) ,$$

$$\sigma_v : (1,2,3,4) \rightarrow (2,1,4,3) , \quad \sigma_v' : (1,2,3,4) \rightarrow (4,3,2,1) ,$$

$$\sigma_d : (1,2,3,4) \rightarrow (1,4,3,2) , \quad \sigma_d' : (1,2,3,4) \rightarrow (3,2,1,4) ,$$

which form a group isomorphic to C_{4v}. The classes are given by

$$\{E\}, \quad \{C_2\}, \quad \{C_4, C_4^3\}, \quad \{\sigma_v, \sigma_v'\}, \quad \{\sigma_d, \sigma_d'\} .$$

The character table is as follows:

	$\{E\}$	$\{C_2\}$	$\{2C_4\}$	$\{2\sigma_v\}$	$\{2\sigma_d\}$
A_1	1	1	1	1	1
A_2	1	1	1	-1	-1
B_1	1	1	-1	1	-1
B_2	1	1	-1	-1	1
E	2	-2	0	0	0 .

$$(4)$$

(iii) From (4), it follows that the projection operator of the representation A_1 is given by

$$\Pi_1 = \frac{1}{8}(O_E + O_{C_2} + O_{C_4} + O_{C_4^3} + O_{\sigma_v} + O_{\sigma_v'} + O_{\sigma_d} + O_{\sigma_d'}) .$$

We recall that $O_p f(x) := f(P^{-1}x)$. We have $C_4^{-1} = C_4^3$, $(C_4^3)^{-1} = C_4$. The other elements of the groups are their own inverse. Thus for $\mathbf{e}_1, \ldots, \mathbf{e}_4 \in \mathbf{C}^2$ we have

$$8\Pi_1(\mathbf{e}_1 \otimes \mathbf{e}_2 \otimes \mathbf{e}_3 \otimes \mathbf{e}_4) = \mathbf{e}_1 \otimes \mathbf{e}_2 \otimes \mathbf{e}_3 \otimes \mathbf{e}_4 + \mathbf{e}_3 \otimes \mathbf{e}_4 \otimes \mathbf{e}_1 \otimes \mathbf{e}_2$$

$$+ \mathbf{e}_4 \otimes \mathbf{e}_1 \otimes \mathbf{e}_2 \otimes \mathbf{e}_3 + \mathbf{e}_2 \otimes \mathbf{e}_3 \otimes \mathbf{e}_4 \otimes \mathbf{e}_1$$

$$+ \mathbf{e}_2 \otimes \mathbf{e}_1 \otimes \mathbf{e}_4 \otimes \mathbf{e}_3 + \mathbf{e}_4 \otimes \mathbf{e}_3 \otimes \mathbf{e}_2 \otimes \mathbf{e}_1$$

$$+ \mathbf{e}_1 \otimes \mathbf{e}_4 \otimes \mathbf{e}_3 \otimes \mathbf{e}_2 + \mathbf{e}_3 \otimes \mathbf{e}_2 \otimes \mathbf{e}_1 \otimes \mathbf{e}_4 .$$

We set

$$|\uparrow\rangle := \begin{pmatrix} 1 \\ 0 \end{pmatrix} \text{ spin up} , \quad |\downarrow\rangle := \begin{pmatrix} 0 \\ 1 \end{pmatrix} \text{ spin down} .$$

These are the eigenfunctions of σ_z. As a basis for the total space we take all Kronecker products of the form $\mathbf{e}_1 \otimes \mathbf{e}_2 \otimes \mathbf{e}_3 \otimes \mathbf{e}_4$, where $\mathbf{e}_1$, $\mathbf{e}_2$, $\mathbf{e}_3$, $\mathbf{e}_4 \in \{|\uparrow\rangle, |\downarrow\rangle\}$. We define

$$|\uparrow\uparrow\uparrow\uparrow\rangle := |\uparrow\rangle \otimes |\uparrow\rangle \otimes |\uparrow\rangle \otimes |\uparrow\rangle, \quad |\downarrow\uparrow\uparrow\uparrow\rangle := |\downarrow\rangle \otimes |\uparrow\rangle \otimes |\uparrow\rangle \otimes |\uparrow\rangle$$

and so on. From the symmetry of Π_1 it follows that we only have to consider the projections of basis elements which are independent under the symmetry operations. We find the following basis for the subspace A_1

$$\Pi_1 |\uparrow\uparrow\uparrow\uparrow\rangle : |1\rangle = |\uparrow\uparrow\uparrow\uparrow\rangle,$$

$$\Pi_1 |\uparrow\uparrow\uparrow\downarrow\rangle : |2\rangle = \frac{1}{2}(|\uparrow\uparrow\uparrow\downarrow\rangle + |\uparrow\uparrow\downarrow\uparrow\rangle + |\uparrow\downarrow\uparrow\uparrow\rangle + |\downarrow\uparrow\uparrow\uparrow\rangle),$$

$$\Pi_1 |\uparrow\uparrow\downarrow\downarrow\rangle : |3\rangle = \frac{1}{2}(|\uparrow\uparrow\downarrow\downarrow\rangle + |\uparrow\downarrow\downarrow\uparrow\rangle + |\downarrow\downarrow\uparrow\uparrow\rangle + |\downarrow\uparrow\uparrow\downarrow\rangle),$$

$$\Pi_1 |\uparrow\downarrow\uparrow\downarrow\rangle : |4\rangle = \frac{1}{\sqrt{2}}(|\uparrow\downarrow\uparrow\downarrow\rangle + |\downarrow\uparrow\downarrow\uparrow\rangle),$$

$$\Pi_1 |\uparrow\downarrow\downarrow\downarrow\rangle : |5\rangle = \frac{1}{2}(|\uparrow\downarrow\downarrow\downarrow\rangle + |\downarrow\downarrow\downarrow\uparrow\rangle + |\downarrow\downarrow\uparrow\downarrow\rangle + |\downarrow\uparrow\downarrow\downarrow\rangle),$$

$$\Pi_1 |\downarrow\downarrow\downarrow\downarrow\rangle : |6\rangle = |\downarrow\downarrow\downarrow\downarrow\rangle.$$

Thus the subspace which belongs to A_1 is six-dimensional. Calculating $\langle j|\hat{H}|k\rangle$, where $j, k = 1, 2, \ldots, 6$, we find the 6×6 symmetric matrix

$$\begin{pmatrix} 4a & 2b & 0 & 0 & 0 & 0 \\ 2b & 0 & 2b & \sqrt{2}b & 0 & 0 \\ 0 & 2b & 0 & 0 & 2b & 0 \\ 0 & \sqrt{2}b & 0 & -4a & \sqrt{2}b & 0 \\ 0 & 0 & 2b & \sqrt{2}b & 0 & 2b \\ 0 & 0 & 0 & 0 & 2b & 4a \end{pmatrix}.$$

(iv) Let $|n\rangle$ be the eigenstates of $\hat{H}$ with eigenvalues E_n. Then we have

$$|\psi(0)\rangle = \sum_{n=1}^{16} c_n(0)|n\rangle.$$

From the Schrödinger equation

$$i\hbar \frac{\partial \psi}{\partial t} = \hat{H}\psi,$$

we obtain

$$|\psi(t)\rangle = e^{-i\hat{H}t/\hbar}|\psi(0)\rangle = \sum_{n=1}^{16} c_n(0)e^{-i\hat{H}t/\hbar}|n\rangle = \sum_{n=1}^{16} c_n(0)e^{-iE_nt/\hbar}|n\rangle \quad (5)$$

as $|n\rangle$ is, by assumption, an eigenstate (eigenvector) of $\hat{H}$ with eigenvalue E_n. Since the dual state of (5) is given by

$$\langle\psi(t)| = \sum_{m=1}^{16} c_m^*(0)e^{iE_mt/\hbar}\langle m|,$$

we obtain (3).

Chapter 25

Number Theory

Problem 1. (i) Show that

$$\frac{\sqrt{5}-1}{2} = \cfrac{1}{1 + \cfrac{1}{1 + \cfrac{1}{1 + \cfrac{1}{\cdots}}}} . \qquad (1)$$

(ii) The resistances of the resistors of the following network

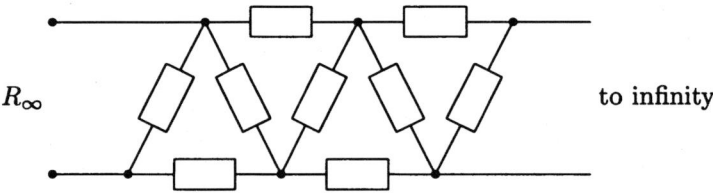

$$R_\infty \qquad\qquad\qquad \text{to infinity}$$

are all equal (say R). Calculate R_∞.

Remark. *The irrational number* $(\sqrt{5}-1)/2$ *is called the* golden mean number. *This is the* worst *irrational number* *due to the representation* (1).

Solution. (i) Since

$$\frac{\sqrt{5}-1}{2} \equiv \frac{(\sqrt{5}-1)(\sqrt{5}+1)}{2(\sqrt{5}+1)} \equiv \frac{4}{2(\sqrt{5}+1)} \equiv \frac{2}{\sqrt{5}+1} \equiv \frac{1}{\sqrt{5}+1/2},$$

we have

$$\frac{\sqrt{5}-1}{2} \equiv \frac{1}{\sqrt{5}+1/2}. \tag{2}$$

Now

$$\frac{\sqrt{5}+1}{2} \equiv \frac{2+\sqrt{5}-2+1}{2} \equiv 1 + \frac{\sqrt{5}-1}{2} \equiv 1 + \frac{(\sqrt{5}-1)(\sqrt{5}+1)}{2(\sqrt{5}+1)}$$

$$\equiv 1 + \frac{4}{2(\sqrt{5}+1)}. \tag{3a}$$

Thus

$$\frac{\sqrt{5}+1}{2} = 1 + \frac{1}{\sqrt{5}+1/2}. \tag{3b}$$

Comparing the right-hand side of (2) and (3b), the continued fraction (1) follows.

(ii) Method 1. The network

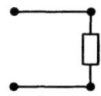

has resistance

$$R_1 = R.$$

The network

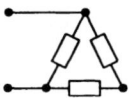

has resistance

$$R_2 = \frac{1}{1/R + 1/2R} \equiv \frac{1}{1+1/2}R. \tag{4}$$

The network

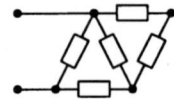

has the resistance

$$R_3 = \frac{1}{1/R} + \frac{1}{R + R_2}$$

where R_2 is given by (4). Obviously, we find the recursion relation

$$R_n = \frac{1}{1/R + 1/R + R_{n-1}}$$

where $(n = 2, 3, \ldots)$. Therefore

$$R_\infty = \frac{1}{1 + \cfrac{1}{1 + \cfrac{1}{1 + \cfrac{1}{\cdots}}}} R = \frac{(\sqrt{5} - 1)}{2} R.$$

Method 2. Since

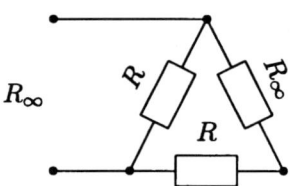

R_∞

we obtain

$$R_\infty = \frac{1}{1/R + 1/R + R_\infty}.$$

It follows that

$$R_\infty = \frac{1}{R_\infty + 2R/R(R + R_\infty)} = \frac{R(R + R_\infty)}{R_\infty + 2R}.$$

Therefore

$$R_\infty^2 + R_\infty R - R^2 = 0.$$

The solution of the quadratic equation is given by

$$R_\infty = \frac{(\sqrt{5}-1)}{2}R.$$

Problem 2.   (i) Show that

$$\sqrt{3}-1 \equiv \cfrac{1}{1+\cfrac{1}{2+\cfrac{1}{1+\cfrac{1}{2+\cfrac{1}{\cdots}}}}}.$$

(ii) The resistances of the resistors of the following network

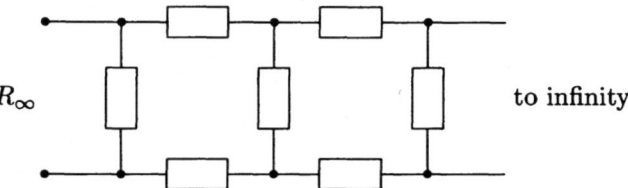

R_∞ to infinity

are all equal (say R). Calculate R_∞.

Solution.   (i) We have

$$\sqrt{3}-1 \equiv \frac{(\sqrt{3}-1)(\sqrt{3}+1)}{\sqrt{3}+1} \equiv \frac{2}{\sqrt{3}+1} = \frac{1}{\sqrt{3}+1/2}. \tag{1}$$

The denominator can be written as

$$\frac{\sqrt{3}+1}{2} \equiv \frac{1+(\sqrt{3}-1)+1}{2} \equiv 1 + \frac{\sqrt{3}-1}{2}.$$

Now

$$\frac{\sqrt{3}-1}{2} \equiv \frac{(\sqrt{3}-1)(\sqrt{3}+1)}{2\sqrt{3}+1} \equiv \frac{1}{\sqrt{3}+1}$$

and

$$\sqrt{3}+1 \equiv 2 + (\sqrt{3}-1). \tag{2}$$

Combining (1) through (2) gives

$$\sqrt{3} - 1 = \frac{1}{\sqrt{3} + 1/2} \equiv \frac{1}{1 + \sqrt{3} - 1/2}$$

$$= \frac{1}{1 + 1/\sqrt{3} + 1} = \frac{1}{1 + 1/2 + (\sqrt{3} - 1)} \, .$$

Thus

$$\sqrt{3} - 1 = \frac{1}{1 + 1/(2 + \sqrt{3} - 1)} \, .$$

(ii) The same methods as in Problem 1 can be used. We apply Method 2. Since

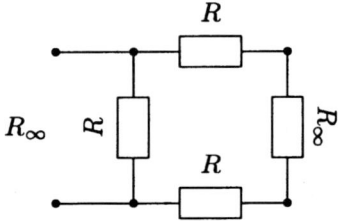

we have

$$R_\infty = \frac{1}{1/R + 1/2R + R_\infty} \equiv \frac{R(2R + R_\infty)}{3R + R_\infty} \, .$$

Consequently,

$$R_\infty^2 + 2RR_\infty - 2R^2 = 0 \, . \tag{3}$$

The solution of (3) is given by

$$R_\infty = (\sqrt{3} - 1)R \, .$$

Problem 3.   Let $x \in \mathbf{R}$. Let $N > 1$ be an integer. Show that there is an integer p and an integer q such that $1 \le q \le N$ and

$$\left| x - \frac{p}{q} \right| \le \frac{1}{qN} \, .$$

Solution.   The proof relies on the *pigeon hole principle* (also called the *Dirichlet box principle*). This principle says that if we put $N + 1$ objects into N boxes, then at least one of the boxes must contain more than one of the objects. We rewrite our statement as follows: There is an integer p and an integer q such that $1 \leq q \leq N$ and

$$|qx - p| \leq \frac{1}{N}.$$

We denote by $[x]$ the largest integer $\leq x$. For example $[\pi] = 3$. Consider the numbers

$$0 - 0 \cdot x, \quad x - [x], \quad 2x - [2x], \quad \ldots \quad Nx - [Nx].$$

This means we consider the numbers $kx - [kx]$ for $k = 0, 1, \ldots, N$. These $N + 1$ numbers will be our objects. They will lie between 0 and 1. We divide the interval $[0, 1]$ into N equal subintervals

$$\left[0, \frac{1}{N}\right), \quad \left[\frac{1}{N}, \frac{2}{N}\right), \quad \left[\frac{2}{N}, \frac{3}{N}\right), \cdots \left[\frac{(N-1)}{N}, 1\right].$$

These N subintervals are the boxes. There are N of them, and so two of the numbers $kx - [kx]$ $(0 \leq k \leq N)$ must lie in the same subinterval; that is, they can be no further apart than $1/N$. Say these two numbers are $nx - [nx]$ and $mx - [mx]$, where $0 \leq n < m \leq N$. Set $q = m - n$ and $p = [mx] - [nx]$. Then

$$|qx - p| = |(nx - [nx]) - (mx - [mx])| \leq \frac{1}{N}.$$

Moreover, $q = m - n \leq m \leq N$, and $q \geq 1$, since $m - n > 0$.

Problem 4. The *Farey sequence* F_N is the set of all fractions in lowest terms between 0 and 1 whose denominators do not exceed N, arranged in order of magnitude. For example, F_6 is given by

$$\frac{0}{1}, \quad \frac{1}{6}, \quad \frac{1}{5}, \quad \frac{1}{4}, \quad \frac{1}{3}, \quad \frac{2}{5}, \quad \frac{1}{2}, \quad \frac{3}{5}, \quad \frac{2}{3}, \quad \frac{3}{4}, \quad \frac{4}{5}, \quad \frac{5}{6}, \quad \frac{1}{1}.$$

N is known as the order of the series.

Let m_1/n_1 and m_2/n_2 be two successive terms in F_N. Then

$$m_2 n_1 - m_1 n_2 = 1. \tag{1}$$

For three successive terms m_1/n_1, m_2/n_2 and m_3/n_3, the middle term is the *mediant* of the other two

$$\frac{m_2}{n_2} = \frac{(m_1 + m_3)}{(n_1 + n_3)}. \tag{2}$$

(i) Find F_4.

(ii) Prove that (1) implies (2).

(iii) Prove that (2) implies (1).

Solution.   (i) We have

$$\frac{0}{1}, \quad \frac{1}{4}, \quad \frac{1}{3}, \quad \frac{1}{2}, \quad \frac{2}{3}, \quad \frac{3}{4}, \quad \frac{1}{1}.$$

(ii) To prove that (1) implies (2), assume we have two identities

$$m_2 n_1 - m_1 n_2 = 1, \quad m_3 n_2 - m_2 n_3 = 1.$$

Subtract one from the other and recombine the terms yields

$$(m_3 + m_1)n_2 = m_2(n_3 + n_1),$$

which leads to (2).

(iii) To prove that (2) implies (1) we use mathematical induction. Assume that we are given a series of fractions m_i/n_i ($i = 1, 2, \ldots$) with the condition that any three consecutive terms satisfy

$$\frac{m_i}{n_i} = \frac{m_{i-1} + m_{i+1}}{n_{i-1} + n_{i+1}}. \tag{3}$$

We show that for such a series (1) also holds. However (1) only contains two fractions. Therefore, let us assume that for $i = 2$, Eq. (1) indeed holds. This is clearly true for F_N which starts with $0/1$, $1/N$. Assume that, for some $i > 2$,

$$m_i n_{i-1} - m_{i-1} n_i = 1. \tag{4}$$

We rewrite (3) as

$$(m_{i+1} + m_{i-1})n_i = m_i(n_{i+1} + n_{i-1}). \tag{5}$$

Subtracting (5) from (4) gives

$$m_{i+1} n_i - m_i n_{i+1} = 1.$$

Equation (5) implies (3) only if $n_i \neq 0$.

Problem 5.   A symmetric ($g \times g$) matrix $B = (B_{jk})$ with negative definite real part $\Re B = (\Re B_{jk})$ is called a *Riemann matrix*. The *Riemann theta function*

$$\theta(z|B) := \sum_{N \in \mathbf{Z}^g} \exp\left(\frac{1}{2}\langle BN, N\rangle + \langle N, z\rangle\right). \tag{1}$$

Here $z = (z_1, \ldots, z_g) \in \mathbf{C}^g$ is a complex vector. The triangular brackets denote the Euclidean scalar product

$$\langle N, z \rangle := \sum_{i=1}^{g} N_i z_i, \quad \langle BN, N \rangle := \sum_{i,j=1}^{N} B_{ij} N_i N_j.$$

The summation in (1) is taken over the lattice of integer vectors $N = (N_1, \ldots, N_g)$. The general term depends only on the symmetric part of the matrix B. From the estimate

$$\Re\langle BN, N \rangle \leq -b\langle N, N \rangle, \quad b > 0$$

where $-b$ is the largest eigenvalue of the matrix $\Re B$. Thus the function is analytic in the whole space $\mathbf{C}^g$. Let $e_1, \ldots, e_g$ be the basis vectors in $\mathbf{C}^g$ with the coordinates $(e_k)_j = \delta_{kj}$. We also introduce vectors $f_1, \ldots, f_g$, setting

$$(f_k)_j = B_{kj}, \quad k, j = 1, \ldots, g.$$

The vector f_k can also be written in the form

$$f_k = Be_k.$$

(i) Show that

$$\theta(z + 2\pi i e_k) = \theta(z). \tag{2}$$

(ii) Show that

$$\theta(z + f_k) = \exp\left(-\frac{1}{2} B_{kk} - z_k\right) \theta(z). \tag{3}$$

Solution. (i) The periodicity of (2) is obvious. The general term of (1) does not change under the shift.

(ii) To prove (3), we change the summation index N in (1) setting $N = M - e_k$, $M \in \mathbf{Z}^g$. We have

$$\theta(z + f_k) = \sum_{N \in \mathbf{Z}^g} \exp\left(\frac{1}{2}\langle BN, N \rangle + \langle N, z + f_k \rangle\right).$$

Thus

$$\theta(z + f_k) = \sum_{M \in \mathbf{Z}^g} \exp\left(\frac{1}{2}\langle BM, M\rangle - \langle M, Be_k\rangle + \frac{1}{2}\langle Be_k, e_k\rangle\right.$$

$$\left. + \langle M, z\rangle + \langle M, f_k\rangle - \langle e_k, z\rangle - \langle e_k, f_k\rangle\right).$$

It follows that

$$\theta(z + f_k) = \exp\left(-\frac{1}{2}B_{kk} - z_k\right) \sum_{M \in \mathbf{Z}^g} \exp\left(\frac{1}{2}\langle BM, M\rangle + \langle M, z\rangle\right).$$

Finally

$$\theta(z + f_k) = \exp\left(-\frac{1}{2}B_{kk} - z_k\right)\theta(z).$$

This proves the assertion.

Problem 6.   The *star-triangle transformation* is given by

$$R_u = \frac{R_2 R_3}{R_1 + R_2 + R_3}, \quad R_1 = \frac{R_u R_v + R_v R_w + R_w R_u}{R_u},$$

$$R_v = \frac{R_1 R_3}{R_1 + R_2 + R_3}, \quad R_2 = \frac{R_u R_v + R_v R_w + R_w R_u}{R_v},$$

$$R_w = \frac{R_1 R_2}{R_1 + R_2 + R_3}, \quad R_3 = \frac{R_u R_v + R_v R_w + R_w R_u}{R_w}.$$

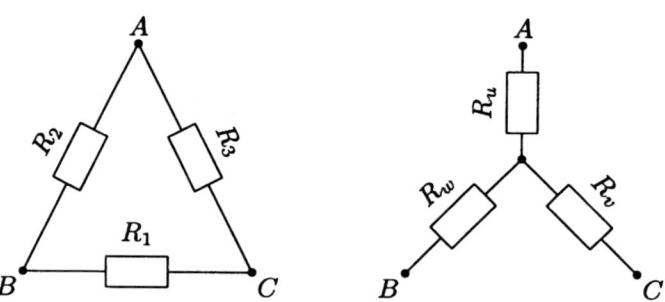

Fig. 25.1.   Start-triangle transformation.

Using the star-triangle transformation find the total resistance of the circuit given below between two opposite corners (say 1 and 7). This means the edges of a cube consist of equal resistors of resistance R, which are joined at the corners. Let a battery be connected to two opposite corners of a face of the cube (say 1 and 7). What is the resistance R_{17}?

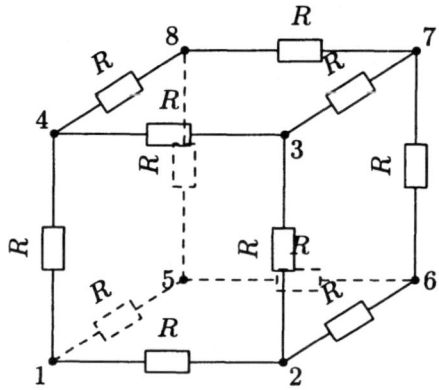

Solution. Drawing the circuit in the plane we have

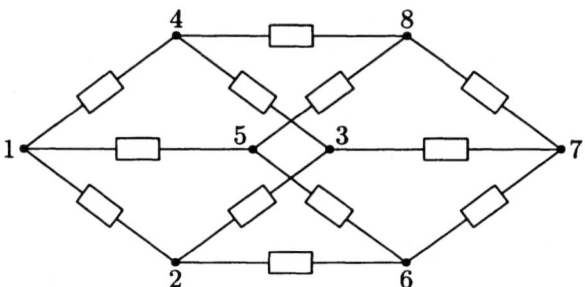

We solve the problem by successively applying the star-triangle transformation. With the star-triangle transformation we can simplify the circuit by using the standard rules for circuits in series

$$R_s = R_1 + R_2 + \cdots + R_n$$

and parallel

$$\frac{1}{R_s} = \frac{1}{R_1} + \frac{1}{R_2} + \cdots + \frac{1}{R_n}.$$

First we note from the star-triangle transformation that if

$$R_u = R_v = R_w = R,$$

we have

$$R_1 = R_2 = R_3 = 3R.$$

In the first step, we convert the star 1, 5, 8, 6 to a triangle. Thus the node 5 disappears. In the second step we convert the star 1, 4, 8, 3 to a triangle. Thus the node 4 disappears. We repeat the step of eliminating the stars. At each step we apply the standard rules for circuits in series and parallel given above.

Finally we arrive at

$$R_{17} = \frac{5}{6}R.$$

Remark.   *Since all the resistors are the same the problem can also be solved by symmetry considerations and Kirchhoff's laws. For example, if we want to find the resistance between 1 and 3, we could argue as follows. Conservation of the current at the corners requires*

$$I = 2I_{14} + I_{15}, \quad I_{15} = 2I_{56} = 2I_{58}$$

where I is the input current and $I_{14} = I_{12}$. The requirement that the voltage between 1 and 3 be independent of path yields the additional equation

$$2I_{14}R = 2(I_{15} + I_{58})R$$

where $I_{58} = I_{56}$. These three equations have the solution

$$I_{14} = \frac{3I}{8}, \quad I_{15} = \frac{I}{4}, \quad I_{58} = \frac{I}{8}.$$

Thus the resistance between 1 and 3 is $2I_{14}R/I$. Therefore

$$R_{13} = \frac{3}{4}R.$$

Chapter 26

Combinatorial Problems

Problem 1. (i) Consider N boxes and n particles with $n \leq N$. In every box we can put a maximum of one particle (*Pauli principle, Fermi particles*). In how many ways can we put the n particles (which are identical) in the N boxes?

(ii) Consider N boxes and an arbitrary number of particles n. We can put an arbitrary number of particles in every box (Bose-particles). Again the particles are identical. In how many ways can we put the n particles in the N boxes?

(iii) The same as case (ii) but now we have distinguishable particles.

Solution. (i) The problem is equivalent to that of determining the total number of ways in which N objects of which n are indistinguishable from type I (say particles) and $N - n$ are indistinguishable from type II (say empty spaces, holes) can be arranged in N possible places. The total number of permutations of N objects is $N!$. From each of these $N!$ arrangements we obtain $n!$ mutually indistinguishable arrangements by permutation of the $n!$ identical particles among themselves. From each of these $n!$ we obtain again $(N - n)!$ mutually indistinguishable arrangements by permutating the $(N - n)!$ identical holes among themselves. The $N!$ possible arrangements can therefore be partitioned into disjoint sets, each containing

$$n!(N - n)!$$

indistinguishable arrangements. The total number of distinguishable arrangements is therefore

$$\frac{N!}{n!(N-n)!} \equiv \binom{N}{n}.$$

Remark.   *We have the recursion relation*

$$\binom{N+1}{n+1} = \binom{N}{n} + \binom{N}{n+1}$$

where

$$\binom{1}{1} = 1.$$

(ii) We select a box to begin with. The number of ways we can choose the
$N-1$ boxes and the n-particles are

$$(N-1+n)!.$$

However the boxes and particles are indistinguishable. Thus we find

$$\binom{n+N-1}{n} \equiv \frac{(n+N-1)!}{n!(n+N-1-n)!} \equiv \frac{(n+N-1)!}{n!(N-1)!}.$$

(iii) For distinguishable particles the first particle can be put in any of the
N boxes, the second one in any of N boxes etc. Thus we find the desired
value to be

$$N^n.$$

Problem 2.   A young man, who lives at location A of the city street
plan shown in the figure, walks daily to the home of his fiancée, who lives
m blocks east and n blocks north of A, at location B. He can only walk
north or east. In how many different ways can he go from A to B? Consider
first the case $n = m = 2$.

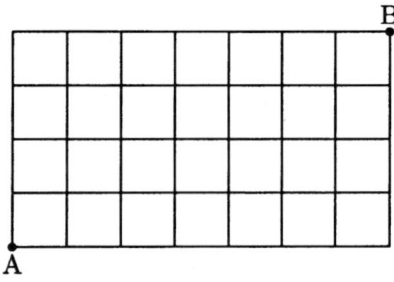

Solution. We consider first the case $n = m = 2$. Each of the paths can be written as vector

$$(u, u, r, r) \quad (u, r, u, r) \quad (u, r, r, u) \quad (r, u, u, r) \quad (r, u, r, u) \quad (r, r, u, u)$$

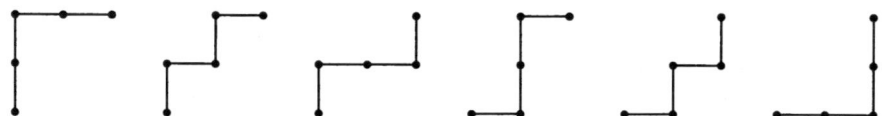

where u stands for one step north ("up") and r stands for one step east ("right"). Thus the number of paths for $n = m = 2$ is given by 6.

The general case is as follows: At each corner the man may walk either up or to the right. Consequently, a particular path is specified by a sequence of the form

$$(u, u, r, u, \ldots, r)$$

where the total number of $u's$ and $r's$ are n and m, respectively. The number of different ways of writing such a sequence and therefore the total number of paths is given by

$$\frac{(m + n)!}{n!\, m!}.$$

Problem 3. (i) Let $\mathbf{Z}$ be the set of integers. Then $\mathbf{Z}^d$ $(d = 1, 2, 3, \ldots)$ denotes the d-tuple of the set of integers, i.e.

$$\mathbf{Z}^d := \mathbf{Z} \times \mathbf{Z} \times \cdots \times \mathbf{Z} \quad d - \text{times}.$$

Find the number of all closed paths A_n with n-steps starting from $(0, 0, \ldots, 0) \in \mathbf{Z}^d$ and returning to $(0, 0, \ldots, 0) \in \mathbf{Z}^d$ (not necessarily the first time).

(ii) Calculate A_4 for $d = 2$.

Solution. (i) Obviously A_n is given by

$$A_n = \sum_{\{\boldsymbol{\Delta}_1, \boldsymbol{\Delta}_2, \ldots, \boldsymbol{\Delta}_n\}} \delta(\boldsymbol{\Delta}_1 + \boldsymbol{\Delta}_2 + \cdots + \boldsymbol{\Delta}_n)$$

where

$$\delta(\boldsymbol{\Delta}_1 + \boldsymbol{\Delta}_2 + \cdots + \boldsymbol{\Delta}_n) := \begin{cases} 1 & \text{for } \boldsymbol{\Delta}_1 + \boldsymbol{\Delta}_2 + \cdots + \boldsymbol{\Delta}_n = \mathbf{0}, \\ 0 & \text{otherwise}. \end{cases}$$

If n is odd we find $A_n = 0$. Now we apply the *completeness relation*

$$\delta(\Delta_1 + \Delta_2 + \cdots + \Delta_n) := \frac{1}{N} \sum_{k \in 1.BZ} e^{ik(\Delta_1 + \Delta_2 + \cdots + \Delta_n)}$$

where N denotes the lattice sites in the first Brillouin zone (1. BZ). It follows that

$$A_n = \sum_{\{\Delta_1, \ldots, \Delta_n\}} \frac{1}{N} \sum_{k \in 1.BZ} e^{ik(\Delta_1 + \cdots + \Delta_n)}$$

$$= \frac{1}{N} \sum_{k \in 1.BZ} \sum_{\{\Delta_1, \ldots, \Delta_n\}} e^{ik(\Delta_1 + \cdots + \Delta_n)} .$$

Thus we can write

$$A_n = \frac{1}{N} \sum_{k \in 1.BZ} \sum_{\{\Delta_1\}} e^{ik\Delta_1} \sum_{\{\Delta_2\}} e^{ik\Delta_2} \cdots \sum_{\{\Delta_n\}} e^{ik\Delta_n} = \frac{1}{N} \sum_{k \in 1.BZ} (\epsilon(k))^n$$

where

$$\epsilon(k) := \sum_{\Delta} e^{ik\Delta} .$$

(ii) In two dimensions ($d = 2$), we have

$$\epsilon(k) = 2(\cos k_1 + \cos k_2)$$

since $\Delta \in \{(1,0), (0,1), (-1,0), (0,-1)\}$. Thus we find

$$A_4 = \frac{1}{N} \sum_{k \in 1.BZ} (\epsilon(k))^4 = \frac{2^4}{4\pi^2} \int_{-\pi}^{\pi} \int_{-\pi}^{\pi} (\cos k_1 + \cos k_2)^4 dk_1 dk_2 = 36 .$$

Remark. *The first Brillouin zone in three dimensions is constructed as follows: Let a_1, a_2 and a_3 be a set of primitive vectors for the direct lattice. Then the reciprocal lattice can be generated by the three primitive vectors*

$$b_1 = 2\pi \frac{a_2 \times a_3}{a_1 \cdot (a_2 \times a_3)}, \quad b_2 = 2\pi \frac{a_3 \times a_1}{a_1 \cdot (a_2 \times a_3)}, \quad b_3 = 2\pi \frac{a_1 \times a_2}{a_1 \cdot (a_2 \times a_3)} .$$

The simple cubic Bravais lattice, with cubic primitive cell of side a, has as its reciprocal a simple cubic lattice with cubic primitive cell of side $2\pi/a$. The construction of the first Brillouin zone is as follows: choose any reciprocal lattice point as the origin. Draw the vectors connecting this point with (all) other lattice points. Next, construct a set of planes that are perpendicular bisectors of these vectors. The smallest solid figure containing the *origin is the first Brillouin zone. Points* **k** *on the surface must satisfy the condition* $\mathbf{k}^2 = (\mathbf{k} - \mathbf{K}_n)^2$ *or* $\mathbf{K}_n^2 - 2\mathbf{k} \cdot \mathbf{K}_n = 0$ *for some reciprocal lattice vector* $\mathbf{K}_n$. *We need only consider those* **k** *values lying within the zone. Any* **k** *vector may be written as*

$$\mathbf{k} = \left(\frac{h_1}{(2N_1 + 1)}\right) \mathbf{b}_1 + \left(\frac{h_2}{(2N_2 + 1)}\right) \mathbf{b}_2 + \left(\frac{h_3}{(2N_3 + 1)}\right) \mathbf{b}_3$$

where the h_i *are integers in the range* $-N_i \leq h_i \leq N_i$. *The values of* **k** *that are given by the above formula may not all be within the first Brillouin zone. However, it is always possible to bring any such* **k** *outside the zone back into it by translation by a reciprocal lattice vector. The values of* **k** *form a uniform and dense distribution. In the limit in which the* N_i *are allowed to become infinite, we may convert sums over possible* **k** *values into integrals over the first Brillouin zone through the relation*

$$\sum_{\mathbf{k} \in 1.BZ} \rightarrow \frac{N}{V} \int_{1.BZ} d\mathbf{k}$$

where $N = (2N_1 + 1)(2N_2 + 1)(2N_3 + 1)$ *is the number of unit cells and* V *is the volume of the first Brillouin zone.*

Problem 4. Find the probability $P_N(\mathbf{R})$ that a random walker reaches a lattice position $\mathbf{R}$,

$$\mathbf{R} = N_1 \mathbf{a}_1 + N_2 \mathbf{a}_2 + N_3 \mathbf{a}_3$$

after

$$N = N_1 + N_2 + N_3$$

steps on a lattice with unit vectors $\mathbf{a}_i$. The lattice is the simple cubic lattice. The lattice is periodic with a periodicity L. The transition probability $W(\mathbf{R})$ for a unit displacement is given by

$$W(\mathbf{R}) := \begin{cases} \dfrac{1}{6} & \text{for } \mathbf{R} = (0, 0, \pm 1), (0, \pm 1, 0), (\pm 1, 0, 0), \\ 0 & \text{otherwise}. \end{cases} \tag{1}$$

Solution. The probability satisfies the recurrence formula

$$P_{N+1}(\mathbf{R}) = \sum_{\mathbf{R}'} W(\mathbf{R} - \mathbf{R}') P_N(\mathbf{R}')$$

where $W(\mathbf{R} - \mathbf{R}')$ is the probability of transition from $\mathbf{R}'$ to $\mathbf{R}$. We introduce a *generating function*

$$\Phi(\mathbf{R}, z) := \sum_{N=0}^{\infty} z^N P_N(\mathbf{R}) \,.$$

Obviously

$$\Phi(\mathbf{R}, z) - z \sum_{\mathbf{R}'} W(\mathbf{R} - \mathbf{R}')\Phi(\mathbf{R}', z) = \delta_{\mathbf{R}, 0} \tag{2}$$

which expresses the fact that the walker starts walking from the origin

$$P_0(\mathbf{R}) = \delta_{\mathbf{R}, 0} = \begin{cases} 1, & \mathbf{R} = 0 \,, \\ 0, & \text{otherwise} \,, \end{cases}$$

where δ is the Kronecker delta. We solve (2) by using the Fourier transform method. Multiplying both sides by

$$\exp\left(\frac{2\pi i \mathbf{k} \cdot \mathbf{R}}{L}\right)$$

and summing over $\mathbf{R}$, we find

$$\tilde{\Phi}(\mathbf{k}, z) - z\lambda(\mathbf{k})\tilde{\Phi}(\mathbf{k}, z) = 1$$

with

$$\tilde{\Phi}(\mathbf{k}, z) := \sum_{\mathbf{R}} \exp\left(\frac{2\pi i \mathbf{k} \cdot \mathbf{R}}{L}\right) \Phi(\mathbf{R}, z) \,,$$

$$\lambda(\mathbf{k}) := \sum_{\mathbf{R}} W(\mathbf{R}) \exp\left(\frac{(2\pi i \mathbf{k} \cdot \mathbf{R})}{L}\right) \,.$$

The Fourier inverse of $\tilde{\Phi}(\mathbf{k}, z)$ is

$$\Phi(\mathbf{R}, z) = \frac{1}{L^3} \sum_{\mathbf{k}} \frac{\exp(-2\pi i \mathbf{k} \cdot \mathbf{R}/L)}{1 - z\lambda(\mathbf{k})}$$

which, in the limit of $L \to \infty$, gives

$$\lim_{L \to \infty} \Phi(\mathbf{R}, z) = \frac{1}{(2\pi)^3} \int_{-\pi}^{\pi} \int_{-\pi}^{\pi} \int_{-\pi}^{\pi} \frac{\exp(-i\boldsymbol{\Theta} \cdot \mathbf{R})}{1 - z\lambda(\boldsymbol{\Theta})} d\boldsymbol{\Theta}$$

where $d\boldsymbol{\Theta} \equiv d\Theta_1 d\Theta_2 d\Theta_3$ and

$$\boldsymbol{\Theta} \cdot \mathbf{R} := \Theta_1 R_1 + \Theta_2 R_2 + \Theta_3 R_3 \,.$$

The probability $P_N(\mathbf{R})$ is given by

$$P_N(\mathbf{R}) = \frac{1}{(2\pi)^3} \int_{-\pi}^{\pi} \int_{-\pi}^{\pi} \int_{-\pi}^{\pi} (\lambda(\Theta))^N \exp(-i\Theta \cdot \mathbf{R}) d\Theta .$$

Owing to (1) we find

$$\lambda(\Theta) = \frac{1}{6}(e^{i\Theta_1} + e^{-i\Theta_1} + e^{i\Theta_2} + e^{-i\Theta_2} + e^{i\Theta_3} + e^{-i\Theta_3})$$

$$= \frac{1}{3}(\cos\Theta_1 + \cos\Theta_2 + \cos\Theta_3) .$$

Thus,

$$P_N(\mathbf{R}) = \frac{1}{(2\pi)^3} \int_{-\pi}^{\pi} \int_{-\pi}^{\pi} \int_{-\pi}^{\pi} \left(\frac{1}{3}(\cos\Theta_1 + \cos\Theta_2 + \cos\Theta_3) \right)^N$$
$$\times \exp(-i\Theta \cdot \mathbf{R}) d\Theta$$

and

$$\Phi(\mathbf{R}, z) = \frac{1}{(2\pi)^3} \int_{-\pi}^{\pi} \int_{-\pi}^{\pi} \int_{-\pi}^{\pi} \frac{\exp(-i\Theta \cdot \mathbf{R})}{1 - 1/3z(\cos\Theta_1 + \cos\Theta_2 + \cos\Theta_3)} d\Theta .$$

Problem 5. Consider a lattice with N lattice sites. Each lattice site can only be occupied by two electrons one with spin up and one with spin down (Pauli principle). Let N_e be the number of electrons we put in the lattice. Obviously, $0 \leq N_e \leq 2N$.

(i) Given N and N_e, find the number of ways to occupy the lattice with electrons.

(ii) Consider the case $N_e = N$.

(iii) Consider the case $N_e = N$ (N = even) and $S_z = 0$, where S_z denotes the total spin.

Solution. (i) Let a be the number of lattice sites occupied by one electron with spin up, b the number of lattice sites occupied by one electron with spin down, c the number of lattice sites occupied by two electrons (one spin up, one spin down) and d the unoccupied lattice sites. Obviously we have the conditions

$$a + b + 2c = N_e , \quad a + b + c + d = N . \tag{1}$$

Thus the number of ways to occupy the lattice is

$$\frac{N!}{a!\,b!\,c!\,d!}$$

with the constraints given by (1).

(ii) If $N = N_e$ we have $d = c$ and thus

$$\frac{N!}{a!\,b!\,c!\,c!}\,.$$

(iii) If $N = N_e$ and if the total spin S_z is equal to 0 ($N =$ even) we find $a = b$ and therefore $a = N/2 - c$. Thus

$$\frac{N!}{(N/2 - c)!(N/2 - c)!\,c!\,c!}\,.$$

Consequently the total number of states with $N = N_e$ and $S_z = 0$ is given by

$$\sum_{c=0}^{N/2} \frac{N!}{(N/2 - c)!(N/2 - c)!\,c!\,c!}\,.$$

For example, for $N = N_e = 4$ and $S_z = 0$ we find 36.

Problem 6.   Consider the problem of randomly placing r balls into n cells. The balls are indistinguishable. Let r_k be the number of balls in the k-th cell. Every n-tuple of integers satisfying

$$r_1 + r_2 + \cdots + r_n = r \tag{1}$$

describes a possible configuration of occupancy numbers. With indistinguishable balls two distributions are distinguishable only if the corresponding n-tuples are not identical. (i) Show that the number of distinguishable distributions (i.e., the number of different solutions of (1)) is

$$A_{r,n} = \binom{n + r - 1}{r} \equiv \frac{(n + r - 1)!}{r!(n - 1)!}\,.$$

(ii) Show that the number of distinguishable distributions, in which no cell remains empty, is

$$\binom{r - 1}{n - 1}\,.$$

Solution.   (i) We represent the balls by stars and indicate the n cells by the n spaces between $n + 1$ bars. Thus, for example,

$$| * * * | * \, \| \, \| * * * * |$$

is used as a symbol for a distribution of $r = 8$ balls in $n = 6$ cells with occupancy number 3, 1, 0, 0, 0, 4. Such a symbol necessarily starts and ends with a bar, but the remaining $n - 1$ bars and r stars can appear in an arbitrary order. Thus the number of distinguishable distributions equals the number of ways of selecting r places out of $n + r - 1$, namely $A_{r,n}$.

(ii) The condition that no cell be empty imposes the restriction that no two bars are adjacent and $r \geq n$. The r stars leave $r - 1$ spaces of which $n - 1$ are to be occupied by bars. Thus we have

$$\binom{r-1}{n-1} \equiv \frac{(r-1)!}{(n-1)!(r-n)!} \, .$$

choices.

Chapter 27

Fermi Operators

Problem 1. For *Fermi operators* we have to take into account the *Pauli principle*. Let $c_j^\dagger$ be Fermi creation operators and let c_j be Fermi annihilation operators, where $j = 1, \ldots, N$, with j denoting the quantum numbers (spin, wave vector, angular momentum, lattice site etc.). Then we have

$$[c_i^\dagger, c_j]_+ \equiv c_i^\dagger c_j + c_j c_i^\dagger = \delta_{ij} I \,,$$

$$[c_i, c_j]_+ = [c_i^\dagger, c_j^\dagger]_+ = 0 \tag{1}$$

where δ_{ij} denotes the Kronecker delta. A consequence of (1) is

$$c_j^\dagger c_j^\dagger = 0$$

which describes the Pauli principle, i.e., two particles cannot be in the same state. The states are given by

$$|n_1, n_2, \ldots, n_N\rangle := (c_1^\dagger)^{n_1} (c_2^\dagger)^{n_2} \cdots (c_N^\dagger)^{n_N} |0, \ldots, 0, \ldots, 0\rangle$$

where, because of the Pauli principle,

$$n_1, n_2, \ldots, n_N \in \{0, 1\} \,.$$

We obtain

$$c_i |n_1, \ldots, n_{i-1}, 0, n_{i+1}, \ldots, n_N\rangle = 0$$

and

$$c_i^\dagger |n_1, \ldots, n_{i-1}, 1, n_{i+1}, \ldots, n_N\rangle = 0 \,.$$

In the following we write

$$|0\rangle \equiv |0, \ldots, 0\rangle .$$

(i) Let $N = 1$. Find the matrix representation of

$$c^\dagger , \quad c, \quad \hat{n}, \quad |0\rangle , \quad c^\dagger |0\rangle$$

where

$$\hat{n} := c^\dagger c$$

is the *number operator*.

(ii) Find the matrix representations for arbitrary N.

(iii) In some cases it is convenient to consider the Fermi operator

$$\{c_{k\sigma}^\dagger, c_{k\sigma}, k = 1, 2, \ldots, N; \sigma \in \{\uparrow, \downarrow\}\}$$

where k denotes the wave vector (or lattice site) and σ denotes the spin. Find the matrix representation.

Solution. (i) A basis is given by

$$\{c^\dagger |0\rangle, |0\rangle\} .$$

The dual basis is

$$\{\langle 0|c, \langle 0|\} .$$

We obtain

$$c^\dagger \to \begin{pmatrix} \langle 0|cc^\dagger c^\dagger |0\rangle & \langle 0|cc^\dagger |0\rangle \\ \langle 0|c^\dagger c^\dagger |0\rangle & \langle 0|c^\dagger |0\rangle \end{pmatrix} .$$

Notice that

$$c|0\rangle = 0 \quad \Leftrightarrow \quad \langle 0|c^\dagger = 0 .$$

Therefore

$$c^\dagger \to \begin{pmatrix} 0 & 1 \\ 0 & 0 \end{pmatrix} = \frac{1}{2}\sigma_+ = \frac{1}{2}(\sigma_x + i\sigma_y) .$$

In an analogous manner we find

$$c \to \begin{pmatrix} 0 & 0 \\ 1 & 0 \end{pmatrix} = \frac{1}{2}\sigma_- \equiv \frac{1}{2}(\sigma_x - i\sigma_y)$$

and

$$c^\dagger c = \hat{n} \rightarrow \begin{pmatrix} 1 & 0 \\ 0 & 0 \end{pmatrix}.$$

The eigenvalues of $\hat{n}$ are $\{0, 1\}$. Next we give the matrix representations of $|0\rangle$ and $c^\dagger|0\rangle$. We find

$$|0\rangle \rightarrow \begin{pmatrix} \langle 0|c|0\rangle \\ \langle 0|0\rangle \end{pmatrix} = \begin{pmatrix} 0 \\ 1 \end{pmatrix}, \quad c^\dagger|0\rangle \rightarrow \begin{pmatrix} \langle 0|cc^\dagger|0\rangle \\ \langle 0|c^\dagger|0\rangle \end{pmatrix} = \begin{pmatrix} 1 \\ 0 \end{pmatrix}.$$

As an example we have

$$c|0\rangle = 0 \Leftrightarrow \begin{pmatrix} 0 & 0 \\ 1 & 0 \end{pmatrix} \begin{pmatrix} 0 \\ 1 \end{pmatrix} = \begin{pmatrix} 0 \\ 0 \end{pmatrix},$$

$$c^\dagger|0\rangle \Leftrightarrow \begin{pmatrix} 0 & 1 \\ 0 & 0 \end{pmatrix} \begin{pmatrix} 0 \\ 1 \end{pmatrix} = \begin{pmatrix} 1 \\ 0 \end{pmatrix}.$$

(ii) For arbitrary N we have

$$c_k^\dagger \rightarrow \overbrace{\sigma_z \otimes \cdots \otimes \sigma_z \otimes \left(\frac{1}{2}\sigma_+\right) \otimes I \otimes \cdots \otimes I}^{N\times} \tag{2}$$

where $\sigma_+/2$ is at the kth place and

$$c_k \rightarrow \overbrace{\sigma_z \otimes \cdots \otimes \sigma_z \otimes \left(\frac{1}{2}\sigma_-\right) \otimes I \otimes \cdots \otimes I}^{N\times} \tag{3}$$

where $\sigma_-/2$ is at the kth place. Recall that

$$[c_k^\dagger, c_q]_+ = \delta_{kq}, \tag{4a}$$

$$[c_k^\dagger, c_q^\dagger]_+ = [c_k, c_q]_+ = 0. \tag{4b}$$

The commutation relations (4) are satisfied by the faithful representations (3) and (2). For $N = 1$ the state $|0\rangle$ is given by

$$|0\rangle \rightarrow \begin{pmatrix} 0 \\ 1 \end{pmatrix}.$$

For arbitrary N the state $|0\rangle \equiv |0, \ldots, 0\rangle$ is given by

$$|0\rangle = \underbrace{\begin{pmatrix} 0 \\ 1 \end{pmatrix} \otimes \begin{pmatrix} 0 \\ 1 \end{pmatrix} \otimes \begin{pmatrix} 0 \\ 1 \end{pmatrix} \otimes \cdots \otimes \begin{pmatrix} 0 \\ 1 \end{pmatrix}}_{N\times}$$

where $|0\rangle$ is called the *vacuum state*. The operator $\hat{n}_k$ takes the form

$$\hat{n}_k := c_k^\dagger c_k \rightarrow I \otimes I \otimes \cdots I \otimes \left(\frac{1}{4}\sigma_+ \sigma_- \right) \otimes I \otimes \cdots \otimes I \,.$$

(iii) We have the commutation relation

$$[c_{k\sigma}^\dagger, c_{q\sigma'}]_+ = \delta_{kq}\delta_{\sigma\sigma'} \,,$$

$$[c_{k\sigma}, c_{q\sigma'}]_+ = [c_{k\sigma}^\dagger, c_{q\sigma'}^\dagger]_+ = 0 \,.$$

The matrix representation is given by

$$c_{k\uparrow}^\dagger \rightarrow \overbrace{\sigma_z \otimes \sigma_z \otimes \cdots \otimes \sigma_z \otimes \left(\frac{1}{2}\sigma_+ \right) \otimes I \otimes \cdots \otimes I}^{2N\times}$$

where $\sigma_+/2$ is at the kth place and

$$c_{k\downarrow}^\dagger \rightarrow \overbrace{\sigma_z \otimes \sigma_z \otimes \cdots \otimes \sigma_z \otimes \left(\frac{1}{2}\sigma_+ \right) \otimes I \otimes \cdots \otimes I}^{2N\times}$$

where $\sigma_+/2$ is at the $(k + N)$th place.

Problem 2. (i) Let $c^\dagger$ be a Fermi creation operator and c a Fermi annihilation operator. Let $\epsilon \in \mathbf{R}$. Show that

$$\exp(\epsilon c^\dagger c)c^\dagger \exp(-\epsilon c^\dagger c) = \exp(\epsilon)c^\dagger$$

$$\exp(\epsilon c^\dagger c)c \exp(-\epsilon c^\dagger c) = \exp(-\epsilon)c \,.$$

(ii) Let

$$\hat{n} := c^\dagger c \,.$$

Show that

$$\exp(-\epsilon \hat{n}) \equiv 1 + \hat{n}(e^{-\epsilon} - 1) \,. \tag{1}$$

(iii) Let

$$\hat{n}_\uparrow \hat{n}_\downarrow := c_\uparrow^\dagger c_\uparrow c_\downarrow^\dagger c_\downarrow \,.$$

Show that

$$\exp(-\epsilon \hat{n}_\downarrow \hat{n}_\uparrow) \equiv 1 + \hat{n}_\uparrow \hat{n}_\downarrow (e^{-\epsilon} - 1) \,. \tag{2}$$

Solution.   (i) We set

$$f(\epsilon) := \exp(\epsilon c^\dagger c) c^\dagger \exp(-\epsilon c^\dagger c)$$

and therefore

$$f(0) = c^\dagger \,. \tag{3}$$

Differentiating both sides with respect to ϵ yields

$$\frac{df}{d\epsilon} = \exp(\epsilon c^\dagger c) c^\dagger c c^\dagger \exp(-\epsilon c^\dagger c) - \exp(\epsilon c^\dagger c) c^\dagger c^\dagger c \exp(-\epsilon c^\dagger c) \,.$$

Since $c^\dagger c^\dagger = 0$ and $cc^\dagger = 1 - c^\dagger c$, we obtain

$$\frac{df}{d\epsilon} = \exp(\epsilon c^\dagger c) c^\dagger \exp(-\epsilon c^\dagger c) = f$$

or

$$\frac{df}{d\epsilon} = f \,. \tag{4}$$

The solution of the linear differential Eq. (4) with the initial condition (3) gives

$$f(\epsilon) = c^\dagger e^\epsilon \,.$$

Analogously, we can prove the second identity.

(ii) We find the identity

$$\hat{n}^2 \equiv c^\dagger c c^\dagger c \equiv c^\dagger (1 - c^\dagger c) c \equiv c^\dagger c \equiv \hat{n} \tag{5}$$

since $cc = 0$. Using identity (5) and

$$\exp(-\epsilon \hat{n}) := \sum_{k=0}^{\infty} \frac{(-\epsilon)^k \hat{n}^k}{k!} \,,$$

we find identity (1).

(iii) We have the identity

$$(\hat{n}_\uparrow \hat{n}_\downarrow)^2 \equiv \hat{n}_\uparrow \hat{n}_\downarrow \hat{n}_\uparrow \hat{n}_\downarrow \equiv \hat{n}_\uparrow \hat{n}_\uparrow \hat{n}_\downarrow \hat{n}_\downarrow \equiv \hat{n}_\uparrow \hat{n}_\downarrow \,. \tag{6}$$

Using identity (6) and

$$\exp(-\epsilon \hat{n}_\uparrow \hat{n}_\downarrow) := \sum_{k=0}^{\infty} \frac{(-\epsilon)^k (\hat{n}_\uparrow \hat{n}_\downarrow)^k}{k!} \,,$$

we find identity (2).

Problem 3. Let $c_{j\uparrow}^\dagger$, $c_{j\uparrow}$, $c_{j\downarrow}^\dagger$, $c_{j\downarrow}$ be Fermi operators, i.e.,

$$[c_{j\sigma}^\dagger, c_{k\sigma'}]_+ = \delta_{jk} \delta_{\sigma,\sigma'} I \,,$$

$$[c_{j\sigma}^\dagger, c_{k\sigma'}^\dagger]_+ = 0 \,, \quad [c_{j\sigma}, c_{k\sigma'}]_+ = 0$$

for all $j, k = 1, 2, \ldots, N$. We define the linear operators

$$\xi_{j\uparrow} := i(c_{j\uparrow} - c_{j\uparrow}^\dagger) \,, \quad \xi_{j\downarrow} := i(c_{j\downarrow} - c_{j\downarrow}^\dagger)(I - 2c_{j\uparrow}^\dagger c_{j\uparrow}) \,,$$

$$\eta_{j\uparrow} := c_{j\uparrow} + c_{j\uparrow}^\dagger \,, \quad \eta_{j\downarrow} := (c_{j\downarrow} + c_{j\downarrow}^\dagger)(I - 2c_{j\uparrow}^\dagger c_{j\uparrow}) \,.$$

(i) Find the properties of these operators.

(ii) Find the inverse transformation.

Solution. (i) By straightforward calculations using the anticommutation relations we find that the operators $\xi_{j\sigma}$ are unitary and self-adjoint and therefore involutions

$$\xi_{j\sigma} = \xi_{j\sigma}^\dagger = \xi_{j\sigma}^{-1} \,, \quad \xi_{j\sigma}^2 = I$$

for all $j = 1, 2, \ldots, N$. The linear operators $\eta_{j\sigma}$ also satisfy

$$\eta_{j\sigma} = \eta_{j\sigma}^\dagger = \eta_{j\sigma}^{-1} \,, \quad \eta_{j\sigma}^2 = I \,,$$

for all $j = 1, 2, \ldots, N$.

(ii) The inverse transformation between $c_{j\sigma}^\dagger$, $c_{j\sigma}$ and $\xi_{j\sigma}$, $\eta_{j\sigma}$ is given by

$$c_{j\uparrow} = \frac{1}{2}(\eta_{j\uparrow} - i\xi_{j\uparrow}) \,, \quad c_{j\uparrow}^\dagger = \frac{1}{2}(\eta_{j\uparrow} + i\xi_{j\uparrow}) \,,$$

$$c_{j\downarrow} = \frac{1}{2}(i\eta_{j\downarrow} + \xi_{j\downarrow})\eta_{j\uparrow}\xi_{j\uparrow} \,, \quad c_{j\downarrow}^\dagger = \frac{1}{2}(i\eta_{j\downarrow} - \xi_{j\downarrow})\eta_{j\uparrow}\xi_{j\uparrow}$$

where we have used that

$$i\eta_{j\uparrow}\xi_{j\uparrow} = I - 2c_{j\uparrow}^\dagger c_{j\uparrow}.$$

Problem 4.   Let $c_{j\sigma}^\dagger$, $c_{j\sigma}$ $(j = 1, \ldots, N)$ be Fermi creation and annihilation operators, where $\sigma \in \{\uparrow, \downarrow\}$. We define

$$\hat{n}_{j\uparrow} := c_{j\uparrow}^\dagger c_{j\uparrow},$$

$$\hat{n}_{j\downarrow} := c_{j\downarrow}^\dagger c_{j\downarrow}.$$

(i) Show that the operators

$$S_{jx} := \frac{1}{2}(c_{j\uparrow}^\dagger c_{j\downarrow} + c_{j\downarrow}^\dagger c_{j\uparrow}),$$

$$S_{jy} := \frac{1}{2i}(c_{j\uparrow}^\dagger c_{j\downarrow} - c_{j\downarrow}^\dagger c_{j\uparrow}),$$

$$S_{jz} := \frac{1}{2}(\hat{n}_{j\uparrow} - \hat{n}_{j\downarrow})$$

form a basis of a Lie algebra under the commutator.

(ii) Show that the operators

$$R_{jx} := \frac{1}{2}(c_{j\uparrow}^\dagger c_{j\downarrow}^\dagger + c_{j\downarrow} c_{j\uparrow}),$$

$$R_{jy} := \frac{1}{2i}(c_{j\uparrow}^\dagger c_{j\downarrow}^\dagger - c_{j\downarrow} c_{j\uparrow}),$$

$$R_{jz} := \frac{1}{2}(\hat{n}_{j\uparrow} + \hat{n}_{j\downarrow} - 1)$$

form a Lie algebra under the commutator.

(iii) Prove the identity

$$\hat{n}_{j\uparrow}\hat{n}_{j\downarrow} \equiv \frac{1}{4}(1 - \alpha_j) + R_{jz} + \frac{1}{3}(\alpha_j - 1)\mathbf{S}_j^2 + \frac{1}{3}(\alpha_j + 1)\mathbf{R}_j^2 \qquad (1)$$

where

$$\mathbf{S}_j^2 := S_{jx}^2 + S_{jy}^2 + S_{jz}^2$$

and

$$\mathbf{R}_j^2 := R_{jx}^2 + R_{jy}^2 + R_{jz}^2.$$

Solution.   (i) Since

$$[c_{j\sigma}^\dagger, c_{j\sigma'}]_+ = \delta_{\sigma\sigma'}\,,$$

$$[c_{j\sigma}^\dagger, c_{j\sigma'}^\dagger]_+ = 0\,,$$

$$[c_{j\sigma}, c_{j\sigma'}]_+ = 0\,,$$

we obtain

$$[S_{jx}, S_{jy}] = iS_{jz}\,,$$

$$[S_{jy}, S_{jz}] = iS_{jx}\,,$$

$$[S_{jz}, S_{jx}] = iS_{jy}\,.$$

Thus the operators S_{jx}, S_{jy}, S_{jz} form a basis of a Lie algebra.

(ii) Analogously, we obtain

$$[R_{jx}, R_{jy}] = iR_{jz}\,,$$

$$[R_{jy}, R_{jz}] = iR_{jx}\,,$$

$$[R_{jz}, R_{jx}] = iR_{jy}\,.$$

Thus the operators R_{jx}, R_{jy}, R_{jz} form a Lie algebra.

Remark.   *The two Lie algebras are isomorphic.*

(iii) We find

$$\mathbf{S}_j^2 = \frac{3}{4}(\hat{n}_{j\uparrow} + \hat{n}_{j\downarrow} - 2\hat{n}_{j\uparrow}\hat{n}_{j\downarrow})$$

and

$$\mathbf{R}_j^2 = \frac{3}{2}\hat{n}_{j\uparrow}\hat{n}_{j\downarrow} - \frac{3}{4}(\hat{n}_{j\uparrow} + \hat{n}_{j\downarrow}) + \frac{3}{4}\,.$$

Thus identity (1) follows.

Problem 5.   Let

$$\hat{H} = \hat{H}_0 + \hat{H}_1$$

be a Hamilton operator describing a many-body Fermi or Bose system. The *grand thermodynamic potential* is given by

$$\Omega := -\frac{1}{\beta}\ln\operatorname{tr}\exp(-\beta(\hat{H} - \mu\hat{N})) = -\frac{1}{\beta}\ln Z$$

where μ is the chemical potential, $\hat{N}$ is the number operator and Z is the grand thermodynamic partition function. Assume that the grand thermodynamic potential for the unperturbed Hamilton operator $\hat{H}_0$

$$\Omega_0 := -\frac{1}{\beta} \ln \operatorname{tr} \exp(-\beta(\hat{H}_0 - \mu\hat{N}))$$

can be calculated.

(i) Let

$$\exp(-\beta(\hat{H} - \mu\hat{N})) \equiv (\exp(-\beta(\hat{H}_0 - \mu\hat{N})))S(\beta).$$

Find $S(\beta)$.

(ii) Calculate

$$Z = \operatorname{tr}(\exp(-\beta(\hat{H}_0 - \mu\hat{N}))S(\beta)). \tag{1}$$

(iii) From (i) we find

$$S(\beta) = 1 + \sum_{n=1}^{\infty}(-1)^n \int_0^\beta d\tau_1 \int_0^{\tau_1} d\tau_2 \cdots \int_0^{\tau_{n-1}} d\tau_n \hat{H}_1(\tau_1)\hat{H}_1(\tau_2)\cdots\hat{H}_1(\tau_n) \tag{2}$$

where

$$\hat{H}_1(\beta) := \exp(\beta(\hat{H}_0 - \mu\hat{N}))\hat{H}_1\exp(-\beta(\hat{H}_0 - \mu\hat{N}))$$

(the so-called *interaction picture*). It can be shown that $S(\beta)$ can be written as

$$S(\beta) = \sum_{n=0}^{\infty}\frac{(-1)^n}{n!}\int_0^\beta d\tau_1 \int_0^\beta d\tau_2 \cdots \int_0^\beta d\tau_n T_\tau[\hat{H}_1(\tau_1)\hat{H}_1(\tau_2)\cdots\hat{H}_1(\tau_n)]$$

where T_τ is the *time-ordering operator*. We have

$$T_\tau[A(\tau_1)B(\tau_2)] := \begin{cases} A(\tau_1)B(\tau_2), & \text{if } \tau_2 < \tau_1, \\ B(\tau_2)A(\tau_1), & \text{if } \tau_1 < \tau_2. \end{cases}$$

We set

$$\left\langle T_\tau \exp\left(-\int_0^\beta \hat{H}_1(\tau)d\tau\right)\right\rangle_0 \equiv \exp\left(\left\langle T_\tau \exp\left(-\int_0^\beta \hat{H}_1(\tau)d\tau\right) - 1\right\rangle_c\right). \tag{3}$$

Express $\langle\cdots\rangle_c$ in terms of $\langle\cdots\rangle_0$.

Remark. *This is the so-called* cumulant expansion. *The expansions given by (1) and (2) cannot be used for calculating* Ω. *The cumulant expansion must be used.*

Solution. (i) We set

$$\exp(-\beta(\hat{H} - \mu\hat{N})) = (\exp(-\beta(\hat{H}_0 - \mu\hat{N})))S(\beta) =: \phi(\beta). \qquad (4)$$

Taking the derivative of (4) with respect to β gives

$$\frac{\partial \phi}{\partial \beta} = -(\hat{H} - \mu\hat{N})\phi.$$

Therefore

$$\frac{\partial S}{\partial \beta} = -\hat{H}_1(\beta)S. \qquad (5)$$

Owing to (4) we have the "initial condition"

$$S(\beta = 0) \equiv S(0) = 1.$$

The integration of (5) with the initial condition gives

$$S(\beta) = 1 - \int_0^\beta \hat{H}_1(\tau)S(\tau)d\tau.$$

By iterating this equation we arrive at

$$S(\beta) = 1 - \int_0^\beta \hat{H}_1(\tau)d\tau + \int_0^\beta d\tau_1 \int_0^{\tau_1} d\tau_2 \hat{H}_1(\tau_1)\hat{H}_1(\tau_2) + \cdots$$

$$+ (-1)^n \int_0^\beta d\tau_1 \int_0^{\tau_1} d\tau_2 \cdots \int_0^{\tau_{n-1}} d\tau_n \hat{H}_1(\tau_1)\hat{H}_1(\tau_2) \cdots \hat{H}_1(\tau_n) + \cdots.$$

(ii) For the grand thermodynamical partition function Z we obtain

$$Z = \text{tr}(\exp(-\beta(\hat{H}_0 - \mu\hat{N}))S(\beta))$$

$$\equiv \frac{[\text{tr}\exp(-\beta(\hat{H}_0 - \mu\hat{N}))][\text{tr}(\exp(-\beta(\hat{H}_0 - \mu\hat{N}))S(\beta))]}{\text{tr}\exp(-\beta(\hat{H}_0 - \mu\hat{N}))}. \qquad (6)$$

Therefore we find

$$Z = \exp(-\beta\Omega_0)\left(1 - \int_0^\beta \langle\hat{H}_1(\tau)\rangle_0 d\tau\right.$$

$$\left. + \int_0^\beta d\tau_1 \int_0^{\tau_1} d\tau_2 \langle\hat{H}_1(\tau_1)\hat{H}_1(\tau_2)\rangle_0 + \cdots\right) \qquad (7)$$

where

$$\langle \cdots \rangle_0 := \frac{\mathrm{tr} \cdots e^{-\beta(\hat{H}_0 - \mu\hat{N})}}{\mathrm{tr} e^{-\beta(\hat{H}_0 - \mu\hat{N})}} \,.$$

Taking the logarithm of both sides of (7), we find

$$\Omega = \Omega_0 - \frac{1}{\beta} \ln \left(1 - \int_0^\beta \langle \hat{H}_1(\tau) \rangle_0 d\tau + \int_0^\beta d\tau_1 \int_0^{\tau_1} d\tau_2 \langle \hat{H}_1(\tau_1) \hat{H}(\tau_2) \rangle_0 - \cdots \right) .$$

(iii) Taking the logarithm of both sides of (3) with $\hat{H}_1 \to \lambda \hat{H}_1$ it follows that we can set

$$\phi(\lambda) = \sum_{n=1}^{\infty} \frac{(-1)^n \lambda^n}{n!} \int_0^\beta d\tau_1 \cdots \int_0^\beta d\tau_n \langle T_\tau \hat{H}_1(\tau_1) \cdots \hat{H}_1(\tau_n) \rangle_c \qquad (8a)$$

and

$$\phi(\lambda) = \ln \left\langle T_\tau \exp \left(-\lambda \int_0^\beta \hat{H}_1(\tau) d\tau \right) \right\rangle_0 \qquad (8b)$$

where λ is a real parameter. Equation (8b) can be written as

$$\phi(\lambda) = \lim_{\alpha \to 0} \ln \left\langle T_\tau \exp \left((\lambda + \alpha) \left(-\int_0^\beta \hat{H}_1(\tau) d\tau \right) \right) \right\rangle_0 \,.$$

Since

$$\exp \left(\lambda \frac{\partial}{\partial \alpha} \right) f(\alpha) \equiv f(\alpha + \lambda) \,,$$

we find

$$\phi(\lambda) = \lim_{\alpha \to 0} \exp \left(\lambda \frac{\partial}{\partial \alpha} \right) \ln \left\langle T_\tau \exp \left(\alpha \left(-\int_0^\beta \hat{H}_1(\tau) d\tau \right) \right) \right\rangle_0 \,.$$

It follows that

$$\phi(\lambda) = \lim_{\alpha \to 0} \sum_{n=1}^{\infty} \frac{\lambda^n}{n!} \left(\frac{\partial}{\partial \alpha} \right)^n \ln \left\langle T_\tau \exp \left(\alpha \left(-\int_0^\beta \hat{H}_1(\tau) d\tau \right) \right) \right\rangle_0 \,. \qquad (9)$$

Comparing powers of λ^n in (8a) and (9), we can express $\langle \cdots \rangle_c$ in terms of $\langle \cdots \rangle_0$. For the first two terms we find

$$\int_0^\beta d\tau \langle \hat{H}_1(\tau) \rangle_c = \int_0^\beta d\tau \langle \hat{H}_1(\tau) \rangle_0 \,,$$

$$\int_0^\beta d\tau_1 \int_0^\beta d\tau_2 \langle T_\tau \hat{H}_1(\tau_1) \hat{H}_1(\tau_2) \rangle_c = \int_0^\beta d\tau_1 \int_0^\beta d\tau_2 \langle T_\tau \hat{H}_1(\tau_1) \hat{H}_1(\tau_2) \rangle_0$$

$$- \left(\int_0^\beta d\tau \langle \hat{H}_1(\tau) \rangle_0 \right)^2.$$

Problem 6. Consider the Hamilton operator (two-point *Hubbard model*)

$$\hat{H} = t(c_{1\uparrow}^\dagger c_{2\uparrow} + c_{1\downarrow}^\dagger c_{2\downarrow} + c_{2\uparrow}^\dagger c_{1\uparrow} + c_{2\downarrow}^\dagger c_{1\downarrow}) + U(n_{1\uparrow} n_{1\downarrow} + n_{2\uparrow} n_{2\downarrow}) \quad (1)$$

where $n_{j\uparrow} := c_{j\uparrow}^\dagger c_{j\uparrow}$, $n_{j\downarrow} := c_{j\downarrow}^\dagger c_{j\downarrow}$. The operators $c_{j\uparrow}^\dagger$, $c_{j\downarrow}^\dagger$, $c_{j\uparrow}$, $c_{j\downarrow}$ are Fermi operators.

(i) Show that the Hubbard Hamilton operator (1) commutes with the total number operator $\hat{N}$ and the total spin operator $\hat{S}_z$, where

$$\hat{N} := \sum_{j=1}^2 (c_{j\uparrow}^\dagger c_{j\uparrow} + c_{j\downarrow}^\dagger c_{j\downarrow}), \quad \hat{S}_z := \frac{1}{2} \sum_{j=1}^2 (c_{j\uparrow}^\dagger c_{j\uparrow} - c_{j\downarrow}^\dagger c_{j\downarrow}).$$

(ii) We consider the subspace with two particles $N = 2$ and total spin $S_z = 0$. A basis in this space is given by

$$c_{1\uparrow}^\dagger c_{1\downarrow}^\dagger |0\rangle, \quad c_{1\uparrow}^\dagger c_{2\downarrow}^\dagger |0\rangle, \quad c_{2\uparrow}^\dagger c_{1\downarrow}^\dagger |0\rangle, \quad c_{2\uparrow}^\dagger c_{2\downarrow}^\dagger |0\rangle.$$

Find the matrix representation of $\hat{H}$ for this basis.

(iii) Find the discrete symmetries of $\hat{H}$ and perform a group-theoretical reduction.

Solution. (i) Using the Fermi anti-commutation relations we obtain $[\hat{H}, \hat{N}] = 0$, $[\hat{H}, \hat{S}_z] = 0$. We also have $[\hat{N}, \hat{S}_z] = 0$.

(ii) Using the Fermi anti-commutation relations and $c_{j\uparrow}|0\rangle = 0$, $c_{j\downarrow}|0\rangle = 0$ we obtain the matrix representation

$$\begin{pmatrix} U & t & t & 0 \\ t & 0 & 0 & t \\ t & 0 & 0 & t \\ 0 & t & t & U \end{pmatrix}.$$

(iii) The Hamilton operator (1) admits the symmetry $1 \to 2$, $2 \to 1$, i.e. swapping the sites 1 and 2 leaves the Hamilton operator invariant. Thus we have a finite group and two elements (identity and the swapping of the sites). There are two conjugacy classes and therefore two irreducible representations. We find the two invariant subspaces

$$\left\{ \frac{1}{\sqrt{2}}(c_{1\downarrow}^\dagger c_{1\uparrow}^\dagger|0\rangle + c_{2\downarrow}^\dagger c_{2\uparrow}^\dagger|0\rangle), \quad \frac{1}{\sqrt{2}}(c_{1\downarrow}^\dagger c_{2\uparrow}^\dagger|0\rangle + c_{2\downarrow}^\dagger c_{1\uparrow}^\dagger|0\rangle) \right\},$$

$$\left\{ \frac{1}{\sqrt{2}}(c_{1\downarrow}^\dagger c_{1\uparrow}^\dagger|0\rangle - c_{2\downarrow}^\dagger c_{2\uparrow}^\dagger|0\rangle), \quad \frac{1}{\sqrt{2}}(c_{1\downarrow}^\dagger c_{2\uparrow}^\dagger|0\rangle - c_{2\downarrow}^\dagger c_{1\uparrow}^\dagger|0\rangle) \right\}.$$

Chapter 28

Bose Operators

Problem 1.   Let b be a Bose annihilation operator. Solve the eigenvalue problem

$$b|\beta\rangle = \beta|\beta\rangle. \tag{1}$$

Hint.   Use the *number representation* $|n\rangle$, where

$$|n\rangle := \frac{(b^\dagger)^n}{\sqrt{n!}}|0\rangle$$

and $n = 0, 1, 2, \ldots$. Therefore

$$b|n\rangle = \sqrt{n}|n-1\rangle, \quad b^\dagger|n\rangle = \sqrt{n+1}|n+1\rangle. \tag{2}$$

Remark.   *The state $|\beta\rangle$ is called a Bose coherent state. The commutation relation for the Bose operators are given by*

$$[b, b^\dagger] = I, \quad [b, b] = [b^\dagger, b^\dagger] = 0$$

where I is the identity operator.

Solution.   To solve the eigenvalue problem we apply the completeness relation of the number representation. This means we expand $|\beta\rangle$ with respect to $|n\rangle$. We obtain

$$|\beta\rangle = \sum_{n=0}^{\infty} |n\rangle\langle n|\beta\rangle \equiv \sum_{n=0}^{\infty} c_n(\beta)|n\rangle \tag{3}$$

291

where $c_n(\beta) := \langle n|\beta \rangle$. If we insert (3) into (1) and use (2) we obtain

$$\sum_{n=1}^{\infty} c_n(\beta)\sqrt{n}|n-1\rangle = \sum_{n=0}^{\infty} \beta c_n(\beta)|n\rangle \,.$$

The first sum goes from 1 to ∞ since the term $n = 0$ vanishes. We can therefore shift indices and put $n \to n+1$. It follows that

$$\sum_{n=0}^{\infty} c_{n+1}(\beta)\sqrt{n+1}|n\rangle = \sum_{n=0}^{\infty} \beta c_n(\beta)|n\rangle \,.$$

If we apply the dual state $\langle m|$ and use

$$\langle m|n\rangle = \delta_{nm}$$

we obtain the linear difference equation

$$c_{n+1}(\beta)\sqrt{n+1} = \beta c_n(\beta) \,.$$

On inspection we see that

$$c_1 = \frac{\beta}{\sqrt{1}}c_0 \,, \quad c_2 = \frac{\beta}{\sqrt{2}}c_1 = \frac{\beta^2}{\sqrt{2!}}c_0 \,, \quad c_3 = \frac{\beta^3}{\sqrt{3!}}c_0 \,, \quad \cdots \quad c_n(\beta) = \frac{\beta^n}{\sqrt{n!}}c_0 \,.$$

Therefore

$$|\beta\rangle = c_0 \sum_{n=0}^{\infty} \frac{\beta^n}{\sqrt{n!}}|n\rangle \,.$$

We normalize $|\beta\rangle$ to determine c_0. From the condition $\langle\beta|\beta\rangle = 1$ we obtain

$$1 = |c_0|^2 e^{|\beta|^2}$$

where from (1) we have

$$\langle\beta|b^\dagger = \langle\beta|\bar{\beta} \,.$$

It follows that

$$|\beta\rangle = e^{-|\beta|^2/2} \sum_{n=0}^{\infty} \frac{\beta^n}{\sqrt{n!}}|n\rangle \equiv e^{-|\beta|^2/2} e^{\beta b^\dagger}|0\rangle$$

where $\beta \in \mathbf{C}$.

Problem 2. Let $b^\dagger$, b be Bose creation and Bose annihilation operators. Define

$$K_0 := \frac{1}{4}(b^\dagger b + bb^\dagger), \quad K_+ := \frac{1}{2}(b^\dagger)^2, \quad K_- := \frac{1}{2}b^2.$$

(i) Show that K_0, K_+, K_- form a basis of a Lie algebra.

(ii) Define

$$L_0 := b_1^\dagger b_1 + b_2 b_2^\dagger, \quad L_+ := b_1^\dagger b_2^\dagger, \quad L_- := b_1 b_2.$$

Show that L_0, L_+, L_- form a basis of a Lie algebra.

(iii) Are the two Lie algebras isomorphic?

(iv) Do the operators

$$\{I, b, b^\dagger, b^\dagger b\}$$

form a Lie algebra under the commutator? Here I denotes the identity operator.

(v) Calculate

$$\mathrm{tr}(b^\dagger b \exp(-\epsilon b^\dagger b))$$

where $\epsilon \in \mathbf{R}$ ($\epsilon > 0$) and $b^\dagger b$ is the infinite-dimensional diagonal matrix

$$b^\dagger b = \mathrm{diag}\,(0, 1, 2, 3, \ldots).$$

(vi) Calculate

$$\exp(-\epsilon b^\dagger)b \exp(\epsilon b^\dagger).$$

(vii) Calculate

$$\exp(-\epsilon b^\dagger b)b \exp(\epsilon b^\dagger b).$$

Solution. (i) Using the commutation relations for Bose operators

$$[b, b^\dagger] = I,$$

$$[b, b] = [b^\dagger, b^\dagger] = 0,$$

we obtain

$$[K_0, K_+] = K_+, \quad [K_0, K_-] = -K_-, \quad [K_-, K_+] = 2K_0. \tag{1}$$

Thus the set $\{K_0, K_+, K_-\}$ forms a basis of a Lie algebra.

(ii) Using the commutation relations for Bose operators

$$[b_j, b_k^\dagger] = \delta_{jk} \,,$$

$$[b_j, b_k] = [b_j^\dagger, b_k^\dagger] = 0 \,,$$

we find

$$[L_0, L_+] = 2L_+ \,, \quad [L_0, L_-] = -2L_- \,, \quad [L_-, L_+] = L_0 \,. \tag{2}$$

(iii) Two Lie algebras, L_1 and L_2, are called *isomorphic* if there is a map $\phi : L_1 \to L_2$ such that

$$\phi([a, b]) = [\phi(a), \phi(b)]$$

where $a, b \in L_1$. Let

$$\phi(K_+) = L_+ \,, \quad \phi(K_-) = L_- \,, \quad \phi(K_0) = \frac{1}{2} L_0 \,.$$

Then we find from (1) and (2) that the Lie algebras are isomorphic.

(iv) Since

$$[I, b] = 0 \,, \quad [I, b^\dagger] = 0 \,, \quad [I, b^\dagger b] = 0 \,,$$

$$[b, b^\dagger] = I \,, \quad [b, b^\dagger b] = b \,, \quad [b^\dagger, b^\dagger b] = -b^\dagger \,,$$

we find that the set of operators $\{I, b, b^\dagger, b^\dagger b\}$ forms a basis of a Lie algebra.

(v) Since

$$\exp(-\epsilon b^\dagger b) = \mathrm{diag}(1, e^{-\epsilon}, e^{-2\epsilon}, \ldots)$$

and

$$b^\dagger b \exp(-\epsilon b^\dagger b) = \mathrm{diag}(0, e^{-\epsilon}, 2e^{-2\epsilon}, 3e^{-3\epsilon}, \ldots) \,,$$

we obtain

$$\mathrm{tr}(b^\dagger b \exp(-\epsilon b^\dagger b)) = \sum_{n=1}^{\infty} n e^{-\epsilon n} \,.$$

Now

$$f(\epsilon) := \sum_{n=1}^{\infty} e^{-\epsilon n} \equiv \frac{1}{e^\epsilon - 1} \quad \text{(geometric series)} \,.$$

The derivative of f yields

$$\frac{df}{d\epsilon} = -\sum_{n=1}^{\infty} n e^{-\epsilon n} = \frac{d}{d\epsilon} \frac{1}{e^\epsilon - 1} = \frac{-e^\epsilon}{(e^\epsilon - 1)^2}.$$

Consequently,

$$\text{tr}(b^\dagger b \exp(-\epsilon b^\dagger b)) = \frac{e^\epsilon}{(e^\epsilon - 1)^2}.$$

(vi) We set

$$f(\epsilon) := \exp(-\epsilon b^\dagger) b \exp(\epsilon b^\dagger).$$

We seek the ordinary differential equation for $f(\epsilon)$, where $f(0) = b$. Taking the derivative with respect to ϵ yields

$$\frac{df}{d\epsilon} = e^{-\epsilon b^\dagger}(-b^\dagger) b e^{\epsilon b^\dagger} + e^{-\epsilon b^\dagger} b b^\dagger e^{\epsilon b^\dagger} = e^{-\epsilon b^\dagger}(-b^\dagger b + b b^\dagger) e^{\epsilon b^\dagger} = I \quad (3)$$

where I is the identity operator. From (3) and the initial condition $f(0) = b$ we find

$$f(\epsilon) = \exp(-\epsilon b^\dagger) b \exp(\epsilon b^\dagger) = b + \epsilon I.$$

(vii) We set

$$g(\epsilon) := \exp(-\epsilon b^\dagger b) b \exp(\epsilon b^\dagger b).$$

We seek the ordinary differential equation for $g(\epsilon)$, where $g(0) = b$. Taking the derivative with respect to ϵ yields

$$\frac{dg}{d\epsilon} = e^{-\epsilon b^\dagger b}(-b^\dagger b) b e^{\epsilon b^\dagger b} + e^{-\epsilon b^\dagger b} b (b^\dagger b) e^{\epsilon b^\dagger b}$$

$$= e^{-\epsilon b^\dagger b}(-b^\dagger b b + b b^\dagger b) e^{\epsilon b^\dagger b} = e^{-\epsilon b^\dagger b} b e^{\epsilon b^\dagger b} = g(\epsilon). \quad (4)$$

From the linear differential Eq. (4) and the initial condition $g(0) = b$ we obtain

$$g(\epsilon) = \exp(-\epsilon b^\dagger b) b \exp(\epsilon b^\dagger b) = b \exp(\epsilon).$$

Problem 3.   Let

$$Q := (b - \epsilon(b + b^\dagger)^2) \otimes c^\dagger \quad (1)$$

be a linear operator, where b is a Bose annihilation operator, $c^\dagger$ is a Fermi creation operator and ϵ is a real parameter.

(i) Show that

$$Q^2 = 0.$$

(ii) We define the Hamilton operator as

$$\hat{H}_\epsilon := [Q, Q^\dagger]_+ \equiv QQ^\dagger + Q^\dagger Q. \tag{2}$$

Find $\hat{H}_\epsilon$.   Calculate $[\hat{H}_\epsilon, Q]$ and $[\hat{H}_\epsilon, Q^\dagger Q]$, where $[,]$ denotes the commutator.

(iii) Give a basis of the underlying Hilbert space.

(iv) Can the Hilbert space be decomposed into subspaces?

Remark.   *The so constructed $\hat{H}_\epsilon$ is called a supersymmetric Hamilton operator.*

Solution.   (i) Since

$$c^\dagger c^\dagger = 0,$$

we find that

$$Q^2 = 0.$$

(ii) From (1) we obtain

$$Q^\dagger = (b^\dagger - \epsilon(b^\dagger + b)^2) \otimes c.$$

Applying $[b, b^\dagger] = I_B$ and $[c, c^\dagger]_+ = I_F$, we arrive at

$$\hat{H}_\epsilon = [Q, Q^\dagger]_+ = QQ^\dagger + Q^\dagger Q$$

$$= I_B \otimes c^\dagger c + b^\dagger b \otimes I_F - 4\epsilon(b + b^\dagger) \otimes c^\dagger c$$

$$+ \epsilon^2(b^4 + b^{\dagger 4} + 6b^2 + 6b^{\dagger 2} + 4b^{\dagger 3}b + 4b^\dagger b^3 + 6b^{\dagger 2}b^2 + 12b^\dagger b + 3) \otimes I_F$$

$$- \epsilon(b^3 + b^{\dagger 3} + b + b^\dagger + 3b^{\dagger 2}b + 3b^\dagger b^2) \otimes I_F$$

where I_B is the identity operator in the Hilbert space $\mathcal{H}_B$ of the Bose operators and I_F is the identity operator in the Hilbert space $\mathcal{H}_F$ of the Fermi operators. Straightforward calculation yields

$$[\hat{H}_\epsilon, Q] = 0,$$

$$[\hat{H}_\epsilon, Q^\dagger Q] = 0.$$

Thus the three operators $\hat{H}_\epsilon$, Q, $Q^\dagger Q$ may be diagonalized simultaneously.

(iii) Let

$$|n\rangle := \frac{(b^\dagger)^n}{\sqrt{n!}}|0\rangle$$

be the number states, where $n = 0, 1, 2, \ldots$. Then a basis in the Hilbert space $\mathcal{H}_B \otimes \mathcal{H}_F$ is given by

$$\{|n\rangle \otimes |0\rangle_F, |n\rangle \otimes c^\dagger|0\rangle_F\}$$

where $n = 0, 1, 2, \ldots$.

(iv) Since

$$c^\dagger c|0\rangle_F = 0\,,$$

$$c^\dagger c c^\dagger|0\rangle_F = c^\dagger|0\rangle_F\,,$$

we find that the Hilbert space $\mathcal{H}_B \otimes \mathcal{H}_F$ decomposes, under the Hamilton operator (2), into two subspaces with the bases

$$\{|n\rangle \otimes |0\rangle\}$$

and

$$\{|n\rangle \otimes c^\dagger|0\rangle\}$$

where $n = 0, 1, 2, \ldots$.

Problem 4. Let

$$\hat{H} = -\Delta I_B \otimes \sigma_z + \frac{k}{2}(b^\dagger \otimes \sigma_- + b \otimes \sigma_+) + \Omega b^\dagger b \otimes I_2$$

where

$$\sigma_- := \begin{pmatrix} 0 & 0 \\ 2 & 0 \end{pmatrix}, \quad \sigma_+ := \begin{pmatrix} 0 & 2 \\ 0 & 0 \end{pmatrix},$$

$$\sigma_z := \begin{pmatrix} 1 & 0 \\ 0 & -1 \end{pmatrix}, \quad I_2 := \begin{pmatrix} 1 & 0 \\ 0 & 1 \end{pmatrix}$$

and Δ, k and Ω are constants. Let

$$\hat{U} := \exp\left[i\pi \left(b^\dagger b \otimes I_2 + \frac{1}{2} I_B \otimes \sigma_z + \frac{1}{2} I_B \otimes I_2 \right) \right],$$

$$\hat{N} := b^\dagger b \otimes I_2 + \frac{1}{2} I_B \otimes \sigma_z,$$

$$I_B \otimes \sigma^2 := I_B \otimes \left[\sigma_z^2 + \frac{1}{2}(\sigma_+ \sigma_- + \sigma_- \sigma_+) \right].$$

(i) Calculate

$$[\hat{H}, \hat{U}], \quad [\hat{H}, \hat{N}], \quad [\hat{H}, I_B \otimes \sigma^2], \quad [\hat{U}, \hat{N}], \quad [\hat{U}, I_B \otimes \sigma^2], \quad [\hat{N}, I_B \otimes \sigma_z].$$

(ii) Find the eigenvalues of $\hat{U}$. Discuss how $\hat{U}$ can be used to simplify the eigenvalue problem.

(iii) Calculate the eigenvalues of $\hat{H}$.

Solution.   (i) We find

$$[\hat{H}, \hat{U}] = 0, \quad [\hat{H}, \hat{N}] = 0, \quad [\hat{H}, I_B \otimes \sigma^2] = 0,$$

$$[\hat{U}, \hat{N}] = 0, \quad [\hat{U}, I_B \otimes \sigma^2] = 0, \quad [\hat{N}, I_B \otimes \sigma_z] = 0.$$

(ii) A basis in the Hilbert space $\mathcal{H}_B \otimes \mathbf{C}^2$ is given by

$$\left\{ |n\rangle \otimes \begin{pmatrix} 1 \\ 0 \end{pmatrix}, \quad |n\rangle \otimes \begin{pmatrix} 0 \\ 1 \end{pmatrix} \right\} \tag{1}$$

where $|n\rangle$ are the Bose number states. Since the commutators of the operators $b^\dagger b \otimes I_2$, $I_B \otimes \sigma_2$ and $I_B \otimes I_2$ vanish, we can write

$$\hat{U} = \exp(i\pi b^\dagger b \otimes I_2) \exp\left(\frac{1}{2} i\pi (I_B \otimes \sigma_z) \right) \exp\left(\frac{1}{2} i\pi (I_B \otimes I_2) \right).$$

We find

$$\hat{U}|n\rangle \otimes \begin{pmatrix} 1 \\ 0 \end{pmatrix} = e^{i\pi(n+1)} |n\rangle \otimes \begin{pmatrix} 1 \\ 0 \end{pmatrix},$$

$$\hat{U}|n\rangle \otimes \begin{pmatrix} 0 \\ 1 \end{pmatrix} = e^{i\pi n} |n\rangle \otimes \begin{pmatrix} 0 \\ 1 \end{pmatrix}.$$

Thus the basis elements (1) are eigenfunctions of $\hat{U}$ and the eigenvalues of $\hat{U}$ are given by $1, -1$, because

$$e^{i\pi m} = 1 \quad \text{for } m \text{ even}, \quad e^{i\pi m} = -1 \quad \text{for } m \text{ odd}.$$

The eigenvalues are infinitely degenerate. Since $\hat{U}$ and $I_B \otimes I_2$ form a finite group (isomorphic to C_2) we can decompose the Hilbert space $\mathcal{H}_B \otimes \mathbf{C}^2$ into two invariant subspaces. We obtain the two subspaces

$$S_1 = \left\{ |0\rangle \otimes \begin{pmatrix} 0 \\ 1 \end{pmatrix}, \quad |1\rangle \otimes \begin{pmatrix} 1 \\ 0 \end{pmatrix}, \quad |2\rangle \otimes \begin{pmatrix} 0 \\ 1 \end{pmatrix}, \quad |3\rangle \otimes \begin{pmatrix} 1 \\ 0 \end{pmatrix}, \dots \right\},$$

$$S_2 = \left\{ |0\rangle \otimes \begin{pmatrix} 1 \\ 0 \end{pmatrix}, \quad |1\rangle \otimes \begin{pmatrix} 0 \\ 1 \end{pmatrix}, \quad |2\rangle \otimes \begin{pmatrix} 1 \\ 0 \end{pmatrix}, \quad |3\rangle \otimes \begin{pmatrix} 0 \\ 1 \end{pmatrix}, \dots \right\}.$$

Remark. *The operator $\hat{U}$ is the so-called parity operator.*

(iii) Since the Hamilton operator $\hat{H}$ commutes with $\hat{N}$, we can write the infinite matrix representation of $\hat{H}$ as the direct sum of 2×2 matrices. This can also be seen when we apply the Hamilton operator to the basis elements (1). We find

$$\hat{H}|n\rangle \otimes \begin{pmatrix} 1 \\ 0 \end{pmatrix} = (-\Delta + \Omega n)|n\rangle \otimes \begin{pmatrix} 1 \\ 0 \end{pmatrix} + k\sqrt{n+1}|n+1\rangle \otimes \begin{pmatrix} 0 \\ 1 \end{pmatrix},$$

$$\hat{H}|n\rangle \otimes \begin{pmatrix} 0 \\ 1 \end{pmatrix} = (\Delta + \Omega n)|n\rangle \otimes \begin{pmatrix} 0 \\ 1 \end{pmatrix} + k\sqrt{n}|n-1\rangle \otimes \begin{pmatrix} 1 \\ 0 \end{pmatrix}.$$

Thus for the subspace S_1 we have

$$\begin{pmatrix} -\Delta & k \\ k & \Delta + \Omega \end{pmatrix} \oplus \begin{pmatrix} -\Delta + 2\Omega & \sqrt{3}k \\ \sqrt{3}k & \Delta + 3\Omega \end{pmatrix}$$

$$\oplus \cdots \oplus \begin{pmatrix} -\Delta + 2n\Omega & \sqrt{2n+1}k \\ \sqrt{2n+1}k & \Delta + (2n+1)\Omega \end{pmatrix} \oplus \cdots.$$

For the subspace S_2 we find

$$\Delta \oplus \begin{pmatrix} -\Delta + \Omega & \sqrt{2}k \\ \sqrt{2}k & \Delta + 2\Omega \end{pmatrix} \oplus \cdots \oplus \begin{pmatrix} -\Delta + (2n+1)\Omega & \sqrt{2n+2}k \\ \sqrt{2n+2}k & \Delta + (2n+2)\Omega \end{pmatrix} \oplus \cdots$$

The eigenvalues of $\hat{H}$ are the eigenvalues of these 2×2 matrices.

Problem 5. Calculate

$$f(\alpha) = \exp(\alpha \sigma_z \otimes (b - b^\dagger)) \sigma_z \otimes (b^\dagger + b) \exp(-\alpha \sigma_z \otimes (b - b^\dagger)) \qquad (1)$$

where α is a real parameter, σ_z is the Pauli matrix and b and $b^\dagger$ are Bose annihilation and creation operators, i.e.

$$[b, b^\dagger] = I.$$

Solution.   We find a differential equation for f and then solve the differential equation. From (1) we find the initial condition

$$f(0) = \sigma_z \otimes (b^\dagger + b). \qquad (2)$$

Differentiating (1) with respect to α yields

$$\frac{df(\alpha)}{d\alpha} = \exp(\alpha\sigma_z \otimes (b - b^\dagger))(\sigma_z \otimes (b^\dagger - b))(\sigma_z \otimes (b^\dagger + b))$$

$$\times \exp(-\alpha\sigma_z \otimes (b - b^\dagger)) - \exp(\alpha\sigma_z \otimes (b - b^\dagger))(\sigma_z \otimes (b^\dagger + b))$$

$$\times (\sigma_z \otimes (b - b^\dagger)) \exp(-\alpha\sigma_z \otimes (b - b^\dagger)).$$

We have

$$\sigma_z^2 = I_2$$

where I_2 is the 2×2 unit matrix and

$$(b - b^\dagger)(b^\dagger + b) - (b^\dagger + b)(b - b^\dagger) = 2I_b$$

where I_b is the unit operator. Thus we find

$$\frac{f(\alpha)}{d\alpha} = 2I_2 \otimes I_b.$$

Integrating this differential equation and inserting the initial condition (2) we obtain

$$f(\alpha) = 2\alpha I_2 \otimes I_b + \sigma_z \otimes (b^\dagger + b).$$

Problem 6.   Consider the Hamilton operator

$$\hat{H} = \hat{H}_0 + g(b^\dagger \otimes S^- + b \otimes S^+) + \gamma b^\dagger b^\dagger bb \otimes I_S + \gamma(I_B \otimes S_z)^2$$

with

$$\hat{H}_0 = \omega b^\dagger b \otimes I_S + \omega_0 I_B \otimes S_z.$$

Here g and γ are real coupling constants, and

$$S^{\pm} := \sum_{n=1}^{N} \sigma_n^{\pm}, \quad S_z := \frac{1}{2} \sum_{n=1}^{N} \sigma_n^z$$

are collective N-atom Dicke operators, i.e. spin operators for which total spin $S \leq N/2$. The operators satisfy the $su(2)$ Lie algebra, i.e.,

$$[S^+, S^-] = 2S_z, \quad [S_z, S^{\pm}] = \pm S^{\pm}.$$

The Bose operators $b^{\dagger}$ and b satisfy the commutation relation

$$[b, b^{\dagger}] = I.$$

The number operator $\hat{M}$ is given by

$$\hat{M} = I_B \otimes S_z + b^{\dagger} b \otimes I_S.$$

(i) Find the commutator $[\hat{H}, \hat{M}]$.

(ii) Find

$$\hat{K} := g^{-1}(\hat{H} + (\gamma - \omega)\hat{M} - \gamma \hat{M}^2).$$

(iii) The total spin operator of the model is given by the Casimir operator

$$\hat{S}^2 = S^+ S^- + S_z(S_z - I_S).$$

Find the commutator $[I_B \otimes \hat{S}^2, \hat{K}]$.

Solution. (i) From the commutation relations for the Bose operators we find

$$[b^{\dagger} b^{\dagger} b b, b^{\dagger} b] = 0$$

and

$$[b^{\dagger}, b^{\dagger} b] = -b^{\dagger}, \quad [b, b^{\dagger} b] = b.$$

Using the commutation relation given above and

$$[\hat{A} \otimes I_S, I_B \otimes \hat{B}] = 0,$$

we find that

$$[\hat{H}, \hat{M}] = 0.$$

(ii) We find

$$\hat{K} = \Delta I_B \otimes S_z + (b^\dagger \otimes S^- + b \otimes S^+) + cb^\dagger b \otimes S_z \qquad (1)$$

where

$$\Delta := g^{-1}(\omega_0 - \omega + \gamma)$$

and

$$c := -2g^{-1}\gamma.$$

The last term on the right-hand side of (1) causes photon number dependent changes in the atomic transitions and describes therefore a *Stark shift*. Since $[\hat{H}, \hat{M}] = 0$ we obtain

$$[\hat{K}, \hat{M}] = 0.$$

(iii) We obtain

$$[I_B \otimes \hat{S}^2, \hat{K}] = 0.$$

Remark.   *Since $[\hat{M}, \hat{K}] = 0$ and $[\hat{S}^2, \hat{K}] = 0$, it is convenient to decompose the Hilbert space*

$$\mathcal{H} = \mathcal{H}_B \otimes \mathbf{C}_1^2 \otimes \cdots \otimes \mathbf{C}_N^2$$

in terms of the irreducible representations of the Lie algebra su(2) with spin S and excitation numbers M.

Problem 7.   Consider the difference equation of first order

$$x_{n+1} = f(x_n), \quad n = 0, 1, 2, \ldots \qquad (1)$$

where $f : \mathbf{R} \to \mathbf{R}$ is an analytic function. Consider the state

$$|x, n\rangle := \exp\left(\frac{1}{2}(x_n^2 - x_0^2)\right) |x_n\rangle \qquad (2)$$

where x_n satisfies (1) and $|x_n\rangle$ is a normalized coherent state. Suppose we are given a Boson operator

$$M := \sum_{k=0}^{\infty} \frac{1}{k!} b^{\dagger k} (f(b) - b)^k$$

where $b^\dagger$ and b are Bose operators. Find

$$M|x, n\rangle,$$

Solution. The *coherent state* $|z\rangle$, where $z \in \mathbf{C}$, is defined as the eigenvectors of the annihilation operator b, i.e.

$$b|z\rangle = z|z\rangle .$$

The normalized coherent state is given by

$$|z\rangle = \exp\left(-\frac{1}{2}|z|^2\right)\exp(zb^\dagger)|0\rangle . \tag{3}$$

On using (1), (2) and (3) we find that

$$|x, n+1\rangle = M|x, n\rangle . \tag{4}$$

Now let $x_n(x_0)$ designate the solution of (1) and let $|x_0, n\rangle$ be the solution of (4). Taking (2) into account we find that the following eigenvalue equation holds true

$$b|x_0, n\rangle = x_n(x_0)|x_0, n\rangle .$$

Thus the solution of (1) is equivalent to the solution of the linear abstract difference Eq. (4).

Chapter 29

Lax Representations and Bethe Ansatz

Problem 1. Let

$$\psi_{m+1} = L_m \psi_m \,, \tag{1}$$

$$\frac{d\psi_m}{dt} = M_m \psi_m \tag{2}$$

where the entries of the square matrices L and M depend on time, λ is the so-called spectral parameter which does not depend on time and $m \in \mathbf{Z}$. Show that

$$\frac{dL_m}{dt} = M_{m+1} L_m - L_m M_m \,. \tag{3}$$

Solution. From (1) we obtain

$$\frac{d\psi_{m+1}}{dt} = \frac{dL_m}{dt} \psi_m + L_m \frac{d\psi_m}{dt} \,. \tag{4}$$

Equation (2) yields

$$\frac{d\psi_{m+1}}{dt} = M_{m+1} \psi_{m+1} \,. \tag{5}$$

Inserting (5) and (2) into (4) gives

$$M_{m+1} \psi_{m+1} = \frac{dL_m}{dt} \psi_m + L_m M_m \psi_m \,. \tag{6}$$

Inserting (1) into (6) leads to

$$M_{m+1} L_m \psi_m = \frac{dL_m}{dt} \psi_m + L_m M_m \psi_m \,.$$

Consequently,

$$\frac{dL_m}{dt}\psi_m = (M_{m+1}L_m - L_m M_m)\psi_m \,.$$

Since ψ_m is arbitrary we obtain (3).

Remark 1. L_m and M_m *are called a* Lax *pair. We can find conservation laws from* L_m.

Remark 2. *For a number of solid state physics-models, the Heisenberg equation of motion*

$$i\hbar\frac{d\hat{A}}{dt} = [\hat{A}, \hat{H}](t)$$

can be written as (3). *An example is the XY Hamilton operator*

$$\hat{H} = -\frac{1}{2}\sum_{j=1}^{N}(J_x S_j^x S_{j+1}^x + J_y S_j^y S_{j+1}^y + h S_j^z)$$

with periodic boundary condition $S_{N+1}^\alpha \equiv S_N^\alpha$, *where* $\alpha = x, y, z$.

Problem 2. The one-dimensional field-quantized nonlinear Schrödinger model is defined by the Hamilton operator

$$\hat{H} = \int_0^L \left(\frac{\partial\psi^\dagger}{\partial x}\frac{\partial\psi}{\partial x} + c\psi^\dagger\psi^\dagger\psi\psi \right) dx \tag{1}$$

where ψ is a nonrelativistic Bose field with canonical equal-time commutation relations

$$[\psi(x,t), \psi^\dagger(x',t)] = \delta(x - x') \,, \tag{2a}$$

$$[\psi(x,t), \psi(x',t)] = 0 \tag{2b}$$

and c is a real constant. The Hamilton operator (1) is in the standard form of a many body problem with the second term corresponding to a two-body delta function potential. The Hamilton operator (1) commutes with the particle number operator

$$\hat{N} := \int_0^L \psi^\dagger\psi dx \,. \tag{2c}$$

We impose cyclic boundary conditions, i.e.

$$\psi(0) = \psi(L) \,.$$

Let

$$|\phi(k_1, k_2)\rangle := \int_0^L \int_0^L dx_1 dx_2 e^{i(k_1 x_1 + k_2 x_2)}(\theta(x_1 - x_2)$$

$$+ S(k_{21})\theta(x_2 - x_1))\psi^\dagger(x_1)\psi^\dagger(x_2)|0\rangle \qquad (3)$$

be a two-particle state, where $k_{21} := k_2 - k_1$ and

$$S(k_{21}) := \frac{k_2 - k_1 - ic}{k_2 - k_1 + ic}. \qquad (4)$$

Moreover $\langle 0|0 \rangle = 1$ and θ denotes the step function with $\theta(0) = 1$.

Remark.   *The right-hand side of (3) is called a Bethe ansatz.*

Calculate

$$\hat{H}|\phi(k_1, k_2)\rangle$$

and discuss.

Hint.   Use

$$\psi|0\rangle = 0$$

where $|0\rangle$ is the *vacuum state*, and integration by parts. Furthermore

$$\frac{\partial}{\partial x_1}\theta(x_1 - x_2) = \delta(x_1 - x_2).$$

Solution.   The Hamilton operator can be written as $\hat{H} = \hat{H}_k + \hat{H}_I$, where $\hat{H}_k$ denotes the kinetic part and $\hat{H}_I$ denotes the interacting part. First we consider the interacting part, i.e.,

$$\hat{H}_I = c \int_0^L \psi^\dagger \psi^\dagger \psi \psi \, dx.$$

We set

$$f(x_1, x_2) := e^{i(k_1 x_1 + k_2 x_2)}(\theta(x_1 - x_2) + S(k_{21})\theta(x_2 - x_1)). \qquad (5)$$

It follows that

$$f(x_1, x_1) = e^{i(k_1 x_1 + k_2 x_1)}(\theta(0) + S(k_{21})\theta(0)). \qquad (6)$$

Since $\theta(0) = 1$, the parenthesis of the right-hand side of (6) can be written as

$$1 + S(k_{21}) = 1 + \frac{k_{21} - ic}{k_{21} + ic} = \frac{2k_{21}}{k_{21} + ic}$$

where we used (4). Using the commutation relation (2) we find

$$H_I|\phi(k_1, k_2)\rangle$$

$$= c \int dx dx_1 dx_2 \psi^\dagger(x)\psi^\dagger(x)\psi(x)\psi(x)f(x_1, x_2)\psi^\dagger(x_1)\psi^\dagger(x_2)|0\rangle$$

$$= c \int dx dx_1 dx_2 f(x_1, x_2)\psi^\dagger(x)\psi^\dagger(x)\psi(x)\delta(x - x_1)\psi^\dagger(x_2)|0\rangle$$

$$+ c \int dx dx_1 dx_2 f(x_1, x_2)\psi^\dagger(x)\psi^\dagger(x)\psi(x)\psi^\dagger(x_1)\psi(x)\psi^\dagger(x_2)|0\rangle .$$

It follows that

$$\hat{H}_I|\phi(k_1, k_2)\rangle$$

$$= c \int dx_1 dx_2 f(x_1, x_2)\psi^\dagger(x_1)\psi^\dagger(x_1)\psi(x_1)\psi^\dagger(x_2)|0\rangle$$

$$+ c \int dx dx_1 dx_2 f(x_1, x_2)\psi^\dagger(x)\psi^\dagger(x)\psi(x)\psi^\dagger(x_1)\psi(x)\psi^\dagger(x_2)|0\rangle$$

$$= c \int dx_1 dx_2 f(x_1, x_2)\psi^\dagger(x_1)\psi^\dagger(x_1)\psi(x_1)\psi^\dagger(x_2)|0\rangle$$

$$+ c \int dx dx_1 dx_2 f(x_1, x_2)\psi^\dagger(x)\psi^\dagger(x)\psi(x)\psi^\dagger(x_1)\delta(x - x_2)|0\rangle$$

$$= c \int dx_1 dx_2 f(x_1, x_2)\psi^\dagger(x_1)\psi^\dagger(x_1)\delta(x_1 - x_2)|0\rangle$$

$$+ c \int dx_1 dx_2 f(x_1, x_2)\psi^\dagger(x_2)\psi^\dagger(x_2)\psi(x_2)\psi^\dagger(x_1)|0\rangle$$

$$= 2c \int dx_1 f(x_1, x_1)\psi^\dagger(x_1)\psi^\dagger(x_1)|0\rangle$$

$$= 2c \int dx_1 e^{i(k_1 x_1 + k_2 x_1)}(\theta(0) + S(k_{21})\theta(0))\psi^\dagger(x_1)\psi^\dagger(x_1)|0\rangle$$

$$= 2c \int dx_1 e^{i(k_1 x_1 + k_2 x_1)} \left(\frac{2(k_2 - k_1)}{k_2 - k_1 + ic} \right) \psi^\dagger(x_1)\psi^\dagger(x_1)|0\rangle$$

$$= 4c \int_0^L dx_1 e^{i(k_1 x_1 + k_2 x_1)} \frac{k_2 - k_1}{k_2 - k_1 + ic} \psi^\dagger(x_1)\psi^\dagger(x_1)|0\rangle .$$

Next we consider the kinetic part. Applying integration by parts we find

$$\hat{H}_k|\phi(k_1,k_2)\rangle = \int dx dx_1 dx_2 \frac{\partial \psi^\dagger}{\partial x}\frac{\partial \psi}{\partial x} f(x_1,x_2)\psi^\dagger(x_1)\psi^\dagger(x_2)|0\rangle\,.$$

Thus

$$\hat{H}_k|\phi(k_1,k_2)\rangle = -\int dx dx_1 dx_2 \frac{\partial^2 \psi^\dagger}{\partial x^2} f(x_1,x_2)\psi(x)\psi^\dagger(x_1)\psi^\dagger(x_2)|0\rangle$$

where we have used the cyclic boundary condition $\psi(0) = \psi(L)$.
It follows that

$$\hat{H}_k|\phi(k_1,k_2)\rangle = -\int dx dx_1 dx_2 \frac{\partial^2 \psi^\dagger}{\partial x^2} f(x_1,x_2)\delta(x-x_1)\psi^\dagger(x_2)|0\rangle$$

$$-\int dx dx_1 dx_2 \frac{\partial^2 \psi^\dagger}{\partial x^2} f(x_1,x_2)\psi^\dagger(x_1)\psi(x)\psi^\dagger(x_2)|0\rangle$$

$$= -\int dx_1 dx_2 \frac{\partial^2 \psi^\dagger}{\partial x_1^2} f(x_1,x_2)\psi^\dagger(x_2)|0\rangle$$

$$-\int dx dx_1 dx_2 \frac{\partial^2 \psi^\dagger}{\partial x^2} f(x_1,x_2)\psi^\dagger(x_1)\delta(x-x_2)|0\rangle$$

$$= -\int dx_1 dx_2 \frac{\partial^2 \psi^\dagger}{\partial x_1^2} f(x_1,x_2)\psi^\dagger(x_2)|0\rangle$$

$$-\int dx_1 dx_2 \frac{\partial^2 \psi^\dagger}{\partial x_2^2} f(x_1,x_2)\psi^\dagger(x_1)|0\rangle$$

$$= -\int dx_1 dx_2 \frac{\partial^2 f}{\partial x_1^2} \psi^\dagger(x_1)\psi^\dagger(x_2)|0\rangle$$

$$-\int dx_1 dx_2 \frac{\partial^2 f}{\partial x_2^2} \psi^\dagger(x_2)\psi^\dagger(x_1)|0\rangle\,.$$

From the function f given by (5) we obtain the derivatives in the sense of generalized functions

$$\frac{\partial f}{\partial x_1} = ik_1 e^{i(k_1 x_1 + k_2 x_2)}(\theta(x_1 - x_2) + S(k_{21})\theta(x_2 - x_1))$$

$$+ e^{i(k_1 x_1 + k_2 x_2)}(\delta(x_1 - x_2) - S(k_{21})\delta(x_2 - x_1))\,,$$

$$\frac{\partial^2 f}{\partial x_1^2} = -k_1^2 e^{i(k_1 x_1 + k_2 x_2)}(\theta(x_1 - x_2) + S(k_{21})\theta(x_2 - x_1))$$

$$+ 2ik_1 e^{i(k_1 x_1 + k_2 x_2)}(\delta(x_1 - x_2) - S(k_{21})\delta(x_2 - x_1))$$

$$+ e^{i(k_1 x_1 + k_2 x_2)}(\delta'(x_1 - x_2) - S(k_{21})\delta'(x_2 - x_1))\,,$$

$$\frac{\partial f}{\partial x_2} = ik_2 e^{i(k_1 x_1 + k_2 x_2)}(\theta(x_1 - x_2) + S(k_{21})\theta(x_2 - x_1))$$

$$+ e^{i(k_1 x_1 + k_2 x_2)}(-\delta(x_1 - x_2) + S(k_{21})\delta(x_2 - x_1)),$$

$$\frac{\partial^2 f}{\partial x_2^2} = -k_2^2 e^{i(k_1 x_1 + k_2 x_2)}(\theta(x_1 - x_2) + S(k_{21})\theta(x_2 - x_1))$$

$$+ 2ik_2 e^{i(k_1 x_1 + k_2 x_2)}(-\delta(x_1 - x_2) + S(k_{21})\delta(x_2 - x_1))$$

$$+ e^{i(k_1 x_1 + k_2 x_2)}(-\delta'(x_1 - x_2) + S(k_{21})\delta'(x_2 - x_1)).$$

We arrive at

$$\hat{H}_k|\phi(k_1, k_2)\rangle = (k_1^2 + k_2^2)|\phi(k_1, k_2)\rangle$$

$$- \int dx_1 2ik_1 e^{i(k_1 x_1 + k_2 x_1)}(1 - S(k_{21}))\psi^\dagger(x_1)\psi^\dagger(x_1)|0\rangle$$

$$- \int dx_1 2ik_2 e^{i(k_1 x_1 + k_2 x_1)}(-1 + S(k_{21}))\psi^\dagger(x_1)\psi^\dagger(x_1)|0\rangle$$

$$= (k_1^2 + k_2^2)|\phi(k_1, k_2)\rangle$$

$$- \int_0^L dx_1 e^{i(k_1 x_1 + k_2 x_1)} 2i(k_1 - k_1 S(k_{21}) - k_2$$

$$+ k_2 S(k_{21}))\psi^\dagger(x_1)\psi^\dagger(x_1)|0\rangle. \tag{7}$$

Since

$$2i(k_1 - k_2 - k_1 S(k_{21}) + k_2 S(k_{21})) = 4c\frac{k_2 - k_1}{k_2 - k_1 + ic},$$

the second term of the right-hand side of Eq. (7) cancels with the term $\hat{H}_I|\phi(k_1, k_2)\rangle$. Consequently we find

$$\hat{H}|\phi(k_1, k_2)\rangle = \hat{H}_k|\phi(k_1, k_2)\rangle + \hat{H}_I|\phi(k_1, k_2)\rangle = (k_1^2 + k_2^2)|\phi(k_1, k_2)\rangle.$$

Consequently, $|\phi(k_1, k_2)\rangle$ is an eigenstate of $\hat{H}$ and the eigenvalue is given by

$$k_1^2 + k_2^2.$$

Remark. *Ansatz (3) can be extended to N particles.*

Problem 3. Let

$$\mathcal{H}_N = \prod_{n=1}^{N} \otimes \eta_n$$

and

$$\hat{H}_N = \frac{J}{4} \sum_{n=1}^{N} (\sigma_n^1 \sigma_{n+1}^1 + \sigma_n^2 \sigma_{n+1}^2 + \sigma_n^3 \sigma_{n+1}^3 - I_N) \tag{1}$$

where $\eta_n = \mathbf{C}^2$, J is a real constant and I_N is the identity operator ($2^N \times 2^N$ unit matrix) in the space $\mathcal{H}_N$ (dim $\mathcal{H}_N = 2^N$). The operators σ_n^a with ($a = 1, 2, 3$) have the following form

$$\sigma_n^a := I_2 \otimes \cdots \otimes I_2 \otimes \sigma^a \otimes I_2 \otimes \cdots \otimes I_2 (N - \text{factors})$$
$$\downarrow \tag{2}$$
$$n - th \text{ place}$$

with

$$\sigma^1 := \begin{pmatrix} 0 & 1 \\ 1 & 0 \end{pmatrix}, \quad \sigma^2 := \begin{pmatrix} 0 & -i \\ i & 0 \end{pmatrix},$$
$$\sigma^3 := \begin{pmatrix} 1 & 0 \\ 0 & -1 \end{pmatrix}, \quad I_2 := \begin{pmatrix} 1 & 0 \\ 0 & 1 \end{pmatrix} \tag{3}$$

(Pauli matrices) and $n = 1, \ldots, N$. Here $\otimes$ denotes the Kronecker product. Let

$$\sigma_{N+1}^a \equiv \sigma_1^a \tag{4}$$

(periodic boundary condition), and

$$L_n(\lambda) := \begin{pmatrix} \lambda I_n + \frac{i}{2} \sigma_n^3 & \frac{i}{2} \sigma_n^- \\ \frac{i}{2} \sigma_n^+ & \lambda I_n - \frac{i}{2} \sigma_n^3 \end{pmatrix} \tag{5}$$

be an operator valued matrix of order 2×2 with

$$\sigma_n^+ := \sigma_n^1 + i \sigma_n^2, \quad \sigma_n^- := \sigma_n^1 - i \sigma_n^2. \tag{6}$$

Here λ is a complex parameter. Thus $L_n(\lambda)$ is a $2^{N+1} \times 2^{N+1}$ matrix. L_n is called a *Lax operator*. The matrix $L_n(\lambda)$ can also be represented in the form

$$L_n(\lambda) = \lambda I_2 \otimes I_N + \frac{i}{2} \sum_{a=1}^{3} \sigma^a \otimes \sigma_n^a. \tag{7}$$

Show that

$$R(\lambda - \mu)(L_n(\lambda) \otimes L_n(\mu)) \equiv (L_n(\mu) \otimes L_n(\lambda))R(\lambda - \mu) \tag{8}$$

where

$$R(\lambda) = \frac{1}{\lambda + i} \left(\left(\frac{\lambda}{2} + i \right) I_2 \otimes I_2 + \frac{\lambda}{2} \sum_{a=1}^{3} \sigma^a \otimes \sigma^a \right).$$

We define

$$(\sigma^1 \otimes \sigma^1)(L_n(\lambda) \otimes L_n(\mu)) := (\sigma^1 L_n(\lambda)) \otimes (\sigma^1 L_n(\mu)) \tag{9}$$

and

$$\begin{pmatrix} 0 & 1 \\ 1 & 0 \end{pmatrix} \begin{pmatrix} \lambda I_N + \frac{i}{2}\sigma_n^3 & \frac{i}{2}\sigma_n^- \\ \frac{i}{2}\sigma_n^+ & \lambda I_N - \frac{i}{2}\sigma_n^3 \end{pmatrix} = \begin{pmatrix} \frac{i}{2}\sigma_n^+ & \lambda I_N - \frac{i}{2}\sigma_n^3 \\ \lambda I_N + \frac{i}{2}\sigma_n^3 & \frac{i}{2}\sigma_n^- \end{pmatrix} \tag{10}$$

and so on. Due to this definition, one says that the identity matrix I_2 and the Pauli matrices σ^a act in the auxiliary space $\mathbf{C}^2$.

Remark. *Equation* (8) *is the so-called Yang–Baxter relation.*

Solution. We put

$$L_n(\lambda) = \begin{pmatrix} a & b \\ c & d \end{pmatrix}, \quad L_n(\mu) = \begin{pmatrix} e & f \\ g & h \end{pmatrix} \tag{11}$$

and

$$D := \frac{1}{\lambda - \mu + i}. \tag{12}$$

Obviously $b = f$ and $c = g$ with

$$b = f = \frac{i}{2}\sigma_n^-, \tag{13a}$$

$$c = g = \frac{i}{2}\sigma_n^+. \tag{13b}$$

Then we find

$$R(\lambda - \mu)(L_n(\lambda) \otimes L_n(\mu))$$

$$= D \begin{pmatrix} \lambda - \mu + i & 0 & 0 & 0 \\ 0 & i & \lambda - \mu & 0 \\ 0 & \lambda - \mu & i & 0 \\ 0 & 0 & 0 & \lambda - \mu + i \end{pmatrix} \begin{pmatrix} ae & af & be & bf \\ ag & ah & bg & bh \\ ce & cf & de & df \\ cg & ch & dg & dh \end{pmatrix}$$

$$= D \begin{pmatrix} (\lambda - \mu + i)ae & (\lambda - \mu + i)af & (\lambda - \mu + i)be & (\lambda - \mu + i)bf \\ iag + (\lambda - \mu)ce & iah + (\lambda - \mu)cf & ibg + (\lambda - \mu)de & ibh + (\lambda - \mu)df \\ (\lambda - \mu)ag + ice & (\lambda - \mu)ah + icf & (\lambda - \mu)bg + ide & (\lambda - \mu)bh + idf \\ (\lambda - \mu + i)cg & (\lambda - \mu + i)ch & (\lambda - \mu + i)dg & (\lambda - \mu + i)dh \end{pmatrix}$$

(14a)

and

$$(L_n(\mu) \otimes L_n(\lambda))R(\lambda - \mu)$$

$$= D \begin{pmatrix} ea & eb & fa & fb \\ ec & ed & fc & fd \\ ga & gb & ha & hb \\ gc & gd & hc & hd \end{pmatrix} \begin{pmatrix} \lambda - \mu + i & 0 & 0 & 0 \\ 0 & i & \lambda - \mu & 0 \\ 0 & \lambda - \mu & i & 0 \\ 0 & 0 & 0 & \lambda - \mu + i \end{pmatrix}$$

$$= D \begin{pmatrix} (\lambda - \mu + i)ea & ieb + (\lambda - \mu)fa & (\lambda - \mu)eb + ifa & (\lambda - \mu + i)fb \\ (\lambda - \mu + i)ec & ied + (\lambda - \mu)fc & (\lambda - \mu)ed + ifc & (\lambda - \mu + i)fd \\ (\lambda - \mu + i)ga & igb + (\lambda - \mu)ha & (\lambda - \mu)gb + iha & (\lambda - \mu + i)hb \\ (\lambda - \mu + i)gc & igd + (\lambda - \mu)hc & (\lambda - \mu)gd + ihc & (\lambda - \mu + i)hd \end{pmatrix}.$$

(14b)

For identity (8) to be true, the matrices on the right-hand sides of (14a) and (14b) must be equal. We number the entries as follows

$$\begin{pmatrix} 1 & 2 & 5 & 6 \\ 3 & 4 & 7 & 8 \\ 9 & 10 & 13 & 14 \\ 11 & 12 & 15 & 16 \end{pmatrix}.$$

Therefore we have to prove identities (14). The first identity to prove is

$$ea = ae.$$

From $ea - ae$ we find

$$\left(\mu I_N + \frac{i}{2}\sigma_n^3\right)\left(\lambda I_N + \frac{i}{2}\sigma_n^3\right) - \left(\lambda I_N + \frac{i}{2}\sigma_n^3\right)\left(\mu I_N + \frac{i}{2}\sigma_n^3\right) = 0$$

since λ and μ are complex parameters. Analogously, we can prove identity (14), i.e.,

$$dh = hd \,.$$

Identity (5) is given by

$$fb = bf \,.$$

This is obviously true, because

$$f = b \,.$$

Analogously, identity (9) is satisfied. Identity (1) is given by

$$(\lambda - \mu + i)af = ieb + (\lambda - \mu)fa$$

or

$$(\lambda - \mu)(af - fa) + i(af - eb) = 0 \,.$$

Using the commutation relation

$$[\sigma_n^-, \sigma_n^3] = 2\sigma_n^- \,,$$

we find that the identity is satisfied. By using the same identity, we can similarly prove (4), (7) and (12). Applying the commutation relation

$$[\sigma_n^+, \sigma_n^3] = -2\sigma_n^+ \,,$$

we can prove (2), (8), (10) and (13). The technique is the same as that used for identity (1). Identity (3) is given by

$$ied + (\lambda - \mu)fc = iah + (\lambda - \mu)cf \,.$$

Using the commutation relation

$$[\sigma_n^+, \sigma_n^-] = 4\sigma_n^3$$

we find that identity (3) holds. By using the same identity, we can prove identity (13) in the same way. Identity (6) is given by

$$(\lambda - \mu)ed + ifc = ibg + (\lambda - \mu)de \,.$$

Thus we have to prove that

$$(\lambda - \mu)(ed - de) + i(fc - bg) = 0 \,.$$

For $ed - de$ we find

$$ed - de = \left(\mu I_N - \frac{i}{2}\sigma_n^3\right)\left(\lambda I_N - \frac{i}{2}\sigma_n^3\right)$$

$$- \left(\lambda I_N - \frac{i}{2}\sigma_n^3\right)\left(\mu I_N - \frac{i}{2}\sigma_n^3\right) = 0.$$

For $fc - bg$ we obtain

$$fc - bg = \frac{i}{2}\sigma_n^- \frac{i}{2}\sigma_n^+ - \frac{i}{2}\sigma_n^- \frac{i}{2}\sigma_n^+ = 0.$$

Identity (9) is proved in the same way. Consequently, we have shown that the corresponding entries of the matrices (14a) and (14b) are equal. Thus identity (8) follows.

Problem 4.   The classical massless *Thirring model* in (1+1) dimensions is given by

$$\frac{\partial\phi}{\partial t} + v\frac{\partial\phi}{\partial x} = -2iJ\chi^*\chi\phi, \tag{1a}$$

$$\frac{\partial\chi}{\partial t} - v\frac{\partial\chi}{\partial x} = -2iJ\phi^*\phi\chi \tag{1b}$$

where v is the constant velocity and J is the coupling constant.

(i) Find the general solution of the initial-value problem

$$\phi(x,0) = a(x)\exp(ib(x)), \quad \chi(x,0) = c(x)\exp(id(x)) \tag{2}$$

where a, b, c and d are real-valued functions.

(ii) Find the Lax pair for (1).

Solution.   (i) Using the new independent variables

$$\xi := x - vt, \quad \eta := x + vt$$

we can write (1) in the form

$$v\frac{\partial\phi}{\partial\eta} = -iJ\chi^*\chi\phi, \tag{3a}$$

$$v\frac{\partial\chi}{\partial\xi} = iJ\phi^*\phi\chi. \tag{3b}$$

Since

$$\frac{\partial(\phi^*\phi)}{\partial\eta} = 0, \quad \frac{\partial(\chi^*\chi)}{\partial\xi} = 0,$$

we set

$$\phi(\xi,\eta) = f(\xi)\exp(ik(\xi,\eta)), \quad \chi(\xi,\eta) = g(\eta)\exp(i\ell(\xi,\eta)) \qquad (4)$$

where f and g are real arbitrary functions. The real functions k and ℓ are determined as follows. Substituting the expressions (4) into (3) yields

$$v\frac{\partial k}{\partial\eta} = -Jg^2(\eta), \quad v\frac{\partial\ell}{\partial\xi} = Jf^2(\xi). \qquad (5)$$

Integrating (5) we obtain

$$k(\xi,\eta) = -\frac{J}{v}\int^\eta g^2(\eta)d\eta + \theta(\xi), \quad \ell(\xi,\eta) = \frac{J}{v}\int^\xi f^2(\xi)d\xi + \epsilon(\eta) \qquad (6)$$

where θ and ϵ are real arbitrary functions. With (4), (6) and (3) we arrive at the general solutions for (1)

$$\phi(x,t) = f(x-vt)\exp\left(-i\frac{J}{v}\int^{x+vt} g^2(\eta)d\eta + i\theta(x-vt)\right),$$

$$\chi(x,t) = g(x+vt)\exp\left(i\frac{J}{v}\int^{x-vt} f^2(\xi)d\xi + i\epsilon(x+vt)\right).$$

Suppose that the initial conditions are given by (2), we obtain

$$f(x) = a(x), \quad \theta(x) = \frac{J}{v}\int^x c^2(\eta)d\eta + b(x),$$

$$g(x) = c(x), \quad \epsilon(x) = -\frac{J}{v}\int^x a^2(\xi)d\xi + d(x).$$

Thus we find

$$\phi(x,t) = a(x-vt)\exp\left(-i\frac{J}{v}\int_{x-vt}^{x+vt} c^2(\eta)d\eta + ib(x-vt)\right),$$

$$\chi(x,t) = c(x+vt)\exp\left(-i\frac{J}{v}\int_{x-vt}^{x+vt} a^2(\xi)d\xi + id(x+vt)\right).$$

(ii) To find the Lax pair we consider

$$\frac{\partial\Psi}{\partial t} = U\Psi, \quad \frac{\partial\Psi}{\partial x} = V\Psi. \qquad (7)$$

Here U and V are 4×4 matrices. The consistency condition for (7) yields

$$\frac{\partial U}{\partial t} - \frac{\partial V}{\partial x} + [U, V] = 0 \tag{8}$$

where $[U, V]$ denotes the commutator. We assume that U and V take the following form

$$U = \lambda U^{(1)} + U^{(0)}, \quad V = \lambda V^{(1)} + V^{(0)}. \tag{9}$$

Substituting the expression (9) into (8) and equating the terms with the same powers of λ, we have

$$[U^{(1)}, V^{(1)}] = 0, \tag{10a}$$

$$\frac{\partial U^{(1)}}{\partial t} - \frac{\partial V^{(1)}}{\partial x} + [U^{(1)}, V^{(0)}] + [U^{(0)}, V^{(1)}] = 0, \tag{10b}$$

$$\frac{\partial U^{(0)}}{\partial t} - \frac{\partial V^{(0)}}{\partial x} + [U^{(0)}, V^{(0)}] = 0. \tag{10c}$$

We take

$$U^{(1)} := \begin{pmatrix} A & 0 \\ 0 & B \end{pmatrix}, \quad U^{(0)} := \begin{pmatrix} C & 0 \\ 0 & D \end{pmatrix}, \tag{11a}$$

$$V^{(1)} := \begin{pmatrix} -vA & 0 \\ 0 & vB \end{pmatrix}, \quad V^{(0)} := \begin{pmatrix} vC & 0 \\ 0 & -vD \end{pmatrix} \tag{11b}$$

where A, B, C and D are 2×2 matrices. With this choice, (10a) is satisfied identically. Equations (10b) and (10c) yield

$$\frac{\partial A}{\partial t} + v\frac{\partial A}{\partial x} + 2v[A, C] = 0, \tag{12a}$$

$$\frac{\partial B}{\partial t} - v\frac{\partial B}{\partial x} - 2v[B, D] = 0, \tag{12b}$$

$$\frac{\partial C}{\partial t} - v\frac{\partial C}{\partial x} = 0, \quad \frac{\partial D}{\partial t} + v\frac{\partial D}{\partial x} = 0. \tag{12c}$$

Thus we can set

$$A = \begin{pmatrix} 0 & \phi \\ \phi^* & 0 \end{pmatrix}, \quad B = \begin{pmatrix} 0 & \chi \\ \chi^* & 0 \end{pmatrix}, \tag{13a}$$

$$C = -\frac{iJ}{2v}\begin{pmatrix} \chi^*\chi & 0 \\ 0 & -\chi^*\chi \end{pmatrix}, \quad D = \frac{iJ}{2v}\begin{pmatrix} \phi^*\phi & 0 \\ 0 & -\phi^*\phi \end{pmatrix}. \tag{13b}$$

Equations (13) with (12) are the massless Thirring model (1).

Problem 5. The quantum three wave interaction equation in one-space dimension is described by the Hamilton operator

$$\hat{H} = \int dx \left(\sum_{j=1}^{3} c_j b_j^* \left(\frac{1}{i} \frac{\partial}{\partial x} \right) b_j + g(b_2^* b_1 b_3 + b_1^* b_3^* b_2) \right)$$

where c_j's ($j = 1, 2, 3$) are constant velocities and g is the coupling constant. We assume that all c_j's are distinct. The three fields b_j's are bosons which satisfy the equal time commutation relations

$$[b_j(x,t), b_k^*(y,t)] = \delta_{jk}\delta(x - y), \tag{1a}$$

$$[b_j(x,t), b_k(y,t)] = [b_j^*(x,t), b_k^*(y,t)] = 0. \tag{1b}$$

We define the *vacuum state* by

$$b_j(x,t)|0\rangle = 0, \quad j = 1, 2, 3. \tag{2}$$

(i) Find the equation of motion.

(ii) Find the commutator of $\hat{H}$ with the operators

$$\hat{N}_a := \int dx (b_1^* b_1 + b_2^* b_2),$$

$$\hat{N}_b := \int dx (b_2^* b_2 + b_3^* b_3).$$

We define the states

$$\|0, 0\rangle\rangle := |0\rangle,$$

$$\|M, 0\rangle\rangle := \int \cdots \int dx_1 \cdots dx_M$$
$$\times \exp(i(p_1 x_1 + \cdots + p_M x_M))b_1^*(x_1, t) \cdots b_1^*(x_M, t)|0\rangle,$$

$$\|0, N\rangle\rangle := \int \cdots \int dx_1 \cdots dx_N$$
$$\times \exp(i(q_1 x_1 + \cdots + q_N x_N))b_3^*(x_1, t) \cdots b_3^*(x_N, t)|0\rangle$$

where $M = 1, 2, \ldots$ and $N = 1, 2, \ldots$.

(iii) Find $\hat{H}\|M, 0\rangle\rangle$ and $\hat{H}\|0, N\rangle\rangle$.

Solution.   (i) The Heisenberg equation of motion is given by

$$\frac{\partial}{\partial t} b_j = i[\hat{H}, b_j(x,t)], \quad j = 1,2,3.$$

Using the commutation relations (1) it follows that

$$\frac{\partial b_1}{\partial t} + c_1 \frac{\partial b_1}{\partial x} = -igb_3^* b_2,$$

$$\frac{\partial b_2}{\partial t} + c_2 \frac{\partial b_2}{\partial x} = -igb_1 b_3,$$

$$\frac{\partial b_3}{\partial t} + c_3 \frac{\partial b_3}{\partial x} = -igb_1^* b_2.$$

(ii) Using the commutation relation (2) we find that the operators $\hat{N}_A$ and $\hat{N}_B$ commute with $\hat{H}$.

(iii) We find

$$\hat{H}\|M,0\rangle = \int dx \int \cdots dx_1 \cdots dx_M \exp(i(p_1 x_1 + \cdots + p_M x_M))$$

$$\times \left(b_1^*(x,t) c_1 \frac{1}{i} \frac{\partial}{\partial x} b_1^*(x,t) \right) b_1^*(x_1,t) \cdots b_1^*(x_M,t)|0\rangle$$

$$= E\|M,0\rangle$$

where the energy E is given by

$$E = c_1(p_1 + \cdots + p_M).$$

Thus $\|M,0\rangle$ is an eigenstate. Similarly we find that $\|0,N\rangle$ is an eigenstate with the energy E, which is

$$E = c_3(q_1 + \cdots + q_N).$$

Chapter 30

Gauge Transformations

Problem 1. Let

$$\omega := \sum_{j=1}^{n} p_j dq_j - H(\mathbf{p}, \mathbf{q}, t)dt \tag{1}$$

be a *Cartan form* of a Hamilton system. Consider the transformation

$$q_j \to q_j, \quad p_j \to p_j + \frac{\partial \Omega(\mathbf{q}, t)}{\partial q_j}, \quad H(\mathbf{p}, \mathbf{q}, t) \to H(\mathbf{p}, \mathbf{q}, t) - \frac{\partial \Omega(\mathbf{q}, t)}{\partial t} \tag{2}$$

where Ω is a smooth function of $\mathbf{q}$ and t.

Remark. *System (2) is called a* gauge transformation.

(i) Show that under the transformation, $d\omega \to d\omega$.

(ii) Calculate $d\omega$.

(iii) Calculate

$$Z \rfloor d\omega = 0$$

where $\rfloor$ denotes the contraction $(\partial/\partial x_j \rfloor dx_k = \delta_{jk})$ and

$$Z := \sum_{j=1}^{n} \left(V_j(\mathbf{p}, \mathbf{q}, t) \frac{\partial}{\partial p_j} + W_j(\mathbf{p}, \mathbf{q}, t) \frac{\partial}{\partial q_j} \right) + \frac{\partial}{\partial t} .$$

Solution. (i) From (1) and transformation (2) we obtain

$$\omega \to \sum_{j=1}^{n} \left(p_j + \frac{\partial \Omega}{\partial q_j} \right) dq_j - \left(H - \frac{\partial \Omega}{\partial t} \right) dt = \omega + \sum_{j=1}^{n} \frac{\partial \Omega}{\partial q_j} dq_j + \frac{\partial \Omega}{\partial t} dt .$$

Now

$$d\left(\sum_{j=1}^{n}\frac{\partial\Omega}{\partial q_j}dq_j + \frac{\partial\Omega}{\partial t}dt\right) = \sum_{j=1}^{n}\sum_{k=1}^{n}\frac{\partial^2\Omega}{\partial q_j\partial q_k}dq_k \wedge dq_j + \sum_{j=1}^{n}\frac{\partial^2\Omega}{\partial t\partial q_j}dt \wedge dq_j$$

$$+ \sum_{j=1}^{n}\frac{\partial^2\Omega}{\partial t\partial q_j}dq_j \wedge dt . \tag{3}$$

Since $dq_j \wedge dq_k = -dq_k \wedge dq_j$ and $dq_j \wedge dt = -dt \wedge dq_j$ we find that the right-hand side of (3) vanishes.

(ii) Straightforward calculation yields

$$d\omega = \sum_{j=1}^{n}\left(dp_j \wedge dq_j - \frac{\partial H}{\partial q_j}dq_j \wedge dt - \frac{\partial H}{\partial p_j}dp_j \wedge dt\right).$$

(iii) The corresponding equations of motion of the vector field Z are given by

$$\frac{dp_j}{d\epsilon} = V_j(\mathbf{p},\mathbf{q},t), \quad \frac{dq_j}{d\epsilon} = W_j(\mathbf{p},\mathbf{q},t), \quad \frac{dt}{d\epsilon} = 1$$

where $t(\epsilon = 0) = 0$. Therefore, we find

$$Z\rfloor d\omega = \sum_{j=1}^{n}\left(V_j dq_j - W_j dp_j - W_j \frac{\partial H}{\partial q_j}dt - V_j \frac{\partial H}{\partial p_j}dt + \frac{\partial H}{\partial q_j}dq_j + \frac{\partial H}{\partial p_j}dp_j\right).$$

Thus the condition $Z\rfloor d\omega = 0$ yields the *Hamilton equations of motion*

$$V_j(\mathbf{p},\mathbf{q},t) = -\frac{\partial H}{\partial q_j}, \quad W_j(\mathbf{p},\mathbf{q},t) = \frac{\partial H}{\partial p_j}.$$

Problem 2.   The equation of motion of a charged particle with charge q in an electromagnetic field is given by

$$m\frac{d\mathbf{v}}{dt} = q(\mathbf{E} + \mathbf{v} \times \mathbf{B}). \tag{1}$$

(i) Show that the equation of motion is invariant under

$$\mathbf{A}(\mathbf{r},t) \rightarrow \mathbf{A}(\mathbf{r},t) + \mathrm{grad}\lambda(\mathbf{r},t), \quad U(\mathbf{r},t) \rightarrow U(\mathbf{r},t) - \frac{\partial\lambda(\mathbf{r},t)}{\partial t}. \tag{2}$$

Remark.   *Transformation (2) is called a* gauge transformation.

(ii) Show that the equation of motion (1) can be derived from the *Lagrange function*

$$L(\mathbf{r}, \mathbf{v}, t) = \frac{m\mathbf{v}^2}{2} + q(-U(\mathbf{r}, t) + \mathbf{A}(\mathbf{r}, t) \cdot \mathbf{v}) \tag{3}$$

where

$$\mathbf{A} \cdot \mathbf{v} = A_1 v_1 + A_2 v_2 + A_3 v_3, \quad \mathbf{v}^2 = v_1^2 + v_2^2 + v_3^2.$$

Is the Lagrange function (3) invariant under the gauge transformation?

Solution. (i) We write

$$\mathbf{A}' = \mathbf{A} + \operatorname{grad} \lambda, \quad U' = U - \frac{\partial \lambda}{\partial t}.$$

Since

$$\mathbf{B} = \operatorname{curl} \mathbf{A}, \quad \mathbf{E} = -\operatorname{grad} U - \frac{\partial \mathbf{A}}{\partial t},$$

we obtain

$$\mathbf{B}' = \operatorname{curl} \mathbf{A}' = \operatorname{curl}(\mathbf{A} + \operatorname{grad} \lambda) = \operatorname{curl} \mathbf{A} = \mathbf{B}$$

where we used

$$\operatorname{curl}(\operatorname{grad} \lambda) = 0.$$

Analogously

$$\mathbf{E}' = \mathbf{E}.$$

Thus $\mathbf{E}$ and $\mathbf{B}$ are invariant under the gauge transformation as well as the equation of motion (1) which only depends on $\mathbf{E}$ and $\mathbf{B}$.

(ii) The *Euler–Lagrange equations* are given by

$$\frac{d}{dt}\frac{\partial L}{\partial v_1} - \frac{\partial L}{\partial x} = 0, \quad \frac{d}{dt}\frac{\partial L}{\partial v_2} - \frac{\partial L}{\partial y} = 0, \quad \frac{d}{dt}\frac{\partial L}{\partial v_3} - \frac{\partial L}{\partial z} = 0. \tag{4}$$

Since

$$\frac{\partial L}{\partial v_1} = m v_1 + q A_1,$$

we obtain

$$\frac{d}{dt}\frac{\partial L}{\partial v_1} = m\frac{dv_1}{dt} + q\frac{\partial A_1}{\partial t} + q\frac{\partial A_1}{\partial x}v_1 + q\frac{\partial A_1}{\partial y}v_2 + q\frac{\partial A_1}{\partial z}v_3.$$

Now

$$\frac{\partial L}{\partial x} = -q\frac{\partial U}{\partial x} + qv_1\frac{\partial A_1}{\partial x} + qv_2\frac{\partial A_2}{\partial x} + qv_3\frac{\partial A_3}{\partial x}.$$

Therefore, from the Euler–Lagrange equations (3), we obtain

$$m\frac{dv_1}{dt} - qE_1 - q(v_2B_3 - v_3B_2) = 0.$$

Analogously we obtain the two other components of (1). Under the gauge transformation the Lagrange function L becomes $L' = L + qd\lambda/dt$, where

$$\frac{d\lambda}{dt} = \frac{\partial\lambda}{\partial t} + \frac{\partial\lambda}{\partial x}v_1 + \frac{\partial\lambda}{\partial y}v_2 + \frac{\partial\lambda}{\partial z}v_3.$$

Problem 3.   Let

$$i\hbar\frac{\partial\psi(\mathbf{x}, t)}{\partial t} = \hat{H}_0\psi(\mathbf{x}, t) \tag{1}$$

be the *Schrödinger equation* in three space dimensions, where

$$\hat{H}_0 = -\frac{\hbar^2}{2m}\triangle$$

with $j = 1, 2, 3$, $\mathbf{x} = (x_1, x_2, x_3)$. ϵ is a real dimensionless parameter, $\hbar$ is Planck's constant divided by 2π and

$$\triangle := \sum_{j=1}^{3}\frac{\partial^2}{\partial x_j^2}.$$

(i) Show that (1) is invariant under the *global gauge transformation*

$$x_j'(\mathbf{x}, t) = x_j, \quad t'(\mathbf{x}, t) = t, \tag{2a}$$

$$\psi'(\mathbf{x}'(\mathbf{x}, t), t'(\mathbf{x}, t)) = \exp(i\epsilon)\psi(\mathbf{x}, t). \tag{2b}$$

(ii) Let

$$i\hbar\frac{\partial\psi}{\partial t} = -\frac{\hbar^2}{2m}\sum_{j=1}^{3}\left(\frac{\partial}{\partial x_j} - i\frac{q}{\hbar}A_j\right)^2\psi + qU\psi \tag{3}$$

where $\mathbf{A}$ is the vector potential and U is the scalar potential. Show that (3) is invariant under the transformation

$$x_j'(\mathbf{x}, t) = x_j, \quad t'(\mathbf{x}, t) = t,$$

$$\psi'(\mathbf{x}'(\mathbf{x}, t), t'(\mathbf{x}, t)) = \exp(i\epsilon(\mathbf{x}, t))\psi(\mathbf{x}, t),$$

$$U'(\mathbf{x}'(\mathbf{x},t),t'(\mathbf{x},t)) = U(\mathbf{x},t) - \frac{\hbar}{q}\frac{\partial\epsilon}{\partial t},$$

$$A'_j(\mathbf{x}'(\mathbf{x},t),t'(\mathbf{x},t)) = A_j(\mathbf{x},t) + \frac{\hbar}{q}\frac{\partial\epsilon}{\partial x_j} \tag{4}$$

where ϵ is a smooth function of $\mathbf{x} = (x_1, x_2, x_3)$ and t, and $j = 1, 2, 3$.

Solution. (i) Since ϵ is a space and time independent parameter we find

$$\frac{\partial^2\psi'}{\partial x_j'^2} = e^{i\epsilon}\frac{\partial\psi}{\partial x_j^2},$$

$$\frac{\partial\psi'}{\partial t'} = e^{i\epsilon}\frac{\partial\psi}{\partial t}$$

where $j = 1, 2, 3$. Thus (1) is invariant under transformation (2).

(ii) Since ϵ is a smooth function of $\mathbf{x} = (x_1, x_2, x_3)$ and t we find

$$\frac{\partial\psi'}{\partial t'} = ie^{i\epsilon}\frac{\partial\epsilon}{\partial t}\psi + e^{i\epsilon}\frac{\partial\psi}{\partial t}, \tag{5a}$$

$$\frac{\partial\psi'}{\partial x_j'} = ie^{i\epsilon}\frac{\partial\epsilon}{\partial x_j}\psi + e^{i\epsilon}\frac{\partial\psi}{\partial x_j} \tag{5b}$$

and

$$\frac{\partial^2\psi'}{\partial x_j'^2} = -e^{i\epsilon}\left(\frac{\partial\epsilon}{\partial x_j}\right)^2\psi + ie^{i\epsilon}\frac{\partial^2\epsilon}{\partial x_j^2}\psi + 2ie^{i\epsilon}\frac{\partial\epsilon}{\partial x_j}\frac{\partial\psi}{\partial x_j} + e^{i\epsilon}\frac{\partial^2\psi}{\partial x_j^2}. \tag{5c}$$

Therefore

$$\frac{\partial\psi}{\partial x_j} = e^{-i\epsilon}\left(\frac{\partial\psi'}{\partial x_j'} - i\frac{\partial\epsilon}{\partial x_j}\psi'\right),$$

$$\frac{\partial\psi}{\partial t} = e^{-i\epsilon}\left(\frac{\partial\psi'}{\partial t'} - i\frac{\partial\epsilon}{\partial t}\psi'\right),$$

$$\frac{\partial^2\psi}{\partial x_j^2} = e^{-i\epsilon}\left(\frac{\partial^2\psi'}{\partial x_j'^2} - \left(\frac{\partial\epsilon}{\partial x_j}\right)^2\psi' - i\frac{\partial^2\epsilon}{\partial x_j^2}\psi' - 2i\frac{\partial\epsilon}{\partial x_j}\frac{\partial\psi'}{\partial x_j'}\right).$$

Since

$$\left(\frac{\partial}{\partial x_j} - i\frac{q}{\hbar}A_j\right)^2\psi \equiv \frac{\partial^2\psi}{\partial x_j^2} - \frac{2iq}{\hbar}A_j\frac{\partial\psi}{\partial x_j} - \frac{iq}{\hbar}\psi\frac{\partial A_j}{\partial x_j} - \frac{q^2}{\hbar^2}A_j^2\psi,$$

we obtain using

$$\frac{\partial A'_j}{\partial x'_j} = \frac{\partial A_j}{\partial x_j} + \frac{\hbar}{q}\frac{\partial^2 \epsilon}{\partial x_j^2}$$

and system (5) that

$$\left(\frac{\partial}{\partial x_j} - i\frac{q}{\hbar}A_j\right)^2 \psi = e^{-i\epsilon}\left(\frac{\partial}{\partial x'_j} - i\frac{q}{\hbar}A'_j\right)^2 \psi'.$$

Moreover

$$i\hbar\frac{\partial \psi}{\partial t} - qU\psi = i\hbar e^{-i\epsilon}\left(\frac{\partial \psi'}{\partial t'} - i\frac{\partial \epsilon}{\partial t}\psi'\right) - qe^{-i\epsilon}\left(U' + \frac{\hbar}{q}\frac{\partial \epsilon}{\partial t}\right)\psi'$$

$$= e^{-i\epsilon}\left(i\hbar\frac{\partial \psi'}{\partial t'} - qU'\psi'\right).$$

Since $e^{-i\epsilon} \neq 0$, we obtain

$$i\hbar\frac{\partial \psi'}{\partial t'} = -\frac{\hbar^2}{2m}\sum_{j=1}^{3}\left(\frac{\partial}{\partial x'_j} - i\frac{q}{\hbar}A'_j\right)^2 \psi' + qU'\psi'.$$

Consequently, (3) is invariant under the gauge transformation (4).

Problem 4.   Consider the linear differential equations

$$\frac{\partial \Phi}{\partial x} = U\Phi, \tag{1a}$$

$$\frac{\partial \Phi}{\partial t} = V\Phi \tag{1b}$$

and

$$\frac{\partial Q}{\partial x} = WQ,$$

$$\frac{\partial Q}{\partial t} = ZQ.$$

Let

$$Q(x,t) = g(x,t)\Phi(x,t) \tag{2}$$

where g is invertible. Show that

$$U = -g^{-1}\frac{\partial g}{\partial x} + g^{-1}Wg, \tag{3}$$

$$V = -g^{-1}\frac{\partial g}{\partial t} + g^{-1}Zg.$$ (4)

Solution. From

$$g^{-1}g = 1,$$

we obtain

$$\frac{\partial g^{-1}}{\partial x}g + g^{-1}\frac{\partial g}{\partial x} = 0, \quad \frac{\partial g^{-1}}{\partial t}g + g^{-1}\frac{\partial g}{\partial t} = 0.$$ (5)

From (2) we obtain

$$\frac{\partial Q}{\partial x} = \frac{\partial g}{\partial x}\Phi + g\frac{\partial \Phi}{\partial x}, \quad \frac{\partial Q}{\partial t} = \frac{\partial g}{\partial t}\Phi + g\frac{\partial \Phi}{\partial t}$$

and

$$\frac{\partial \Phi}{\partial x} = \frac{\partial g^{-1}}{\partial x}Q + g^{-1}\frac{\partial Q}{\partial x}, \quad \frac{\partial \Phi}{\partial t} = \frac{\partial g^{-1}}{\partial t}Q + g^{-1}\frac{\partial Q}{\partial t}.$$ (6)

Inserting (5) through (6) into (1a) yields

$$\frac{\partial g^{-1}}{\partial x}Q + g^{-1}WQ = U\Phi = Ug^{-1}Q.$$

Thus

$$\left(-g^{-1}\frac{\partial g}{\partial x}g^{-1} + g^{-1}W - Ug^{-1}\right)Q = 0.$$ (7)

It follows that the expression in the parenthesis of the left-hand side of (7) must vanish. Multiplying this expression with g on the right-hand side, we obtain (3). Analogously we can prove (4).

Problem 5. Consider the derivative nonlinear Schrödinger equation in one-space dimension

$$\frac{\partial \phi}{\partial t} - i\frac{\partial^2 \phi}{\partial x^2} + \frac{\partial}{\partial x}(|\phi|^2\phi) = 0$$ (1)

and another derivative nonlinear Schrödinger equation in one-space dimension

$$\frac{\partial q}{\partial t} - i\frac{\partial^2 q}{\partial x^2} + |q|^2\frac{\partial q}{\partial x} = 0$$ (2)

where ϕ and q are complex-valued functions. Both equations are integrable by the inverse scattering method. Both arise as consistency conditions of a system of linear differential equations

$$\frac{\partial \Phi}{\partial x} = U(x, t, \lambda)\Phi \,, \tag{3a}$$

$$\frac{\partial \Phi}{\partial t} = V(x, t, \lambda)\Phi \,, \tag{3b}$$

$$\frac{\partial Q}{\partial x} = W(x, t, \lambda)Q \,, \tag{4a}$$

$$\frac{\partial Q}{\partial t} = Z(x, t, \lambda)Q \tag{4b}$$

where λ is a complex parameter. The consistency conditions have the form

$$\frac{\partial U}{\partial t} - \frac{\partial V}{\partial x} + [U, V] = 0 \,,$$

$$\frac{\partial W}{\partial t} - \frac{\partial Z}{\partial x} + [W, Z] = 0 \,.$$

For (1) we have

$$U = -i\lambda^2 \sigma_3 + \lambda \begin{pmatrix} 0 & \phi \\ -\phi^* & 0 \end{pmatrix} \tag{5a}$$

$$V = -(2i\lambda^4 - i|\phi|^2 \lambda^2)\sigma_3 + 2\lambda^3 \begin{pmatrix} 0 & \phi \\ -\phi^* & 0 \end{pmatrix}$$

$$+ \lambda \begin{pmatrix} 0 & i\phi_x - |\phi|^2 \phi \\ i\phi_x^* + |\phi|^2 \phi^* & 0 \end{pmatrix} \tag{5b}$$

where σ_3 is the Pauli spin matrix. For (2) we have

$$W = \left(-i\lambda^2 + \frac{i}{4}|q|^2\right)\sigma_3 + \lambda \begin{pmatrix} 0 & q \\ -q^* & 0 \end{pmatrix} \,, \tag{6a}$$

$$Z = \left(-2i\lambda^4 - i|q|^2\lambda^2 + \frac{1}{4}(qq_x^* - q^*q_x) - \frac{i}{8}|q|^4\right)\sigma_3 + 2\lambda^3 \begin{pmatrix} 0 & q \\ -q^* & 0 \end{pmatrix}$$

$$+ \lambda \begin{pmatrix} 0 & iq_x - |q|^2 q/2 \\ iq_x^* + |q|^2 q^*/2 & 0 \end{pmatrix} \,. \tag{6b}$$

Show that (1) and (2) are gauge-equivalent.

Solution. We look for a transformation which converts (4a) with (6a) into the form of (3a) with (5a). For this purpose the transformation which eliminates

$$\frac{i|q|^2}{4}\sigma_3$$

in (6a) is introduced, i.e.,

$$Q = g\Phi \tag{7}$$

where

$$g = \begin{pmatrix} f & 0 \\ 0 & f^{-1} \end{pmatrix}, \tag{8a}$$

$$f(x,t) := \exp\left(\frac{i}{4}\int^x |q(s,t)|^2 ds\right). \tag{8b}$$

Note that

$$g(x,t) = Q(x,t,\lambda = 0).$$

Equations (3a), (3b), (4a), (4b) and (7) yield (see Problem 4)

$$U = -g^{-1}\frac{\partial g}{\partial x} + g^{-1}Wg, \tag{9}$$

$$V = -g^{-1}\frac{\partial g}{\partial t} + g^{-1}Zg.$$

Substituting (5a), (6a) and (8) into (9) we obtain

$$-i\lambda^2\sigma_3 + \lambda\begin{pmatrix} 0 & \phi \\ -\phi^* & 0 \end{pmatrix} = -i\lambda^2\sigma_3 + \lambda\begin{pmatrix} 0 & qf^{-2} \\ -q^*f^2 & 0 \end{pmatrix}.$$

We find that the transformation of eigenfunction (10) and the transformation of the field variables

$$\phi = qf^{-2}$$

where f is given by (8), keep the inverse scattering method invariant. We call the transformation which changes an eigenvalue problem into another eigenvalue problem the *gauge transformation*. Here g is the gauge transformation.

Problem 6. (i) Consider the general linear group $GL(n, \mathbf{R})$ with the usual coordinates. We write

$$X = (x_{ij}), \quad dX = (dx_{ij}).$$

Consider the matrix

$$\Omega := X^{-1}dX \tag{1}$$

whose entries are one-forms on G. Show that

$$d\Omega + \Omega \wedge \Omega = 0 \tag{2}$$

where $\wedge$ denotes the wedge product of matrices.

(ii) Show that

$$d\Omega + \Omega \wedge \Omega = 0 \tag{3}$$

is invariant under the transformation

$$\Omega \to R^{-1}dR + R^{-1}\Omega R \tag{4}$$

where R is an invertible matrix. This is called a gauge transformation.

Solution.   (i) From (1) we obtain

$$dX = X\Omega. \tag{5}$$

Taking the exterior derivative of (5) we obtain

$$0 = (dX) \wedge \Omega + X \wedge d\Omega \tag{6}$$

since

$$ddX = 0.$$

Inserting (5) into (6) we obtain

$$X(\Omega \wedge \Omega + d\Omega) = 0.$$

Since X is invertible ($X \in GL(n, \mathbf{R})$), we find (2).

(ii) From

$$R^{-1}R = I$$

where I is the $n \times n$ unit matrix, we obtain

$$(dR^{-1})R + R^{-1}dR = 0$$

since

$$dI = 0.$$

Thus

$$dR^{-1} = -R^{-1}(dR)R^{-1}. \tag{7}$$

Inserting (4) into (3) we obtain

$$d(R^{-1}dR + R^{-1}\Omega R) + (R^{-1}dR + R^{-1}\Omega R) \wedge (R^{-1}dR + R^{-1}\Omega R) = 0.$$

From

$$ddR = 0,$$

$$d(R^{-1}dR) = dR^{-1} \wedge dR,$$

$$d(R^{-1}\Omega R) = dR^{-1} \wedge \Omega R + R^{-1}d\Omega R - R^{-1}\Omega \wedge dR$$

and (7) we find

$$R^{-1}(d\Omega + \Omega \wedge \Omega)R = 0.$$

Thus (2) follows.

Problem 7.   The dynamics of a free Dirac particle with rest mass m is given by the *Dirac equation*

$$(i\gamma^\mu \partial_\mu - m)\psi = 0 \tag{1}$$

where ψ is a four-component spinor and γ^μ are the usual 4×4 Dirac matrices satisfying

$$\gamma^\mu \gamma^\nu + \gamma^\nu \gamma^\mu = 2g^{\mu\nu}. \tag{2}$$

The particle's probability density is then given by $\rho = \psi^\dagger \psi$, which is unchanged if we multiply ψ by any constant 4×4 unitary matrix, i.e.,

$$\psi \to U^\dagger \psi, \quad U^\dagger U = I. \tag{3}$$

Then

$$\rho \to (U^\dagger \psi)^\dagger U^\dagger \psi = \psi^\dagger U U^\dagger \psi = \rho.$$

The transformation of the 4×4 matrices γ^μ is given by

$$\gamma^\mu \to U^\dagger \gamma^\mu U. \tag{4}$$

The transformations (3) and (4) also leaves invariant the Dirac Eq. (1), the anticommutation relations (2) and the free particle Lagrangian density

$$\mathcal{L}_F = \bar{\psi}(i\gamma^\mu\partial_\mu - m)\psi\,.$$

The set of all unitary 4×4 matrices comprise the sixteen parameter Lie group $U(4)$. An arbitrary transformation may be expressed in the form

$$U = \exp(i(\epsilon^0 I_4 + \epsilon^a T_a))$$

where I_4 is the 4×4 unit matrix, T_a $(a = 1, 2, \ldots, 15)$ are the hermitian, traceless generators of the Lie algebra $SU(4)$ and ϵ^0, ϵ^a are sixteen real parameters.

Discuss the case where ϵ^0 and ϵ^a become smooth functions of space-time.

Solution.   The invariance of $\mathcal{L}_F$ under the transformations (2) and (4) leads to the Lagrangian density

$$\mathcal{L} = \bar{\psi}(i\gamma^\mu D_\mu - m)\psi - \frac{1}{4}F_{\mu\nu}F^{\mu\nu} - \frac{1}{4}G_{\mu\nu}G^{\mu\nu}$$

where

$$D_\mu := \partial_\mu - ig_0 B_\mu - ig_1 W_\mu\,,$$

$$W_\mu = T_a W_\mu^a\,,$$

$$F_{\mu\nu} = \partial_\mu B_\nu - \partial_\nu B_\mu\,,$$

$$G_{\mu\nu} = \partial_\mu W_\mu - \partial_\nu W_\mu + ig_1[W_\mu, W_\nu]\,.$$

We find that the Lagrangian density $\mathcal{L}$ is invariant under the gauge transformation

$$\psi \to U^\dagger\psi\,, \quad U^\dagger U = I\,,$$

$$\gamma^\mu \to U^\dagger\gamma^\mu U\,,$$

$$B_\mu \to B_\mu - \frac{1}{g_0}\partial_\mu\epsilon^0\,,$$

$$W_\mu \to U_0^\dagger W_\mu U_0 + \frac{i}{g_1}U_0^\dagger\partial_\mu U_0\,,$$

with

$$U_0 = \exp(i\epsilon^a(x)T_a)\,,$$

$$U = \exp(i\epsilon^0(x))U_0\,.$$

Since

$$U(4) = U(1) \times \mathrm{SU}(4)\,,$$

the coupling constants need not be the same.

Chapter 31

Chaos, Fractals and Complexity

Problem 1.   Let

$$x_{t+1} = 4x_t(1 - x_t) \tag{1}$$

where $t = 0, 1, 2, \ldots$ and $x_0 \in [0, 1]$.

Remark.   *This nonlinear difference equation (map) is the so-called lo-gistic equation or logistic map.   It can also be considered as a map $f : [0, 1] \to [0, 1]$, $f(x) = 4x(1 - x)$ and $x \in [0, 1]$.*

(i) Show that $x_t \in [0, 1]$ for $t = 0, 1, 2, \ldots$

(ii) Find the fixed points.

(iii) Give the variational equation.

(iv) Study the stability of the fixed points.

(v) Show that the general solution is given by

$$x_t = \frac{1}{2} - \frac{1}{2} \cos(2^t \cos^{-1}(1 - 2x_0)) \tag{2}$$

where x_0 is the initial value.

(vi) Find the periodic orbits.

(vii) Find the maximal Ljapunov exponent.

(viii) Find the autocorrelation function.

Solution. (i) Let $x \in [0,1]$. Then $x(1-x) \leq 1/4$. Consequently, $4x(1-x) \leq 1$. If $x \neq 1/2$, then $x(1-x) < 1/4$. In other words, the function $g(x) = x(1-x)$ has one maximum at $x = 1/2$ with $g(1/2) = 1/4$.

(ii) The fixed points x^* are solutions of the algebraic equation

$$4x^*(1-x^*) = x^*.$$

We obtain

$$x_1^* = 0, \quad x_2^* = \frac{3}{4}.$$

Remark. *The fixed points are time-independent solutions.*

(iii) Let

$$x_{t+1} = f(x_t)$$

be a one-dimensional difference equation. Assume that f is differentiable. Then

$$y_{t+1} = \frac{df}{dx}(x_t)y_t$$

is called the (difference) *variational equation* or *linearized equation*. Since

$$\frac{df}{dx} = 4 - 8x,$$

we obtain the variational equation

$$y_{t+1} = (4 - 8x_t)y_t \tag{3}$$

where $t = 0, 1, 2, \ldots$.

(iv) Inserting the fixed point $x_1^* = 0$ into the variational Eq. (3) we obtain

$$y_{t+1} = 4y_t.$$

The solution is given by $y_t = 4^t y_0$. Therefore the fixed point $x_1^* = 0$ is unstable, since $y_t \to \infty$ as $t \to \infty$. Inserting the fixed point $x_2^* = 3/4$ into the variational equation yields

$$y_{t+1} = -2y_t.$$

Therefore this fixed point is also unstable, since $|y_t| \to \infty$ as $t \to \infty$.

(v) Let

$$\alpha = \cos^{-1}(1 - 2x_0).$$

Then

$$x_t = \frac{1}{2} - \frac{1}{2}\cos(2^t\alpha).$$

It follows that

$$x_{t+1} = \frac{1}{2} - \frac{1}{2}\cos(2^{t+1}\alpha) = \frac{1}{2} - \frac{1}{2}(2\cos^2(2^t\alpha) - 1) = 1 - \cos^2(2^t\alpha).$$

The left-hand side of (1) is given by

$$4x_t(1 - x_t) = 4x_t - 4x_t^2$$

$$= 2 - 2\cos(2^t\alpha) - 1 - \cos^2(2^t\alpha) + 2\cos(2^t\alpha) = 1 - \cos^2(2^t\alpha).$$

This proves that (2) is the general solution of (1), where x_0 is the initial value.

(vi) The periodic orbits are given by the initial values

$$x_0 = \frac{1}{2} - \frac{1}{2}\cos\left(\frac{r\pi}{2^s}\right)$$

where r and s are positive integers. We have

$$\arccos(1 - 2x_0) = \arccos\left(\cos\left(\frac{r\pi}{2^s}\right)\right) = \frac{r\pi}{2^s}.$$

It follows that

$$x_t = \frac{1}{2} - \frac{1}{2}\cos\left(\frac{2^t r\pi}{2^s}\right).$$

(vii) The *Ljapunov exponent* λ is given by

$$\lambda(x_0) := \lim_{T \to \infty} \frac{1}{T} \sum_{t=0}^{T-1} \ln\left|\frac{df(x)}{dx}\right|_{x=x_t}$$

where x_0 is the initial value, i.e., the Ljapunov exponent depends on the initial value. The Ljapunov exponent measures the exponential rate at which the derivative grows. Since

$$\frac{df}{dx} = 4 - 8x,$$

we find that the Ljapunov exponent is given by

$$\lambda(x_0) = \ln 2$$

for almost all initial values, i.e., the periodic orbits given in (vi) are excluded.

(viii) The *time average* is defined by

$$\langle x_t \rangle := \lim_{T \to \infty} \frac{1}{T} \sum_{t=0}^{T} x_t \,.$$

For almost all initial values we find $\langle x_t \rangle = 1/2$. The *autocorrelation function* is defined by

$$C_{xx}(\tau) := \lim_{T \to \infty} \frac{1}{T} \sum_{t=0}^{T} (x_t - \langle x_t \rangle)(x_{t+\tau} - \langle x_t \rangle)$$

where $\tau = 0, 1, 2, \ldots$. Using the exact solution (2) we find that the autocorrelation function is given by

$$C_{xx}(\tau) = \begin{cases} \dfrac{1}{8} & \text{for } \tau = 0 \,, \\ 0 & \text{otherwise} \,. \end{cases}$$

Problem 2. (i) Let $f : [-1, 1] \mapsto [-1, 1]$ be defined by

$$f(x) := 1 - 2x^2 \,.$$

Let $-1 \leq a \leq b \leq 1$ and

$$\mu([a, b]) := \frac{1}{\pi} \int_a^b \frac{dx}{\sqrt{1 - x^2}} \,. \tag{1}$$

Calculate $\mu([-1, 1])$.

(ii) Show that

$$\mu(f^{-1}([a, b])) = \mu([a, b]) \tag{2}$$

where $f^{-1}([a, b])$ denotes the set S which is mapped under f to $[a, b]$, i.e., $f(S) = [a, b]$. The quantity μ is called the *invariant measure* of the map f.

(iii) Let $g : [0, 1] \mapsto [0, 1]$ be defined by

$$g(x) := \begin{cases} 2x & 0 \leq x \leq \dfrac{1}{2} \,, \\ 2(1 - x) & \dfrac{1}{2} < x \leq 1 \,. \end{cases} \tag{3}$$

This map is called the *tent map*. Let $0 \le a \le b \le 1$ and

$$\nu([a, b]) := \int_a^b dx \,.$$

Show that

$$\nu(g^{-1}([a, b])) = \nu([a, b])$$

where $g^{-1}([a, b])$ is the set S which is mapped under g to $[a, b]$, i.e., $g(S) = [a, b]$.

(iv) Find the Ljapunov exponent of the tent map (3).

Solution.   (i) Since

$$\int_a^b \frac{dx}{\sqrt{1 - x^2}} = \arcsin(b) - \arcsin(a) \qquad (4)$$

and $\arcsin(1) = \pi/2$, $\arcsin(-1) = -\pi/2$, we obtain $\mu([-1, 1]) = 1$.

(ii) The inverse function f^{-1} is not globally defined. We set $f_1 : [-1, 0] \mapsto [-1, 1]$ with $f_1(x) = 1 - 2x^2$. Then $f_1^{-1} : [-1, 1] \mapsto [-1, 0]$,

$$f_1^{-1}(x) = -\sqrt{\frac{1 - x}{2}}$$

with $f_1^{-1}(-1) = -1$ and $f_1^{-1}(1) = 0$. Analogously, we set $f_2 : [0, 1] \mapsto [-1, 1]$ with $f_2(x) = 1 - 2x^2$. Then $f_2^{-1} : [-1, 1] \mapsto [0, 1]$,

$$f_2^{-1}(x) = \sqrt{\frac{1 - x}{2}}$$

with $f_2^{-1}(-1) = 1$ and $f_2^{-1}(1) = 0$. It follows that

$$f_1^{-1}(a) = -\sqrt{\frac{1 - a}{2}}, \quad f_1^{-1}(b) = -\sqrt{\frac{1 - b}{2}},$$

$$f_2^{-1}(a) = \sqrt{\frac{1 - a}{2}}, \quad f_2^{-1}(b) = \sqrt{\frac{1 - b}{2}} \,.$$

Thus f maps the intervals

$$\left[-\sqrt{\frac{1 - a}{2}}, -\sqrt{\frac{1 - b}{2}} \right]$$

and

$$\left[\sqrt{\frac{1-b}{2}}, \sqrt{\frac{1-a}{2}} \right]$$

into the interval $[a, b]$. From (1) and (4) we notice that

$$\pi\mu([a, b]) = \arcsin(b) - \arcsin(a).$$

Furthermore $\arcsin(-x) \equiv -\arcsin(x)$. Now

$$\int_{-\sqrt{(1-a)/2}}^{-\sqrt{(1-b)/2}} \frac{dx}{\sqrt{1-x^2}} + \int_{\sqrt{(1-b)/2}}^{\sqrt{(1-a)/2}} \frac{dx}{\sqrt{1-x^2}}$$

$$= 2\left(\arcsin\sqrt{\frac{1-a}{2}} - \arcsin\sqrt{\frac{1-b}{2}} \right)$$

$$= \arcsin(b) - \arcsin(a).$$

Remark. *Condition (2) can also be written as*

$$\frac{1}{\pi}\frac{1}{\sqrt{1-x^2}} = \frac{d}{dx}\int_{f^{-1}([-1,x])} \frac{ds}{\pi\sqrt{1-s^2}}.$$

(iii) *Obviously*

$$\nu([a, b]) = \int_a^b dx = b - a.$$

The set $g^{-1}([a, b])$ is given by

$$g^{-1}([a, b]) = \left[\frac{a}{2}, \frac{b}{2} \right] \cup \left[1 - \frac{b}{2}, 1 - \frac{a}{2} \right]$$

where $0 \le a \le b \le 1$. Therefore

$$\nu(g^{-1}([a, b])) = \int_{a/2}^{b/2} dx + \int_{1-b/2}^{1-a/2} dx = b - a = \nu([a, b]).$$

(iv) *Since g is not differentiable at $x = 1/2$ we define the* Ljapunov exponent
as

$$\lambda(x_0) := \lim_{T\to\infty} \lim_{\epsilon\to 0} \frac{1}{T} \ln\left| \frac{g^{(T)}(x_0 + \epsilon) - g^{(T)}(x_0)}{\epsilon} \right| \qquad (5)$$

where x_0 is the initial value and $g^{(T)}$ is the Tth iterate of the function g. Inserting the tent map (3) into (5) we obtain

$$\lambda(x_0) = \ln 2$$

for almost all initial values, i.e., the initial conditions

$$x_0 \in \mathbf{Q} \cap [0, 1]$$

are excluded, where $\mathbf{Q}$ *denotes the rational numbers. Thus the Ljapunov exponent is equal to* $\ln 2$ *if the initial value* $x_0 \in [0, 1]$ *is an irrational number. If* $x_0 \in \mathbf{Q} \cap [0, 1]$, *then we obtain a periodic orbit. For example, if* $x_0 = 1/7$, *then we find*

$$x_1 = \frac{2}{7}, \quad x_2 = \frac{4}{7}, \quad x_3 = \frac{6}{7}, \quad x_4 = \frac{2}{7}, \dots .$$

Remark.   *The logistic map can be transformed into the tent map (see Volume I, Chapter 8, Problem 1). The transformation is*

$$g = \phi \circ f \circ \phi^{-1}$$

where $\phi : [0, 1] \to [0, 1]$ *is given by*

$$\phi(x) = \frac{2}{\pi} \arcsin(\sqrt{x})$$

and $\circ$ *denotes the composition of maps. Both maps have the same Ljapunov exponent.*

Problem 3.   The *Cantor set* is constructed as follows: We set

$$E_0 = [0, 1],$$

$$E_1 = \left[0, \frac{1}{3}\right] \cup \left[\frac{2}{3}, 1\right],$$

$$E_2 = \left[0, \frac{1}{9}\right] \cup \left[\frac{2}{9}, \frac{1}{3}\right] \cup \left[\frac{2}{3}, \frac{7}{9}\right] \cup \left[\frac{8}{9}, 1\right],$$

$$\cdots = \cdots .$$

In other words, delete the middle third of the line segment $[0, 1]$ and then the middle third from all the resulting segments and so on, ad infinitum.

The set defined by

$$C := \bigcap_{k=0}^{\infty} E_k$$

is called the *standard Cantor set* (or *Cantor ternary set*).

(i) Show that the Cantor set is of *Lebesgue measure* zero.

(ii) Show that there is a bijective mapping $f : C \mapsto [0, 1]$.

(iii) Let X be a subset of $\mathbf{R}^n$. Let $N(\epsilon)$ be the number of n-dimensional cubes (boxes) of side ϵ to cover the set X. The *capacity D* of X is defined as

$$D := \lim_{\epsilon \to 0} \frac{\ln N(\epsilon)}{\ln(1/\epsilon)}. \tag{1}$$

Find the capacity D of the interval $I = [0, 2]$. Find the capacity of the standard Cantor set C.

Remark. *The capacity is a so-called* fractal dimension.

(iv) Let X be a subset of $\mathbf{R}^n$. A *cover* of the set X is a (possibly infinite) collection of balls, the union of which contains X. The diameter of a cover $\mathcal{A}$ is the maximum diameter of the balls in $\mathcal{A}$. For $d, \epsilon > 0$, we define

$$\alpha(d, \epsilon) := \inf_{\substack{\mathcal{A}=\text{cover of } X \\ \text{diam } \mathcal{A} \le \epsilon}} \sum_{A \in \mathcal{A}} (\operatorname{diam} A)^d$$

and

$$\alpha(d) := \lim_{\epsilon \to 0} \alpha(d, \epsilon).$$

There is a unique d_0 such that

$$d < d_0 \Rightarrow \alpha(d) = \infty,$$

$$d > d_0 \Rightarrow \alpha(d) = 0.$$

This d_0 is defined to be the *Hausdorff dimension* of the set X, written $HD(X)$. Show that for the standard Cantor set $HD(C) \le (\ln 2)/(\ln 3)$.

Solution. (i) We define

$$\chi_{E_k} := \begin{cases} 1 & x \in E_k, \\ 0 & x \notin E_k. \end{cases}$$

Owing to the construction of the Cantor set the sets E_k are the union of 2^k pairwise disjoint intervals with length $1/3^k$. Thus

$$\int_0^1 \chi_{E_k} d\mu = 2^k \frac{1}{3^k} = \left(\frac{2}{3}\right)^k$$

and

$$\lim_{k \to \infty} \int_0^1 \chi_{E_k} d\mu = 0.$$

On the other hand,

$$\chi_{E_{k+1}} \le \chi_{E_k}$$

and

$$\lim_{k \to \infty} \chi_{E_k} = \chi_C.$$

Therefore χ_C is integrable and

$$\mu(C) = \int_0^1 \chi_C d\mu = 0.$$

(ii) Every number $x \in [0, 1]$ can be represented as (*binary representation*)

$$x = \sum_{j=1}^{\infty} a_j 2^{-j}, \quad a_j \in \{0, 1\}.$$

Every $y \in C$ can be written in the form (*ternary representation*)

$$y = \sum_{j=1}^{\infty} c_j 3^{-j}, \quad c_j \in \{0, 2\}.$$

Let $f : C \mapsto [0, 1]$ be defined by

$$f\left(\sum_{j=1}^{\infty} c_j 3^{-j}\right) = \sum_{j=1}^{\infty} a_j 2^{-j} \tag{2}$$

with

$$a_j = \begin{cases} 0 & \text{for } c_j = 0, \\ 1 & \text{for } c_j = 2. \end{cases}$$

A mapping $g : A \mapsto B$ is called *surjective* if $g(A) = B$. Obviously f, given by (2), is surjective. A mapping g is called *injective* (one-to-one) when

$$\text{for all } a, a' \in A \quad g(a) = g(a') \Rightarrow a = a'.$$

Since different numbers have different binary and ternary representations, the mapping f is injective. Since the mapping f is surjective and injective we find that the mapping f is bijective.

(iii) Let $I = [0, 2]$. Let $\epsilon > 0$ and $N \cdot \epsilon = 2$. Then $N(\epsilon) = 2/\epsilon$. Inserting this into definition (1) we find

$$D = \lim_{\epsilon \to 0} \frac{\ln(2/\epsilon)}{\ln(1/\epsilon)} = 1.$$

To calculate the capacity of the Cantor set we note the following: In the first step of the construction we have $\epsilon = 1/3$ and $N = 2$. In the second step of the construction we have $\epsilon = 1/9$ and $N = 4$ and at the pth step of the construction we find

$$\epsilon = \left(\frac{1}{3}\right)^p, \quad N = 2^p.$$

Consequently, the capacity of the Cantor set is given by

$$D = \frac{\ln 2}{\ln 3}.$$

(iv) Owing to the construction of the standard Cantor set the following cover is obvious. For $n = 1, 2, \ldots$, let $\mathcal{A}_n$ be the cover of C consisting of 2^n intervals of length $1/3^n$ each. We set

$$\sum_{A \in \mathcal{A}_n} (\text{diam } A)^d = 1.$$

It follows that

$$2^n \left(\frac{1}{3^n}\right)^d = 1.$$

This leads to

$$d = \frac{\ln 2}{\ln 3}.$$

We conclude that with d given by (3) we have

$$\alpha\left(d, \frac{1}{3^n}\right) \leq \sum_{A \in \mathcal{A}_n} (\text{diam } A)^d = 1$$

for every n. Therefore

$$\alpha\left(\frac{\ln 2}{\ln 3}\right) \leq 1.$$

This implies that

$$HD(C) \leq \frac{\ln 2}{\ln 3}.$$

Remark.   *In fact one can show that*

$$HD(C) = \frac{\ln 2}{\ln 3}.$$

Thus we have

$$D(C) = HD(C)$$

for the ternary Cantor set. In general we have the inequality

$$HD(X) \leq D(X).$$

The capacity of a set does not distinguish between a set and its closure, while the Hausdorff dimension of the two is different. For example, the set of rationals has

$$HD = 0$$

because it is a countable union of points, and has capacity $= 1$ because its closure $= \mathbf{R}$.

Remark.   *Both the capacity and the Hausdorff dimension are so-called fractal dimensions. Another fractal dimension is the similarity dimension. The topological dimension can only take integer values. Within the topological dimension a point has dimension 0, curves have dimension 1, surfaces have dimension 2 and so on.*

Remark.   *If $X \subset Y \subset \mathbf{R}^n$, then*

$$HD(X) \leq HD(Y).$$

The Hausdorff dimension of a set does not change under smooth reparametrizations. That is, if $f : \mathbf{R}^n \rightarrow \mathbf{R}^n$ is a C^1 diffeomorphism and $X \subset \mathbf{R}^n$, then

$$HD(X) = HD(f(X)).$$

Problem 4. (i) Show that the two-dimensional map

$$\mathbf{f} : x_{k+1} = \lambda x_k + \cos\theta_k, \quad \theta_{k+1} = 2\theta_k \bmod(2\pi) \tag{1}$$

when $2 > \lambda > 1$, has two attractors, at $x = \pm\infty$. Show that the map has no finite attractor.

(ii) Find the *boundary of basins of attractions*.

Solution. (i) There are no finite attractors because the eigenvalues of the Jacobian matrix of the two-dimensional map (1) are 2 and λ, where $\lambda > 1$. Thus

$$\mathbf{f}^{(n)}(x_0, \theta_0) = (x_n, \theta_n \bmod(2\pi))$$

and x_n either tends to $+\infty$ or $-\infty$ as $n \rightarrow \infty$, except for the (unstable) boundary set $x = g(\theta)$, for which x_n remains finite.

(ii) To find this boundary set, we note that

$$\theta_k = 2^k \theta_0 \bmod(2\pi).$$

The map $\mathbf{f}$ is noninvertible, but we can select any x_n and find one orbit that ends at (x_n, θ_n), by using the above θ_k and taking

$$x_{k-1} = \lambda^{-1} x_k - \lambda^{-1} \cos(2^{k-1}\theta_0).$$

For the given (x_n, θ_0) we find that this orbit started at

$$x_0 = \lambda^{-n} x_n - \sum_{l=0}^{n-1} \lambda^{-l-1} \cos(2^l \theta_0).$$

The boundary between the two basins are those (x_0, θ_0) such that x_n is finite as $n \rightarrow \infty$. Thus the x and θ are related by

$$x = -\sum_{l=0}^{\infty} \lambda^{-l-1} \cos(2^l \theta) \equiv g(\theta). \tag{2}$$

Since $\lambda > 1$, this sum converges absolutely and uniformly. On the other hand

$$\frac{dg(\theta)}{d\theta} = \frac{1}{2} \sum_{l=0}^{\infty} \left(\frac{2}{\lambda}\right)^{l+1} \sin(2^l \theta)$$

and the sum diverges, since $\lambda < 2$. Hence $g(\theta)$ is nondifferentiable.

Remark.   *The curve (2) has a fractal dimension*

$$d_c = 2 - (\ln \lambda)(\ln 2)^{-1} .$$

Problem 5.   Consider the *tent map* $f : \mathbf{R} \to \mathbf{R}$ defined by

$$f(x) := \begin{cases} 3x & \text{if } x \le \dfrac{1}{2} , \\[2mm] 3 - 3x & \text{if } x > \dfrac{1}{2} . \end{cases} \tag{1}$$

Discuss the dynamics of the map f on the set Σ of all points x whose iterates stay in the unit interval $[0, 1]$. For any other point the iterates approach $-\infty$.

Solution.   First we note that $f(x) \le 1$ for $x \in [1/3, 2/3]$ and $f(0) = 0$, $f(1) = 0$. The fixed points are $x^* = 0$ and $x^* = 3/4$. If $x \in (1/3, 2/3)$ then the orbit escapes to $-\infty$. We show that Σ is the ternary Cantor set. To prove that Σ is the ternary Cantor set we use the representation of the Cantor set in ternary numbers. We adopt the convention that, where two alternative representations of a given number exist, we choose the form that ends with an infinite string of 2's, rather than the form that ends with a 1 followed by an infinite string of 0's. The Cantor set then consists of the numbers in $[0, 1]$ whose ternary expansions contain only 0's and 2's. Let

$$x = 0 \cdot x_1 x_2 x_3 \cdots$$

in ternary form and introduce the notation

$$\bar{x}_i := 2 - x_i .$$

Multiplication by 3 shifts x one digit to the left, and $3 = 2.222 \cdots$ in ternary form. Hence

$$f(x) = \begin{cases} x_1 \cdot x_2 x_3 x_4 \cdots & \text{if } 0 \le x \le \dfrac{1}{2} , \\[2mm] \bar{x}_1 \cdot \bar{x}_2 \bar{x}_3 \bar{x}_4 \cdots & \text{if } \dfrac{1}{2} < x \le 1 . \end{cases}$$

Note that $\bar{1} = 1$. Hence $f(x) \notin [0, 1]$ if $x_1 = 1$. Otherwise, since $\bar{0} = 2$ and $\bar{2} = 0$,

$$f(x) = \begin{cases} 0 \cdot x_2 x_3 x_4 \cdots & \text{if } x_1 = 0, \\ 0 \cdot \bar{x}_2 \bar{x}_3 \bar{x}_4 \cdots & \text{if } x_1 = 2. \end{cases}$$

Hence $f(x) \in [0, 1]$ if and only if $x_1 = 0$ or 2. An easy inductive argument shows that for all $i \in \mathbf{N}$,

$$f^{(i)}(x) \in [0, 1] \quad \text{if and only if } x_i = 0 \text{ or } x_i = 2.$$

Thus $x \in \Sigma$ if and only if x is in the ternary Cantor set and so Σ is the ternary Cantor set. Hence inspection of the formula (1) shows that f maps Σ into itself.

Problem 6. Let M be a manifold or simply $\mathbf{R}^n$. Assume the map $f : M \to M$ is continuous. μ is called a *Borel probability measure* if μ assigns to every Borel subset $E \subset M$ a certain probability or measure that we denote by $\mu(E)$. A property is said to hold a.e. (almost everywhere) if it is enjoyed by every $x \in M$ except possibly on a set E with $\mu(E) = 0$. Given a probabilistic dynamical system

$$f : (M, \mu) \to (M, \mu).$$

Such a system is said to be *ergodic* if

$$f^{-1}(E) = E \Rightarrow \mu(E) = 0 \text{ or } \mu(E) = 1.$$

That is, an ergodic system cannot be decomposed into two nontrivial subsystems that do not interact with each other.

Birkhoff Ergodic Theorem. *For any integrable function* $\phi : (M, \mu) \to \mathbf{R}$, *the* time average

$$\frac{1}{n} \sum_{i=0}^{n-1} \phi(f^{(i)}(x))$$

converges for a.e. x. *Denote this limit by* $\phi^*(x)$. *If the system is ergodic, then* $\phi^*(x)$ *equals a.e. the* space average

$$\int_M \phi(x) d\mu(x).$$

Apply the theorem to the logistic map $f : [0, 1] \to [0, 1]$,

$$f(x) = 4x(1 - x)$$

where

$$\phi(x) = \ln \left| \frac{df}{dx} \right| .$$

Solution.   It can be shown that

$$d\mu(x) = \frac{1}{\pi \sqrt{x(1 - x)}} dx$$

is an ergodic invariant measure and that

$$\ln \left| \frac{df}{dx} \right|$$

is integrable. Substituting $\phi(x) = \ln |df/dx|$ in the ergodic theorem, we have for a.e. x

$$\frac{1}{n} \ln \left| \frac{f^{(n)'}(x)}{dx} \right| = \frac{1}{n} \ln \prod_{i=0}^{n-1} |f'(f^{(i)}(x))| = \frac{1}{n} \sum_{i=0}^{n-1} \phi(f^{(i)}(x)) .$$

For $n \to \infty$ we obtain

$$\int_0^1 \ln |4 - 8x| \frac{1}{\pi \sqrt{x(1 - x)}} dx = \ln 2$$

where we used that

$$\frac{df}{dx} = 4 - 8x .$$

This says that for a.e. x, $|(f^{(n)})'(x)|$ is in the order of 2^n for large enough n. The number $\ln 2$ is called the *Ljapunov exponent*.

Problem 7.   Given a compact set X embedded in $\mathbf{R}^n$. The *capacity* of X can be calculated by counting the number of grid boxes of side $\epsilon = B^{-\ell}$ that cover X. $B > 1$ and $\ell > 0$ are integers. Let $N(\ell)$ be the number of such boxes. The capacity dimension is assumed to satisfy

$$\lim_{\ell \to \infty} \frac{\ln N(\ell)}{\ell \cdot \ln B} = D_c . \tag{1}$$

The definition of the capacity dimension is

$$D_c := \overline{\lim_{\epsilon \to 0}} \frac{\ln(\bar{N}/\epsilon)}{\ln(1/\epsilon)}$$

where $\bar{N}(\epsilon)$ is the minimum number of boxes of side ϵ of any type that covers the set X. If the limit in the definition exists the limit in (1) will also exist. Discuss how D_c could be calculated numerically.

Solution. Denote by G_ℓ the set of all grid boxes of side $B^{-\ell}$. Numerical calculations of the capacity are based on the heuristic idea that the set X has finite D_c-dimensional measure V. The quantity V satisfies

$$V \approx (\bar{N}(B^{-\ell})) \cdot (B^{-\ell})^{D_c}$$

for sufficiently large ℓ. Since

$$N(B^{-\ell}) \leq N(\ell) \leq N(B^{-\ell}) \cdot c$$

for $c > 1$, a constant independent of X or ℓ, implies that

$$V \approx O(1) \cdot N(\ell) \cdot [B^{-\ell}]^{D_c}$$

where $O(1)$ holds as $\ell \to \infty$. Thus if V is constant and $O(1)$ is also assumed to be a constant then

$$\ln N(\ell) \approx \ln(\text{const} \cdot V) + \ell \cdot \ln B \cdot D_c,$$

that is, for large ℓ a plot of $\ln N$ versus $\ell \cdot \ln B$ is approximately a straight line with slope D_c. The replacement of $O(1)$ by a constant introduces a possible error in the computation that results from the use of boxes to cover X.

Problem 8. Consider the *Rössler model*

$$\frac{dx}{dt} = -y - \epsilon z, \tag{1a}$$

$$\frac{dy}{dt} = x + \epsilon r y, \tag{1b}$$

$$\frac{dz}{dt} = 1 + (x - C)z \tag{1c}$$

where ϵ, r and C are positive constants. There is numerical evidence that the system shows chaotic behaviour for $\epsilon = 0.2$, $r = 1$ and $C = 4$. The system can be considered as a harmonic oscillator in x and y nonlinearly coupled to a third variable. Find a *Poincaré map* for the system with $\epsilon \ll 1$.

Solution.   For $\epsilon = 0$ we have a harmonic oscillator

$$\frac{dx}{dt} = -y, \quad \frac{dy}{dt} = x$$

with the solution

$$x(t) = a_0 \cos(\omega_0 t),$$

$$y(t) = a_0 \sin(\omega_0 t)$$

where $\omega_0 = 1$. For $\epsilon \ll 1$ we consider the expansion

$$x(t) = x_0(t) + \epsilon x_1(t) + \cdots,$$

$$y(t) = y_0(t) + \epsilon y_1(t) + \cdots,$$

$$z(t) = z_0(t) + \epsilon z_1(t) + \cdots$$

where the zeroth solution of x and y is a harmonic oscillator

$$x_0(t) = a_0 \cos(\omega t), \quad y_0(t) = a_0 \sin(\omega t) \tag{2}$$

with

$$\omega = \omega_0 + \epsilon \omega_1 + \cdots.$$

We still have to find ω_1. The terms of first order are given by

$$x_1(t) = a_1(t) \cos(\omega t) + b_1(t) \sin(\omega t),$$

$$y_1(t) = a_1(t) \sin(\omega t) - b_1(t) \cos(\omega t).$$

Inserting this ansatz into (1) we find up to the zeroth order that

$$(\omega_0 - 1)x_0 = 0, \quad (\omega_0 - 1)y_0 = 0, \quad \frac{dz_0}{dt} = 1 + (x_0 - C)z_0.$$

The first two equations are satisfied for $\omega_0 = 1$. The third equation can be integrated. We obtain

$$z_0(t) = \frac{z_0(0) + \int_0^t f(s)ds}{f(t)} \tag{3}$$

where $z_0(0)$ is the initial value of z_0 and the function f is given by

$$f(t) = \exp\left(-\int_0^t (x_0(s) - C)ds\right) = \exp\left(-\frac{a_0}{\omega} \sin(\omega t) + Ct\right).$$

Since we are only interested in the attractor, we have to find $z_0(t \to \infty)$. At time $t + n\tau$ with $0 \le t \le \tau$ the quantity z_0 is given by

$$z_0(t + n\tau) = \frac{z_0 + \int_0^{t+n\tau} f(s)ds}{f(t + n\tau)}.$$

The function f has the property

$$f(t + n\tau) = f(t) \exp(nC\tau).$$

Using this property we find

$$z_0(t + n\tau) = \frac{(z_0 - z_C(\tau)) \exp(-nC\tau) + z_C(\tau) + \int_0^t f(s)ds}{f(t)}$$

where

$$z_C(\tau) = \frac{\int_0^\tau f(s)ds}{\exp(C\tau) - 1}.$$

For $n \to \infty$ the function $z_0(t + n\tau)$ approaches the asymptotic function $\tilde{z}_0(t)$

$$z_0(t + n\tau) \to \tilde{z}_0(t) = \frac{z_C(\tau) + \int_0^t f(s)ds}{f(t)}.$$

We also find

$$\tilde{z}_0(t + n\tau) = \tilde{z}_0(t).$$

In first-order we find

$$\frac{da_1}{dt} = (\omega_1 y_0 - z_0) \cos(\omega t) + (-\omega_1 x_0 + r y_0) \sin(\omega t), \qquad (4a)$$

$$\frac{db_1}{dt} = (\omega_1 y_0 - z_0) \sin(\omega t) - (-\omega_1 x_0 + r y_0) \cos(\omega t) \qquad (4b)$$

where x_0 and y_0 are given by (2) and z_0 is given by (3). Equations (4) can be integrated to give

$$a_1(\tau) = \frac{\tau}{2} a_0 \tau - \int_0^\tau \tilde{z}_0(t) \cos(\omega t)dt,$$

$$b_1(\tau) = \omega_1 a_0 \tau - \int_0^\tau \tilde{z}_0(t) \sin(\omega t)dt$$

with the initial values $a_1(0) = b_1(0) = 0$ and

$$x(0) = a_0, \quad y(0) = 0, \quad z(0) = z_C(\tau).$$

After one period τ, we have

$$x(\tau) = a_0 + \epsilon a_1(\tau) + O(\epsilon^2) = a(\tau),$$

$$y(\tau) = -\epsilon b_1(\tau) + O(\epsilon^2) = b(\tau),$$

$$z(\tau) = z_C(\tau) + O(\epsilon).$$

The initial values a_0 and $b_0 = 0$ are mapped, after one period, into

$$a(\tau) = a_0 + \epsilon\tau\left(\frac{r}{2}a_0 - \frac{1}{\tau}\int_0^\tau \tilde{z}_0(t)\cos(\omega t)dt\right),$$

$$b(\tau) = \epsilon\tau\left(\omega_1 a_0 - \frac{1}{\tau}\int_0^\tau \tilde{z}_0(t)\sin(\omega t)dt\right).$$

This is the discrete Poincaré map.

Problem 9.   Consider the one-dimensional map $f : [0,1] \to [0,1]$

$$x_{t+1} = f(x_t)$$

where $t = 0, 1, 2, \ldots$ and $x_0 \in [0,1]$. The nth moment of the time evolution of x_t is defined as

$$\langle x_t^n \rangle := \lim_{T \to \infty} \frac{1}{T} \sum_{t=0}^{T} x_t^n$$

where $n = 1, 2, \ldots$. The moments depend on the initial conditions. In the case of ergodic systems, the moments and the probability density ρ are related by

$$\langle x_t^n \rangle = \int_0^1 y^n \rho(y)dy$$

where $\rho > 0$ for $x \in [0,1]$ and

$$\int_0^1 \rho(x)dx = 1.$$

We can calculate the Ljapunov exponent as follows

$$\lambda = \int_0^1 \rho(x) \ln\left|\frac{df}{dx}\right| dx.$$

The missing information function (entropy) of a probability density ρ is defined as

$$I = - \int_0^1 \rho(x) \ln \rho(x) dx.$$

Apply the *maximum entropy formalism* to obtain the probability density approximately using as information, N moments. Apply it to the logistic map $f(x) = 4x(1-x)$ with $N = 2$.

Solution. In the maximum entropy formalism, one maximizes the missing information subject to the constraints of the available information and to the normalization of the probability density. We assume that we have the N lowest moments. The constraints are introduced via the method of the *Lagrange multipliers* $\lambda_1, \lambda_2, \ldots, \lambda_N$. Our aim is to find the approximate probability density ρ_{app} which minimizes

$$I' = - \int_0^1 \rho_{app} \ln \rho_{app} dx + \lambda_0 \left(1 - \int_0^1 \rho_{app} dx \right)$$

$$+ \sum_{n=1}^N \lambda_n \left(\langle x^n \rangle - \int_0^1 x^n \rho_{app} dx \right)$$

where λ_n, $n = 0, 1, 2, \ldots, N$ are the Lagrange multipliers. Performing the minimization we obtain

$$\rho_{app}(x) = \exp \left(-1 - \sum_{n=0}^N \lambda_n x^n \right) \equiv \frac{1}{Z} \exp \left(- \sum_{n=1}^N \lambda_n x^n \right)$$

where $Z = \exp(1 + \lambda_0)$. From the normalization condition, we obtain

$$1 = \frac{1}{Z} \int_0^1 \exp \left(- \sum_{n=1}^N \lambda_n x^n \right) dx.$$

The remaining Lagrange multipliers are obtained by solving the following set of N coupled nonlinear equations for λ_m, $m = 1, 2, \ldots, N$,

$$\langle x^m \rangle = \frac{1}{Z} \int_0^1 x^m \exp \left(- \sum_{n=1}^N \lambda_n x^n \right) dx, \quad m = 1, 2, \ldots, N.$$

For the logistic map $f(x) = 4x(1-x)$, the moments are given by

$$\langle x^n \rangle = \frac{1}{2^{2n}} \binom{2n}{n}.$$

Thus $\langle x \rangle = 1/2$ and $\langle x^2 \rangle = 3/8$. Thus we have to solve

$$1 = \int_0^1 \exp(-1 - \lambda_0 - \lambda_1 x - \lambda_2 x^2)dx \,,$$

$$\frac{1}{2} = \int_0^1 \exp(-1 - \lambda_0 - \lambda_1 x - \lambda_2 x^2)dx \,,$$

$$\frac{3}{8} = \int_0^1 \exp(-1 - \lambda_0 - \lambda_1 x - \lambda_2 x^2)dx \,.$$

We solve this system numerically and find

$$\lambda_0 = 2.69242 \,, \quad \lambda_1 = -6.76825 \,, \quad \lambda_2 = -\lambda_1 \,.$$

Obviously $\lambda_1 = -\lambda_2$.

Problem 10.   Given a binary string

"100001010110010101110 $\cdots\cdot$ 10111"

of finite length n. Let A^* denote the set of all finite-length sequences (strings) over a finite alphabet A (in our case $\{0, 1\}$). The quantity $S(i, j)$ denotes the substring $S(i, j) := s_i s_{i+1} \cdots s_j$. A *vocabulary* of a string S, denoted by $v(S)$, is the subset of A^* formed by all the substrings, or words, $S(i, j)$ of S. The *complexity*, in the sense of Lempel and Ziv, of a finite string is evaluated from the point of view of a simple self-delimiting learning machine which, as it scans a given n digit string $S = s_1 s_2 \cdots s_n$ from left to right, adds a new word to its memory every time it discovers a substring of consecutive digits not previously encountered. Thus the calculation of the complexity $c(n)$ proceeds as follows. Let us assume that a given string $s_1 s_2 \cdots s_n$ has been reconstructed by the program up to the digit s_r and that s_r has been newly inserted, i.e., it was not obtained by simply copying it from $s_1 s_2 \cdots s_{r-1}$. The string up to s_r will be denoted by $R := s_1 s_2 \cdots s_r \circ$, where the $\circ$ indicates that s_r is newly inserted. In order to check whether the rest of R, i.e., $s_{r+1} s_{r+2} \cdots s_n$ can be reconstructed by simple copying or whether one has to insert new digits, we proceed as follows: First, one takes $Q \equiv s_{r+1}$ and asks whether this term is contained in the vocabulary of the string R so that Q can simply be obtained by copying a word from R. This is equivalent to the question of whether Q is contained in the vocabulary $v(RQ\pi)$ of $RQ\pi$ where $RQ\pi$ denotes the string which is composed of R and Q (concatenation) and π means that the last digit has to be deleted. This can be generalized to situations where Q also contains two (i.e., $Q =$

$s_{r+1}s_{r+2}$) or more elements. Let us assume that s_{r+1} can be copied from the vocabulary of R. Then we next ask whether $Q = s_{r+1}s_{r+2}$ is contained in the vocabulary of $RQ\pi$ and so on, until Q becomes so large that it can no longer be obtained by copying a word from $v(RQ\pi)$ and one has to insert a new digit. The number c of production steps to create the string S, i.e., the number of newly inserted digits (plus one if the last copy step is not followed by inserting a digit), is used as a measure of the complexity of a given string.

(i) Find the complexity of a string which contains only zeros.

(ii) Find the complexity of a string which is only composed of units of 01, i.e.,

$$01010101\cdots01.$$

(iii) Find the complexity of the string 0010.

(iv) Write a C++ program that finds the complexity $c(n)$ for a given binary string of length n.

Solution. (i) If we have a sequence which contains only zeros we could say that it should have the smallest possible complexity of all strings (equivalent to a string consisting only of 1's). One has only to insert the first zero and can then reconstruct the whole string by copying this digit, i.e.,

$$00000\cdots \rightarrow 0 \circ 000\cdots.$$

Thus the complexity of this string is $c = 2$.

(ii) Similarly one finds for a sequence which is only composed of units 01, i.e.,

$$010101\cdots01 \rightarrow 0 \circ 1 \circ 0101\cdots01$$

the value $c = 3$.

(iii) The complexity c of the string $S = 0010$ can be determined as follows:

(1) The first digit has always to be inserted $\rightarrow 0\circ$;
(2) $R = 0$, $Q = 0$, $RQ = 00$, $RQ\pi = 0$, $Q \in v(RQ\pi) \rightarrow 0 \circ 0$;
(3) $R = 0$, $Q = 01$, $RQ = 001$, $RQ\pi = 00$, $Q \notin v(RQ\pi) \rightarrow 0 \circ 01\circ$;
(4) $R = 001$, $Q = 0$, $RQ = 0010$, $RQ\pi = 001$, $Q \in v(RQ\pi) \rightarrow 0 \circ 01 \circ 0$.

Now c is equal to the number of parts of the string that are separated by $\circ$, i.e., $c = 3$.

Remark.   *The set of the rational numbers in the interval* $[0,1]$ *are of Lebesgue measure zero. This implies that for almost all numbers in* $[0,1]$ *(i.e., for all irrationals) the string of zeros and ones which represents their binary decomposition is not periodic. Therefore we expect that almost all strings which correspond to a binary representation of a number* $x \in [0,1]$ *should be random and have maximal complexity. For almost all* $x \in [0,1]$ *the complexity* $c(n)$ *(of the string which represents the binary decomposition) tends to the same value, namely*

$$\lim_{n \to \infty} c(n) = b(n) \equiv \frac{n}{\log_2 n}. \tag{1}$$

$b(n)$ gives, therefore, the asymptotic behaviour of $c(n)$ for a random string. One often normalizes $c(n)$ via this limit.

(iv) A computer program in C++ for finding the complexity uses a do-while loop, a for loop and a while loop as follows:

```
// ziv.cppa

#include <iostream>    // for cout, cin
using namespace std;

long complexity(const long* S,long n)   // n:  length of string
{                                       // S: string
   long c = 1, l = 1;
   do
   {
      long k = 0, kmax = 1;
      for(long i=0; i<l; i++)
      {
      while(S[i+k] == S[l+k])
      {
      ++k;
      if(l+k >= n-1) return (++c);
      }
      if(k >= kmax) kmax = k + 1; k = 0;
      }
      ++c; l += kmax;
   } while(l < n);
   return c;
}
```

```
void main()
{
   long n = 0;
   cout << "enter length of string:  "; cin >> n;
   long *S;   S = new long[n];          // allocating memory
   for string
   cout << "enter string:  " << endl;
   for(int i=0; i<n; i++)
   {
   cout << "S[" << i << "] = ";    cin >> S[i];
   }
   cout << "complexity of the string:  " << endl << "[";
   for(i=0; i<n; i++)   cout << S[i];
   cout << "]" << endl;
   cout << "is:  " << complexity(S,n) << endl;
   delete[] S;       // deallocating memory
}
```

Bibliography

E. J. Barbeau, *Polynomials*, Springer-Verlag, New York (1989).

M. Berger, P. Pansu, J. P. Berry and X. Saint-Raymond, *Problems in Geometry*, Springer-Verlag, New York (1984).

F. Constantinescu and E. Magyari, *Problems in Quantum Mechanics*, Pergamon Press, Oxford (1971).

J. A. Cronin, D. F. Greenberg and V. L. Telegdi, *Graduate Problems in Physics*, Addison Wesley, Reading (1967).

P. N. de Souza and J.-N. Silva, *Berkeley Problems in Mathematics*, Springer-Verlag, New York (1998).

S. Flügge, *Practical Quantum Mechanics*, Springer-Verlag, Berlin (1974).

B. Gelbaum, *Problems in Analysis*, Springer-Verlag, New York (1982).

K. Knoop, *Problem Book in the Theory of Functions*, Volume I, Dover, New York (1952).

J. G. Krzyz, *Problems in Complex Variables Theory*, Elsevier, New York (1971).

L. C. Larson, *Problem Solving Through Problems*, Springer-Verlag, New York (1983).

J. M. Rassias, *Counter Examples in Differential Equations and Related Topics*, World Scientific, Singapore (1991).

M. R. Spiegel, *Advanced Calculus*, Schaum's Outline Series, McGraw Hill, New York (1974).

M. R. Spiegel, *Finite Differences and Difference Equations*, Schaum's Outline Series, McGraw Hill, New York (1971).

M. R. Spiegel, *Complex Variables*, Schaum's Outline Series, McGraw Hill, New York (1971).

I. Tomescu, *Problems in Combinatorics and Graph Theory*, Wiley, New York (1985).

Index